PUBLIC HEALTH IN THE 21ST CENTURY SERIES

HEALTH AND ENVIRONMENT: SOCIAL SCIENCE PERSPECTIVES

PUBLIC HEALTH IN THE 21ST CENTURY SERIES

Family History of Osteoporosis
Afrooz Afghani (Editor)
2009. ISBN: 978-1-60876-190-6

Cross Infections: Types, Causes and Prevention
Jin Dong and Xun Liang (Editors)
2009. ISBN: 978-1-60741-467-4

Swine Flu and Pig Borne Diseases
Viroj Wiwanitkit
2009. ISBN: 978-1-60876-291-0

Home Fire Safety: Preventive Measures and Issues
Cornelio Moretti (Editor)
2009. ISBN: 978-1-60741-651-7

Biological Clocks: Effects on Behavior, Health and Outlook
Oktav Salvenmoser and Brigitta Meklau (Editors)
2010. ISBN: 978-1-60741-251-9

Sexual Risk Behaviors
Luigi Passero and Cecilia Sgariglia (Editors)
2010. ISBN: 978-1-60741-227-4

MRSA (Methicillin Resistant Staphylococcus aureus) Infections and Treatment
Manal M. Baddour
2010. ISBN: 978-1-61668-038-1

MRSA (Methicillin Resistant Staphylococcus aureus) Infections and Treatment
Manal M. Baddour
2010. ISBN: 978-1-61668-450-1
(Online Book)

Health-Related Quality of Life
Erik C. Hoffmann (Editor)
2009. ISBN: 978-1-60741-723-1

Infectious Disease Modelling Research Progress
Jean Michel Tchuenche and C. Chiyaka (Editors)
2010. ISBN: 978-1-60741-347-9

Expedited Partner Therapy in the Management of STDs
H. Hunter Handsfield, Matthew Hogben, Julia A. Schillinger, Matthew R. Golden, Patricia Kissinger and P. Frederick Sparling
2010. ISBN: 978-1-60741-571-8

Firefighter Fitness: A Health and Wellness Guide
Ernest L. Schneider
2010. ISBN: 978-1-60741-650-0

The Making of a Good Doctor
B. D. Kirkcaldy, R. J. Shephard and R. G. Siefen
2010. ISBN: 978-1-60876-449-5

Handbook of Disease Outbreaks: Prevention, Detection and Control
Albin Holmgren and Gerhard Borg (Editors)
2010. ISBN: 978-1-60876-224-8

Avian Influenza: Etiology, Pathogenesis and Interventions
Salomon Haugan and Walter Bjornson (Editors)
2010. ISBN: 978-1-60741-846-7

Overweightness and Walking
Caleb I. Black (Editor)
2010. ISBN: 978-1-60741-298-4

Overweightness and Walking
Caleb I. Black (Editor)
2010. ISBN: 978-1-61668-516-4
(Online Book)

Smoking Relapse: Causes, Prevention and Recovery
Johan Egger and Mikel Kalb (Editors)
2010. ISBN: 978-1-60876-580-5

Medicare Advantage: The Alternate Medicare Program
Charles V. Baylis (Editor)
2010. ISBN: 978-1-60876-031-2

Diabetes in Women
Eliza I. Swahn (Editor)
2010. ISBN: 978-1-61668-692-5

Diabetes in Women
Eliza I. Swahn (Editor)
2010. ISBN: 978-1-61668-801-1
(Online Book)

Sepsis: Symptoms, Diagnosis and Treatment
Joseph R. Brown (Editor)
2010. ISBN: 978-1-60876-609-3

Sepsis: Symptoms, Diagnosis and Treatment
Joseph R. Brown (Editor)
2010. ISBN: 978-1-61668-448-8
(Online Book)

Breaking Down Barriers to Care: Treating Tobacco Dependence in Vulnerable Populations
John E. Snyder and Megan J. Engelen
2010. ISBN: 978-1-60876-976-6

Strategic Implications for Global Health
Maria Labonté (Editor)
2010. ISBN: 978-1-60741-660-9

Health and Environment: Social Science Perspectives
Helen Kopnina and Hans Keune (Editors)
2010. ISBN: 978-1-60876-216-3

The Anti-Inflammatory Effects of Exercise
Pedro Tauler and Antoni Aguiló
2010. ISBN: 978-1-60876-886-8

Controlling Disease Outbreaks: The Changing Role of Hospitals
Raoul E. Nap, Nico E.L. Meessen, Maarten P.H.M. Andriessen and Tjip S. van der Werf
2010. ISBN: 978-1-61668-314-6

Controlling Disease Outbreaks: The Changing Role of Hospitals
Raoul E. Nap, Nico E.L. Meessen, Maarten P.H.M. Andriessen and Tjip S. van der Werf
2010. ISBN: 978-1-61668-739-7
(Online Book)

The H1N1 Influenza Pandemic of 2009
Charles R. Bartolotti (Editor)
2010. ISBN: 978-1-61668-357-3

The H1N1 Influenza Pandemic of 2009
Charles R. Bartolotti (Editor)
2010. ISBN: 978-1-61668-392-4
(Online Book)

Heavy Metals: A Rapid Clinical Guide to Neurotoxicity and other Common Concerns
Kenneth R. Spaeth, Antonios J. Tsismenakis and Stefanos N. Kales
2010. ISBN: 978-1-60876-634-5

Scientific and Ethical Approaches for Observational Exposure Studies
Alain E. Hughes
2010. ISBN: 978-1-60876-034-3

Physical Activity in Rehabilitation and Recovery
Holly Blake
(Editor)
2010. ISBN: 978-1-60876-400-6

Treadmill Exercise and its Effects on Cardiovascular Fitness, Depression and Muscle Aerobic Function
Nuno Azóia and Pedra Dobreiro
(Editors)
2010. ISBN: 978-1-60876-857-8

COPD Is/Is Not a Systemic Disease?
Claudio F. Donner
(Editor)
2010. ISBN: 978-1-60876-051-0

Research Methodologies in Public Health
David A. Yeboah
2010. ISBN: 978-1-60876-810-3

Diarrhea: Causes, Types and Treatments
Hannah M. Wilson (Editor)
2010. ISBN: 978-1-61668-449-5

Diarrhea: Causes, Types and Treatments
Hannah M. Wilson (Editor)
2010. ISBN: 978-1-61668-874-5
(Online Book)

Strategic Implications for Global Health
Maria Labonté (Editor)
2010. ISBN: 978-1-60741-660-9

Influenza Pandemic - Preparedness and Response To A Health Disaster
Emma S. Brouwer (Editor)
2010. ISBN: 978-1-60692-953-7

PUBLIC HEALTH IN THE 21ST CENTURY SERIES

HEALTH AND ENVIRONMENT: SOCIAL SCIENCE PERSPECTIVES

HELEN KOPNINA
AND
HANS KEUNE
EDITORS

Nova Science Publishers, Inc.
New York

LIBRARY OF CONGRESS CATALOGING-IN-PUBLICATION DATA

Health and environment : social science perspectives / [edited by] Helen Kopnina, Hans Keune.
p. ; cm.
Includes bibliographical references and index.
ISBN 978-1-60876-216-3 (hardcover)
1. Environmental health--Social aspects. 2. Social medicine. I. Kopnina, Helen. II. Keune, Hans.
[DNLM: 1. Environmental Health. 2. Public Policy. 3. Social Conditions. 4. World Health. WA 30.5 H4334 2010]
RA566.H425 2010
362.1--dc22

2010012227

Published by Nova Science Publishers, Inc. ✢ New York

DEDICATION

To the memory of Lucy Galbenstein (1917-2008)
who loved both people and nature

CONTENTS

PREFACE

Picture 1. Picking Apples.

In this book we examine the contribution of social scientists to the topics of health and environment. We present diverse perspectives on classical and contemporary debates by focusing on social scientific framing of environment and health, as well as on the potential contribution of social science to problem solving strategies in these fields. We trace environment and health debates to the classical scholars and examine the evolution of contemporary academic disciplines concerned with health and environment. We address current complex relationships between human health and environmental problems, and between human and non-human interests.

Scientific as well as public discourse has shifted back and forth between 'humanist' and 'environmental' approaches, often focusing either on their duality or mutual complementarity. After examining the roots of these approaches, we turn to contemporary environmental issues with the potential to affect public health, such as climate change, air, soil and water pollution, food contamination, and availability of natural resources. Using the social scientific lens, we examine human impact on the environment, particularly the consequences of population growth and industrialization. Popular notions like ‘responsible consumption’, ‘ecological production’, ‘renewable energy’, ‘sustainable enterprise’, and ‘traditional medicine’ are discussed in turn. Environmental risks for human health perceived as priority issues by policy-makers are critically evaluated leading to reflection on what role the social scientists can play in addressing the issues of societal and environmental importance.

The combined perspective of this collection of essays exemplifies the social relevance of a social scientific focus on health and environment. Not only do the socal aspects of environment and health legitimize our choice of a social scientific perspective, but also the social complexity involved in the problem solving strategies warrants such a point of view. This complexity relates not only to the different interests and divergent social perspectives, but also to the social process of environmental and health knowledge development.

We do not pretend to be exhaustive in our presentation of social scientific perspectives on environment and health related issues. We present a selection of diverse perspectives that will give readers the opportunity to get acquainted with these socially and politically significant topics from other than a merely natural scientific perspective. This volume combines insights from different sub-disciplines dealing with both health and environmental topics, such as medical anthropology, ecological sociology, and environmental health. This volume is an example of the complementarity between different social scientific views on the fields of health and environment. This makes *Health and Environment* useful for scholars and students of social sciences, health and environment, as well as for the general audience.

In: Health and Environment: Social Science Perspectives ISBN 978-1-60876-216-3
Editors: Helen Kopnina and Hans Keune

INTRODUCTION

Helen Kopnina and Hans Keune

ETERNAL QUESTIONS AND PRESENT CONCERNS

What is actually meant by health and environment? There are many possible answers, depending on which academic discipline or interest group is engaged in discussing the topic. Scientific and public discourses, examined through space and time, provide us with diverging perspectives on what health and environment may mean.

Can health be seen as related to the state of the (human) body, while environment to the state of nature? Are human beings part of nature? Can they be seen as part of the same continuum or do they comprise irreconcilable universes? How does health relate to sickness and environment to culture, development or civilization? For many academics and laymen, disease will negate health, and culture (in a sense of human civilization, progress, activity) will negate nature. For others, health and sickness, and environment and human culture are part of the same continuum.

Public discourse in the twenty-first century embraces thoughts, feelings and aspirations of all people for whom health and environment are not merely an intellectual debate, but an everyday experience. Unlike social scientists, known for their (claims of) detachment and attempted neutrality, the public (which includes social scientists with their "academic hats off") may feel passionate about health and/or environmental issues. Be it through business (as in the case of commercializing health and sustainability), through the media (which usually presents the most nationally dominant of the battling debates), or through the public discourse (ranging from "street talk" to charity groups), health and environment are far from emotionally neutral or scientifically objective issues.

In this volume, objectivity of health and environmental problems are critically discussed and examined embedded in social worlds. Social scientists who contributed to this volume attempt to rise above confounds of the academic disciplines and stress the importance of an interdisciplinary perspective on health and environmental issues. It is the engagement of social scientists with their non-scientific or (natural) scientific counterparts that enables contributors to this volume to reflect on inter-relatedness of expert, public, and policy debates and render fruitful analyses to help find viable solutions to both medical and environmental problems. Social scientists contributing to this volume have been as "engaged" with the issues of health and environment as their (natural and medical) scientific colleagues, policymakers, interest groups and the media. Throughout this book, the contributors illustrate how social scientists' engagement with socially and environmentally significant issues could be more fruitful.

Two approaches may be distinguished when reflecting on the issue of engagement with health and environment: humanistic (anthropocentric) focus and nature (environmental or conservational) focus. Modern society is largely influenced by political and social agendas, prioritizing either human interest or conservation or both. While during the Middle Ages in Europe, Man was often placed at the center of Creation by the Judeo-Christian doctrine, present "humanist" or "anthropocentric" approach derives from the ideals of individualism and rationality inspired by Renaissance and Enlightenment, as well as the logic of advanced capitalism. Interest in human health and promotion of (Western) medicine across the world has become one of the trademarks of globalization. "Environmental" interest, on the other hand, refers to centrality of non-manmade nature and viewing humans as part of (pristine) nature. Interest in nature preservation is often expressed independently of human interests by political and public pressure groups.

Picture 2. Jumping over ditch.

"Charity paradox," which will be discussed in more detail further in this introduction, reveals potential conflict between human and environmentalist interests. This paradox may be illustrated by an example of charities that focus on providing health services to the poor countries, which leads to population growth, which, in turn, leads to destruction of wild habitats. Environmental charities, struggling with effects of population growth, rapid urbanization and appropriation of wilderness areas for farming, may find themselves at odds with health-promoting groups. Health-promoting charities may have little to do with the "environment" when "environmental health" is at stake.

ENGAGEMENT OF SOCIAL SCIENTISTS WITH THE TOPICS OF HEALTH AND ENVIRONMENT

The relationship between health and environment is quite complex, not least because the terms health and environment themselves are all encompassing and yet elusive, socially defined and morally loaded, precluding easy definitions. As fascinating as it would be to bring together medical, ecological and other specialist perspectives, such a project would likely fail due to the great scope and complexity of the topics involved. By placing health and environment in a social scientific context rather than providing the reader with medical, scientific or technical expertise, we achieve three aims. At the theoretical level, we explore the disciplinary links on the subject of health and environment between social sciences and public discourse, on the one hand, and social sciences and natural sciences on the other. At the practical level, we address the complexity of the issues from a social point of view, which is a useful contribution to the process of policy development. Finally, in articulating the social complexity of health and environment, and differences of opinions, interests, and knowledge, we ponder our own role as social scientists in the contemporary debates.

Social scientists have often engaged with issues of human interest without giving explicit attention to the place of human actors within the physical environment and, indeed, the influence that the human actors have on the environments they live in. Contributors to this volume examine not only anthropocentric, but also interactive (human-environment and environment-human) perspectives. They demonstrate that engagement with the issues of health and environment is contingent on moral and ethical, as well as expert, opinions.

In this volume, the view that the environment (in a sense of "wilderness" or "nature") can be seen objectively and outside the human sphere is critically discussed. The environmental issues, such as the rapid loss of biodiversity triggered by human activities, are presently addressed by environmental activists and scientists. These issues also belong to the scope of social scientific inquiry. The environmental-minded social scientists suggest that objects in natural environment (such as tigers) certainly do exist within a shared human domain, and human activity clearly contributes to their destruction (through land development leading to the loss of habitat, poaching, etc.). What matters for tigers (if we were to assume the "tiger's perspective" and think outside the Homo sapiens' frame of mind) is the threat to their objective existence and not the way humans (such as social scientists) perceive this problem. In other words, an argument can be made that aside from social (scientific) views, perceptions, theories and discourses, there is a simple practical urgency to deal with the issue of threatened extinction of other species (if we, people, claim to care about them). This kind of activist or moral engagement is perhaps very different in nature to that of the social scientific discourse the social scientists are used to. Yet refusal to engage with such pressing issues may lead to a marginalization of social scientists in relation to socially or environmentally significant issues.

Contributors to this volume argue that social scientists, with their desire to weigh different interests, views and arguments, could indeed play a very significant role in revealing the underlying social mechanisms behind the problems and solutions to the problems. Social scientists need to continue with their good practice of empirical inquiry into social reality, which shapes the issues, problems and solutions of human and environmental significance. Part of this engagement is the recognition of potential conflicts between human and conservationist interests, as well as between different stakeholder interests.

For the most part, it is precisely the connection between environment and human health that deserves social scientific attention. Remarkably, humanistic focus on health brings to light the topic of environment, and environmental focus on pristine nature has a large relevance to the concern for human health. Contributors to this volume perceive the urgency of engagement with both human-interest and environmental issues. Engaging in contemporary debates on health and environment implies that we need to address social values and understandings surrounding these topics before we dive into discussions from our unique sub-disciplinary perspectives. We shall now briefly examine both health and environment from a bird's eye point of view to reflect upon what is meant by these terms in contemporary academic and laymen's discourse. Interconnections as well as potential conflicts between health and environment will be briefly examined. After this broad overview, we shall examine ways in which our social scientific approaches could help understanding of health and environment in the context of living society and perhaps even provide us with practical insights.

Health

Health, defined in its broadest sense, is the lack disease, sickness, illness, impairment, disability or handicap. The World Health Organization (WHO) defined health as a "state of complete physical, mental and social well-being and not merely the absence of disease and infirmity" (WHO, 1946). Some commonly used definitions stem from the International Statistical Classification of Diseases and Related Health Problems (ICD) formulated by the WHO, which distinguishes between some of the categories of diseases linked by common and preventable biological risk factors, notably high blood pressure, high blood cholesterol and obesity, and by related major behavioral risk factors.

"Health" may be seen as a multidimensional and holistic concept, which includes physical health, psychological health, social health and, inevitably, environmental health (Helman, 2007:127). Besides medical and physical manifestations of disease, there are also social, cultural, psychological and environmental [*sic*] factors that play a significant role in how illness is experienced by sufferers and perceived by others (Verbrugge and Jette, 1994:2). The accent may better lie on the quality of life of the individual within society or at work, rather than on assumed objective manifestations of illness. The International Classification of Functioning, Disability and Health (ICF), developed by the WHO, classifies health-related domains from body, individual and societal perspectives. The severity of symptoms but also other characteristics, such as age or gender, may also play a role in how health and illness are experienced. In line with the Weberian concept of rationality, applied to both Western and non-Western ideas about the cause and cure of the disease, not only the diagnosis, but also treatment of the disease may be subjective. According to the school of symbolic interactionism, people's reactions to drugs are mediated by the beliefs, expectations, and knowledge they have acquired through interaction with other people and through previous experiences. Whether we define health expansively (as did the World Health Organization [WHO] in 1946) or narrowly (like the World Bank did in 1993), health conceptions have a significant impact on the political economy of health care provision.

Berkman et al. (2000) note that the scope of external factors, such as social networks and health behaviors, as well as psychological attributes, play a large role in how our perception of risk, either for health or nature, is constructed. Social influence or supportive functions influence health-promoting or health-damaging behaviors such as tobacco and alcohol consumption, physical activity, dietary patterns, sexual practices, or illicit drug use. Second, social networks via any number of pathways influence cognitive and emotional states such a self-esteem, social competence, self-efficacy, depression and affect. Third, networks may have direct effects on health outcomes by influencing a series of physiologic pathways largely related to stress responses (Berkman et al., 2000).

The recent volume on health issues in the global world (Kirton, 2009) examines the link between global health and international relations. Contributors to the volume assert that the spread of communicable diseases, the challenge of migrating health workers and the development of new technologies and medicines have all contributed to the ever-expanding issue of global health. International organizations, the utilization of techniques such as the creation of the framework convention on tobacco control and the development of civil society organizations such as the Gates Foundation, have all changed the face and framework of global health (Kirton, 2009).

Environment

Environment, interpreted in the most common contemporary (Western) sense, may mean anything from "nature" to "surroundings." When we speak of "nature," we may refer to what in the words of the philosopher and poet Ralph Waldo Emerson, consists of "essences unchanged by man; space, the air, the river, the leaf." When we refer to environment as surroundings, we often speak of influences, contexts and conditions surrounding human existence and activity. While the former definition conceives of nature as independent of the human place in it, the latter definition refers to a dependent construct connected to the human perception of it. To rephrase David Hume's famous dictum, if the tree falls in the forest and nobody hears the sound—has the tree really fallen? In other words, does human perception of nature or surroundings not influence in what is meant (by us humans) by environment?

If we do assume some form of objective reality outside of our human perception of it, we may distinguish between at least four major types of environment: the biophysical, natural, built and social environment. The biophysical environment is conceived as the world outside humans. Natural environment is similar to biophysical one, with the greater emphasis on "ecosystem," including all living and non-living organisms that occur naturally on Earth. The founding fathers of environmental sociology, William Catton and Riley Dunlap (1978), distinguished between built, modified and natural (physical) environments and social environments. The biophysical and natural environments in common discourse are mostly associated with nature or wilderness. The other two types of environment, built and social, refer to, respectively, human-constructed surroundings, including public and private homes and urban landscapes; and to the culture and society that people develop and interact in. In Durkheimean terms, man depends upon only three kids of environment: the organism, the external world and society (Durkheim [1893] 1984:285-286). Recently, sociologists debated relation of different typologies of environment in relation to human society and ecology (Murphy, 2002; Carolan, 2005).

While some see environment as a neutral category, others as a (business) opportunity, yet others feel personally or socially involved with it. Current public discourse often associates environment with "issues" or "problems" that are often lumped together under one umbrella. These include climate change, aridification and desertification (drying up of regions, often associated with erosion), risks involved in nanotechnology, air and water pollution, and many others. While the scope of this book does not have the intention to discuss in detail each of those issues, we shall discuss the controversy around some of these perceived problems in the chapters of our contributors.

Placing environment in the context of globalization, we may note that environment is also presently seen as a commodity or as a public good. Environment in the narrow sense of "nature" (pristine wilderness), which exists independently of the humans (or to use the judicial or moral jargon, having a "right to exist"), is often relegated to the domain of environmentalist groups. Policymakers, developers, and (social) scientists often have divergent perspectives that are sometimes at odds with environmental activists. But it is the interplay between both human and environmental interests that earns a focal point in this volume.

TRACING LINKAGES

As some contributors to this volume have argued, addressing only health problems and their so-called objective basis creates a long-term vicious circle (greater impact of population on the environment, which creates worse health conditions for the populations). But understanding social context that connects both health and environment can help both better health of populations (greater development) AND better environment. One of the many things that social scientists can contribute to the debate and possible solutions is to show how stakeholders and policy actors deal with the uncertainties and complexities that surround the issues of health and the environment (individually and in conjunction).

As Ney argues in his contribution to this volume, at the heart of the definitions and framings of health and environment are divergent normative commitments that shape the way actors produce, interpret and deploy scientific knowledge. There are good grounds to believe that an authoritative pronouncement on the precise relationship between health and the environment is likely to remain elusive. All we ever are likely to get from experts are some pieces of the overall puzzle. The real question is who or what is to shape or determine the process of putting the pieces together.

In order to understand the interdependency between health and environment, the following questions may be addressed: In which ways are health and environment (seen as) connected in the contemporary discourses? Would modern humans care about environmental problems if they did not affect their health? Does "better" environment lead to "better" health? Could addressing environmental problems to treat illness and using illness to address environmental problems be a good example of positive feedback?

We may remember the four types of environment mentioned in the previous section: the biophysical, natural, built and social environment. Contributors to the Dunlup et al. (2002) volume, *Sociological Theory and the Environment: Classical Foundations, Contemporary Insights,* argue that environmental issues are often related to human concerns and are socially constructed. Environment is rarely viewed as an issue of "pure nature" but involves human social processes, such as beliefs and values of social environmental movements (Harper, 2004; Rosa and Richter, 2008). McElroy and Townsend (2003) examine cultural variants in perceiving and interpreting environmental and health topics and discover that there are also some universal features that most societies share in their view of health and environment. The vital basics of life are warmth, food security, freedom from disease, and long life. These basics require a high standard of living, and people are prepared to ignore the environmental impacts of industrialization until the basics are achieved.

Damaging environmental effects are associated with production: heavy industry and large-scale, intensive agriculture necessary to feed large numbers of people. But also with lifestyle: e.g., over consumption in the West creating enormous waste problems and excessive personal hygiene in relation to causing allergies. No form of agriculture is really environmentally friendly because wilderness is eliminated and diversity is largely replaced by crop monocultures (Trawivas, 2001). Recent concerns in the Western countries about the food safety (particularly genetically modified foods, pesticides, color- and taste-enhancing additives and other chemicals in agrarian produce) have led to wide-spread preoccupation with "biological," "ecological" and "health" foods. While debates about the health and environmental benefits of "ecological" or "biological" crops abound, general public, largely informed by the media and "popular science" believe such crops to be the panacea for both environment and humans.

At the level of daily life, we may think of examples when health and environment can "work together." Taking a stairs instead of taking an elevator can be seen as being beneficial to both health (movement, exercise) and environment (saving electricity, conserving energy). Innovative, sustainable production processes replacing unsustainable production processes benefit both health and environment. Recycling schemes might be good for both tackling urban degradation and "helping" the environment (Gutberlet, 2008). Yet, health's relation to environment is often complicated by the multidimensionality of the very definitions of health and environment.

Driving might be said to be bad for health because of the lack of movement of the driver (which may affect their weight as well as muscular and cardio-vascular systems) and inhalation of possibly harmful fumes (which may effect their respiration or even cause cancer). Alternatively, taking a walk in the neighborhood can have a positive effect on cardiovascular and muscular systems and weight loss (Demers, 2005). Driving a car might be said to be bad for the environment for a number of reasons. For one, the very process of car production involves the use of natural resources and is polluting in the process. The even greater consequence of having many cars is that the increasing territory of the planet Earth gets covered in tar (where pristine nature once reigned). Then, there is an issue of non-renewable (as of yet) fuel; and last and not least, the car produces harmful fumes that are absorbed through the air.

Proponents of technological solutions (as the discussion in the following chapter will reveal), believe that human ingenuity can solve most problems. Optimists argue that a new energy source for the cars will be much less harmful (less pollution and fine particles in the air). Car manufacturers and automobilist organizations introduce various schemes encouraging "ecological driving" (http://www.ecodrive.org/), alternative fuels, CO2 taxation, noise reduction, etc. (European Automobile Manufacturers Association).

Environmental activists argue, however, that this new energy will come at a cost and will not solve the problem of continuously increasing production and (over)consumption of cars. Besides, having nuclear energy as an alternative energy source might not be that "safe." Planting monocultures from which oil or whatever other "renewable" fuel can be derived may also have a negative effect on biodiversity (as pristine areas will be cleared to make space for harvested mono-crops). Health proponents would note that the negative effects associated with driving a car, particularly its effect on cardio-vascular and muscular system, will not disappear, either. Technical optimists may (and probably will) come up with the car design that will allow the driver continuous movement and muscle stimulation while driving. It is the more pressing question here, however, in how far these public and expert debates affect decisive policies.

The effects of exhaust fumes (carbon dioxide, fine particles released in the air) ignite both public and policy discussions in industrialized countries. However, the discussions as to what is exactly so harmful about the car waste products is contested among the health, environmental and policy experts (to say nothing of the lay public and the media).

The issue of the environmental illness, however, is highly contested. Not all expert decisions on the nature of the supposedly environmental factors determining disease should be believed without critical and continuous empirical investigation. *Contested environmental illness* thesis is explored in the work of Phil Brown (2007) and Tom Shriver (2008). Health knowledge is often constructed by social actors (Levin and Browner, 2005) and perceived environmental risks to health are subject to subjective interpretation (Brown, 2002). The work of scholars in the field of environmental health, medical sociology and anthropology concentrates on strategies that sick people employ to achieve public recognition of somatic complaints and disabilities. The authors also examine resulting political, legal, and medical conflicts, related to "manufactured environments" (Brown, 2008; Shriver and Cable, 2008; Moss and Teghtsoonian, 2008). Collected volume by Kroll-Smith et al. (2000) examines different instances of illnesses that people attribute to environmental factors:

"Just as environments have adapted in response to human activity, so, too, is the human body now, in turn, forced to adapt to these altered conditions. Today, mysterious illnesses, from chronic fatigue to Gulf War Syndrome, meet us at every turn. Yet, even as an increasing number of people attribute ailments to environmental problems, the suspected relationships between illness and environment remain unclear" (Kroll-Smith et al, 2008).

Not only as social scientists, but also as regular citizens, we can come up with many examples of how both health and environment may be effected by the same threat, such as the "change of climate." However, the objectivity of the existence, the origin and the potential consequences of the threat may be highly contested. We may ask, for example, *Is* there really a climate change and *if* so, who (*if* anybody) is responsible for it (political or ethical question), and what can be done about it (technical question) and why (ethical or philosophical question) and by who (practical question involving all of the above questions)? Similar questions can be asked in relation to cars.

Another example of such contested arguments is the food safety discussion, related to organic and genetically modified (GM) crops. Chemicals, pesticides, hormones, color additives and the like often cause public concerns about the safety of food in developed countries. Expert debates range between natural scientists, genetic engineers, governments and the public about what potential benefits and dangers of such crops may be. Besides these expert debates, there are also philosophical and spiritual concerns lying behind environmental and health opposition to GM foods. Another disputed issue is the effect of carbon dioxide on plants. Some argue that C02 has been proven to make plants grow more rapidly, yet, at the same time, they grow with less nutritional value (Future Print http://www.futurepundit.com/archives/001938.html). Others argue that high levels of CO2 in plants being eaten by animals within human food chains are also lower in nutrition; therefore, most of what we are currently consuming is of less nutritional value. Some attribute the increase in current cases of ADHD, ADD, Alzheimer's, lupus, fibromyalgia and cancer, not to mention the increase in celiac diseases and food intolerances to the failure of human immune system.

Yet another related example of contested illness is that of air pollution and its relation to such diseases as cancer and asthma. Studies in Ontario show how the effects of smog can relate to the higher rates of asthma-related illnesses. The highest of the rates of cancer appear to be concentrated in industrious areas of Northern Ontario, where coal-fired plants are still in operation (http://www.climatechange.gc.ca). Similar findings related to air-way diseases related to air pollution in studies of the effects of poor environmental conditions, including air pollution on children in particular (Dexler, 2008). Some studies show health and environmental affects of concentrated air traffic (Randles and Mander, 2009). Other studies show that air pollution may trigger asthma in some groups (such as young girls) and not in others (Sahsuvaroglu et al., 2009). Yet, other studies show no clear correlation or objectively proven relationship between air pollution and the incidence of cancer or asthma (Brown et al., 2003).

As some contributors to this volume argue (Baralt, Dépelteau, Potrata and Akobirshoeva, Zavestovski), "scientific truths" about environmental illness need to be carefully examined as part of social processes before any policy action is to take place. Sometimes, the very cause of disease is widely disputed by different stakeholders (the illness victims, doctors, policymakers, environmental and health "experts," industrial developers, etc.) and even, at times, consciously misconstrued to make the research results fit within certain agendas. The reluctance to recognize causation may be caused by a number of political considerations and commercial interests (when polluting industries, for example, finance the research into the cause of disease). Contributors to this volume place their chapters in the context of global industrial development and consider health and environment as connected to capitalist system and development concerns.

The Sustainability Paradox

If we consider modern capitalist society to be driven to a large extent by economic growth as the main indicator of human well-being, we can explain part of the feel of panic in the current system crisis also known as the financial crisis. Interestingly though, some of the challenges from a sustainability perspective—less pollution by cars and industry, less waste—seem to go hand in hand with this crisis. So we may wonder, is there a paradoxical relation between the ruling paradigm of economic growth and prosperity and the sustainability of human life on this planet when also care for the environment and nature are to be taken into account (see e.g., Latouche 2003)?

The Brundlandt report (1987) already pointed to the importance of this tension between current economic growth and the future planet's well-being, sometimes labeled as the sustainability paradox (Krueger and Gibbs, 2007; Latouche, 2003). The Brundlandt report defines sustainable development as: "development that meets the needs of the present without compromising the ability of future generations to meet their own needs." The report deals with the change of politics needed for achieving that. It contains within it two key concepts: the concept of "needs," in particular the essential needs of the world's poor, to which overriding priority should be given; and the idea of limitations imposed by the state of technology and social organization on the environment's ability to meet present and future needs.

Proponents of "natural capitalism" which has sustainability at the core of their philosophy, recognize the critical interdependency between the production and use of human-made capital and the maintenance and supply of natural capital:

The traditional definition of capital is accumulated wealth in the form of investments, factories, and equipment. Actually, an economy needs four types of capital to function properly: human capital, in the form of labor and intelligence, culture, and organization; financial capital, consisting of cash, investments, and monetary instruments; manufactured capital, including infrastructure, machines, tools, and factories; natural capital, made up of resources, living systems, and ecosystem services. The industrial system uses the first three forms of capital to transform natural capital into the stuff of our daily lives: cars, highways, cities, bridges, houses, food, medicine, hospitals, and schools (Hawkens et al., 1999: 2).

Literature related to the effect of industrial capitalism on natural environment and health are univocal in the realization that, at the present rate, the "global economy is crashing against the earth: for all the material blessings economic progress has provided, for all the disease and destitution avoided, for all the glories that shine in the best of our civilization, the costs to the natural world, the costs to the glories of nature, have been huge and must be counted in the balance of tragic loss" (Speth, 2008:1).

Proponents of "natural capitalism" (Speth, 2008) or "cradle-to-cradle" principle (McDonough and Braungart, 2002) or "sustainable business" (Kolk and Pinske, 2008) would undoubtedly salute the current global financial crisis (2009) as an opportunity to rethink the very foundations of the capitalist system. Capitalism, as it used to exist up until the present, "is a financially profitable, non-sustainable aberration in human development. What might be called 'industrial capitalism' does not fully conform to its own accounting principles. It liquidates its capital and calls it income" (Hawkes et al., 1999:2). Communism or developing countries' political and economic systems proved to be quite similar to that kind of capitalism.

The awareness of human fragility and the crisis-induced unsuspected critical introspection of the richest global players may yet lead to rethinking of the present situation and turn to much more sustainable practices. Many multinational companies and businesses have been responsive to environmental policies and public pressure groups (particularly, consumer boycotts) since the late nineteen fifties. Corporate social-responsibility-related issues (CSR), such as environment, worker safety and consumer rights, moved from informal social pressure group sphere into national policies (Carrol, 1999: 275; Forman & Jørgenson, 2001, 2005). Innovation and compensation strategies, such as trade in CO2 emissions, have been also popular in the past decennia (Kolk and Pinske, 2008).

Picture 3. Gipsy Girl.

CHARITY PARADOX

Unlike "sustainability paradox," the "charity paradox" outlines conflict rather than complementarity in human and conservationist interests. Within public and media discourse the "charity paradox" is rarely explicit as both environmental and human-related NGOs are seen as doing good, charitable work and commonly contributing to the "better world." Social AND environmental problems are often presented as something that can be solved by the same formula. On the surface of it, the main relevance of stakeholder organizations like Amnesty International, Greenpeace or World Wide Fund for Nature (WWF) is the fact that they represent social movements with specific world views and are dealing with problems in society.

There is a significant percentage of donors donating to "mixed" charities (such as "Friends of the Earth"), which take both human and ecological interests into account. Rockefeller and Ford Foundations claim to have helped to fund the "Green Revolution" through the programs of agricultural research, extension, and infrastructural development. However, health and environment may be seen as polarized in mutually opposing camps in their competition for the state or public attention and funding. Recent research (Wilhelm, 2002) demonstrates that most citizens and philanthropists choose to donate to either human-interest charities (Amnesty International, War Child, Food for Africa, etc.) or environmental charities (World Wide Fund for Nature, Greenpeace, etc.).

Yet, the substantial differences between "human-interest" and "nature-centered" charities lie at the level of their less explicit but quite fundamental goals and objectives. These derive from divergent priorities given to either human OR environmental (independent of human) interests, expressing themselves in either anthropomorphic or environmental bias in the choice of problem-solving strategies. While on the surface of it, NGOs' societal contributions may be seen from a complementary point of view, the "paradox" arises from the often conflicting problem-solving strategies. The "interest groups" or the "stakeholders" involved in creation, inspiration and funding of NGOs may employ problem-solving strategies that may potentially create problems for another group and counteract others' objectives. We may consider an example of the factors affecting the survival of a mere 3,000 tigers in the wild, noting that environmental factors affecting the health (and survival) of tigers (such as poaching and destruction of natural habitat) are often quite different from those affecting human health in the same region. A charity organization that supports human health by combating AIDS, for example, will have very different priorities from the charity concerned with survival of tigers. The growth of the human population as the result of having more people surviving AIDS might be celebrated by human-focus charities and regretted by conservationist ones.

To illustrate the point, we may consider The Bill & Melinda Gates Foundation (B&MGF), the financially largest operated private foundation in the world, founded in 2000, and doubled in size by Warren Buffett in 2006. The primary aims of the foundation are to enhance health care and reduce extreme poverty. Despite recent economic downturn (2009), it presently spends more money on AIDS research and treatment and poverty reduction in the Third World than some of the European countries together. Environmental groups may complain, however, that keeping the (Third) word population healthy (and thereby sizeable) may have devastating effects on the ecology (including the irreversible loss of biodiversity) of the same countries when this goes hand in hand with Western type of unsustainable development. While cautious not to sound callous to the problems of the ailing homo sapiens, many environmental charities might have wished that Bill Gates have chosen other priorities in his charitable spending. Significantly, charity fights as a metaphor could be applied to government spending and shifts in state priorities.

However, we may note that the charities and NGOs also differ within their own "camps." Each charity has a vision and a mission that construct the notions of "health" and "environment" in various ways. The way "nature," for example, is conceived by Greenpeace may be much broader than that of World Wide Fund for Nature (WWF). While the Greenpeace agenda may include opposition to genetically modified crops and nuclear testing, WWF may prioritize protection of pristine environment and biodiversity as its prerogative. The two objectives may overlap, as in the case of planting mono-crops that will affect pristine forests, yet priorities of both groups are often placed differently.

A similar case may be made for charities such as Amnesty International or War Child. While both may be inspired by general principles of protection of human rights, their focuses and priorities differ greatly. Also, the way charities work in regulated, financed and implemented areas differs from organization to organization. The types of donors also differ, in accordance with the income levels, social class and occupation. Webb and Abzug (2008) find evidence that individuals in managerial and military occupations are more likely tc donate to charities than are individuals in other occupational categories. This relates to the relative number of donors rather than the amount of money given by each category. The choice of charity is also dependent on socio-demographic characteristics of the donors (Mesch et al., 2006).

The fundamental difference remains, however, that most human-interest charities may be described as anthropocentric in conceiving of environment in rather functional (resource-related) terms, while environmental charities generally place human interests outside the sphere of their objectives. Environmental charities work to preserve and protect the environment and to promote environmental research, conservation and appreciation. Charity Navigator, America's largest independent charity evaluator, defines two causes for environment charities: environmental protection and conservation, and botanical gardens, parks and nature centers. Health charities can be defined as charities targeted at curing diseases, treating and supporting the sick and disabled, seeking improvements in medical treatments, and promoting public understanding and awareness of particular health risks, diseases and disabilities. Charity Navigator classifies health charities in four causes: diseases, disorders and disciplines; patient and family support; Treatment and Prevention services; and medical research. (http://www.charitynavigator.org/index.cfm?bay=search.categories&category id=5).

Some charities attempt to embrace both humanity and nature by projects such as native people's involvement in maintaining the forest (rather than chopping it down) for eco-tourists, or by encouraging a "return to a traditional way of life" of the indigenous people by allowing them to keep and maintain their lands in their pristine states. Sadly though, many of such idealistic projects partially failed, as eco-tourism turned into mass tourism, or as those "traditional hunters and gatherers" preferred both to hunt and to sell their game to those with less "traditional values." Most "traditional" or "indigenous" folks have often abandoned bow and arrow in favor of Western tools and weapons. Well-meaning charities or NGO workers often fail to recognize the extent to which globalization and, to a certain degree

Westernization, has shaped and altered values, beliefs and ways of life of the so-called traditional cultures. What often shapes human decisions throughout the world is not "traditional culture" and "ancient myths of the forest" but quite prosaic economic interests (Dietrich, 1992). As we shall further discuss in the section on globalization and as some of our contributors reflect, there are hard to find peoples on earth who still live or want to live in "harmony with nature."

To return to the role that the social scientists (may) play in the socially and environmentally significant debates, we need to note that similar to the case of charity or sustainability paradox, the stakeholders within one or the other "camp" are rarely aware of the inherent conflicts as well as possible connections among their viewpoints. Yet, it is precisely through the interaction between humanistic and environmental interests that more holistic understanding of the phenomena as well as solutions to problems can be achieved. As our contributors demonstrate, though, social scientists can help disentangle very complex social issues by deconstructing health and environment framings and interpretations. Social scientists may potentially not only contribute to the (social) scientific debate but also to societal and political debates by structuring deliberation on health and/or environmental problems.

Structure of the Book

Chapters of this volume are divided into three parts: "problem-framing and debates," "social sciences perspectives and "policy and society: in search of solutions." Part one of the volume contains four chapters, addressing the roots of social scientific engagement with health and environment, the main social science disciplines addressing the topics of health and environment, current environment and health problems, as well as various approaches possible in dealing with these topics.

Chapter 1 briefly addresses the history of social sciences' engagement with topics of health and environment. These topics range from a more general discussion about the place of humanity and society within natural environment to more specific topics within sub-disciplines of social science. Examining interpretations of "health" and "environment" concepts and their significance in the social science throws light on the evolution of both medical and environmental knowledge. We argue that inspiration can be found in the works of the founding fathers of social science that may inspire us today when thinking critically or pragmatically about health and environment. Classical works possibly hint at social and cultural solutions to some of health and environment problems. Having discussed some of the classical theories, we briefly address contemporary sub-disciplines of Environmental Health, Environmental and Ecological Sociology, Medical Anthropology and similarities and differences between our own approaches as social scientists.

Helen Kopnina and Hans Keune trace main arguments about interdependency between health and environment. The classical theories discussed in the beginning of the chapter are complemented by the reflection on two main changes occurring both in the world and in the history of thought. In the past hundred years, there has been an advance of medical science and technology as well as shift in the geography of those dispossessed of access to this technology. Assuming that health and environment can "work together," in which ways are they interdependent? Could health and environment be operationalized in the first place, since their definitions vary in time and space? The difficulty in answering these questions lies in the fact that divergent perspectives are often colored by moral, economic and political considerations that change over time.

Hans Keune and Nik Van Larebeke focus on the question: Why is scientific information on serious environment and health problems not easily recognized in our society? In part one, they focus on environment and health science. First, they introduce current environment and health problems based on scientific knowledge. They then look into the issue of scientific proof and its problems related to traditional statistical analysis. Finally, they introduce Physics Chemical Hygiene as an alternative approach. In part, two they focus on the complicated relation of environment and health science and society. They discuss the problem of economic and social consequences of recognizing health problems caused by environmental hazards. They present a Flemish example in which mysterious health problems may have environmental causes, but where scientific research seems complicated both from a scientific and policy perspective. They also present a case in which scientific problem knowledge does not easily lead to policy action. Finally, they pose the problem of social choice related to environment and health: wealth or health. In part three, they focus on the social fabrication of environment and health. They present the issue of biased expertise. They illustrate this with the example of scientists working secretly for industry and the example of growing influence of industry experts in official risk assessment. Finally, they focus on how complicated it is to apply expert knowledge on environment and health to personal daily life.

By relying on the case study of the Sudbury Soils Study (Canada) and the works of U. Beck and B. Latour, F. Depelteau explains that the level of public trust towards risk assessments is related to definitional struggles and democratic problems. In brief, he shows that definitional struggles and the low level of public trust towards this risk assessment comes are related to two important democratic issues: i) the central role played by private interests (two mining companies in this case) in the realization of the risk assessment, which create a situation of conflict of interest, and ii) the lack of key resources available to citizens who cannot evaluate the scientific validity of the risk assessment by themselves due to their limited scientific skills and knowledge. In this context, these two democratic problems reinforce each other in terms of definitional struggles and lack of trust: since they cannot evaluate the risks by themselves, residents of Sudbury are asked to *believe* the conclusions of experts, which are not seen as being independent from political and economic interests.

Ney's chapter discusses how stakeholders define and frame the relationship between health and environment. Few people would argue that health and the environment are intricately related. It stands to reason, one may think, that as long as we depend on the environment or, better, environments, for our sustenance, both physical and social, then the state of these environments is likely to have an impact on our health. And yet, the precise ways in which the environment affects health—and vice versa—remains shrouded in uncertainty. One thing, however, does seem clear: whatever the relationship, it is unlikely to be straightforward.

Given this degree of uncertainty and complexity, how do stakeholders define and frame the relationship between health and environment? Using narrative analysis, this chapter systematically compares the stories contending policy actors tell about health and the environment. The chapter compares contending policy narratives in terms of how divergent underlying assumptions of the relationship between health and the environment structures leads to very different and, ultimately, policy recommendations.

By analysing policy positions of key stakeholders at national and international levels, the chapter identifies three contending policy-story archetypes. Each narrative is based on a specific set of assumptions, identifies a particular causal relationship between health and the environment, as well as recommends specific policy measures. The first policy story, one told by market-oriented stakeholders, argues that human health— infinitely adaptable and flexible as it is—thrives when we efficiently and effectively make use of the environment. The second narrative, associated with environmental groups, tells us that human health is an integral and irretrievable part of the environment: in this view, it is the degradation of the natural world that makes us sick. The third policy discourse tells a story of how human health and the environment are in a delicate balance that can only be maintained by rational and expert-driven management of the sort only well-organised institutions can provide.

In the final section, the chapter explores the implications for policymaking of this argumentative and conflictual policy process. In particular, the chapter explores ways of harnessing the creative and innovative energies of this policy conflict while avoiding the tendency for policy impasse as debate deteriorates into an unresponsive shouting match.

In the second part of the book, social sciences' perspectives on health and environment are examined from the points of view of different sub-disciplines, such as environmental health, medical anthropology, environmental and ecological sociology.

Gary Bowden's chapter traces the rise and institutionalization of environmental sociology in order to answer the following questions: A) Has the emergence of environmental sociology forced a fundamental restructuring of sociological theory? B) If not, what have been the blockages to such a development, and do they still exist? C) What would an ecological (as opposed to environmental) sociology look like? And D) what are the implications of ecological sociology for our understanding of environment and health issues?

The rise and institutionalization of environmental sociology represents one of the most significant changes to the discipline of the past quarter century. William Catton and Riley Dunlap, the sub-discipline's founders and most tireless advocates, developed a concept of New Ecological Paradigm (NEP) (Catton and Dunlap, 1978). NEP stressed that humans were still as ecologically interdependent as other species was a departure from "Human Exceptionalism Paradigm" (HEP). Underlying this success, however, is a deeper question: Do these developments represent the first step down a longer path toward more fundamental theoretical change or have the changes to date delivered us to sociological acceptance or at least the recognition of NEP? The first three sections of this chapter combine to argue the former—that fundamental theoretical change is still required. The first section distinguishes between environmental and ecological sociology and advocates the latter. The second section examines current sociological theorizing about the environment and concludes that the label "environmental sociology" accurately describes the current state of affairs. The third section identifies three conceptual blockages that have, in the past, prevented the development of ecological sociology. Recent scholarship has removed each of these blockages, thus paving the way for the development of ecological sociology, the characteristics of which are briefly described in the fourth section. The fifth and final section examines the implications of ecological sociology for our understanding of health.

Lori Baralt's chapter describes the complexity of dealing with "environmental" determinants in the case of breast cancer. Baralt explores the dominant biomedical paradigm and the challenges that social scientific environmental health perspectives pose to this model. The two predominant charity and advocacy approaches to breast cancer are used as examples to demonstrate the conflict between the two perspectives outside of the realm of social science and the potential role of social scientists in furthering environmental health perspectives. While both approaches to breast cancer are anthropocentric in that they maintain the ultimate goal of the eradication of breast cancer as an illness, the dominant charity form of breast cancer advocacy relies on a biomedical model of health, which ignores the environment altogether, while the alternative environmental breast cancer model is a hybrid advocacy model that highlights the inextricable connections between human health and human-induced environmental harms. Embedded within a sociological framework, this chapter begins with an overview of the main tenets of the biomedical paradigm of health and illness. The author then describes challenges posed to the biomedical paradigm by the sociological field of environmental health studies, which has developed at the intersection of medical and environmental sociology. The divergent charity and advocacy approaches to breast cancer are then explored within this context. Finally, the role of social scientists in contributing to positive social change for the well-being of the environment and human health are then discussed.

Alexander Stingl reflects on similar issues associated with the difficulty of discussing the objectivity of environmental factors supposed to be responsible for Attention Deficit Disorder (ADHD). This chapter focuses on "health and environment"-aspects in a genealogical analysis of the discursive production of the concept *Attention* and its implications as a public, medical and psychological problem. Attention itself is, however, a vague concept, despite its frequent use as a medical and psychological category. It has a two-hundred year history, beginning with Kantian psychology/biology. In psychology, it occupies different roles, and so do its pathologies. Neurologically, it is still unclear what attention actually is. Currently, it is assumed that there is a balance of different neurotransmitters that seems to play a role in the neurochemistry of attention. But while the concept itself is vague, doctors, parents, journalists, etc. use it in statements concerning normal and pathological forms of attention. They treat it as if it were self-evident and clear what attention is.

Barbara Potrata and Anzurat Akobirshoeva provide a medical anthropology perspective, asking how anthropology can throw light on health and environment topics. Their chapter addresses the question of whether debates within the areas of health and environment can inform the process of social inquiry? How can methodology from social science aid our understanding of environmental and health topics? Traditional medicine and native knowledge are discussed in the context of harsh political and social environment of Central Asia. The poorest of Central Asia cannot help themselves because the "nature" (for traditional medicine) is no longer there, and because of ideological hegemonism (which claims that biomedicine is superior to traditional medicine, and if TM is tolerated, it is only as a repository of potential products with active ingredients). The western model of the "science of healing" is critically examined and debatable "myths," "facts" and aspirations involved in health and environmental problems are discussed. The authors argue that religions can contribute to environmental preservation.

In the third part of the book, policy-related debates are brought to light. Climate change, renewable energy, population growth, threats to biodiversity, ecological agriculture and biological nutrition, air and water pollution, and (chemical) waste are examined in turns. We note that public, as well as policy, debates differ greatly among the so-called industrial, industrializing or developing countries. The geography and distribution of health and environment debates shift constantly in accordance to global power balances and various issues that affect the complex system of trans-national interdependent actors.

Often guided by economic and political interests, many Western countries have seen a shift in policies in favor of sustainable use of natural resources. The chapters in the third part of the book address the following questions: In which ways are health and environment connected in the public and policy discourses? Would social actors care about environmental problems if they did not affect their health? Does "better" environment" lead to "better" health? Addressing environmental problems to treat illness and using illness to address environmental problems are provided as examples of the positive feedback.

The Hans Keune et al. chapter provides a social scientific view on the complexity of health and environment topics. The authors discuss the relevance of a complexity approach to environment and health research. They discuss two cases of complexity in relation to environment and health in order to illustrate what complexity is and how we may deal with complexity. They introduce the case of climate change by focussing on the work of the Intergovernmental Panel on Climate Change and their practical experience as social scientists in Centre of Expertise for Environment and Health in Flanders (Belgium). They want to illustrate how a complexity approach is helpful to the field of environment and health research and that social science can make a relevant contribution to solving some of the problems. The approach presented is characterized by modesty, respect and transparency with regard to complexity of reality, and openness to different relevant forms of knowledge and opinion. The approach criticizes the traditional interpretation of objectivity and emphasizes the inter-subjectivist nature of science. The discussion exemplifies the need of a social science contribution. Moreover, it illustrates the complexity of such an endeavor in practice. The approach will not make work easier in all respects. But it will enhance the quality by respecting the complexity of real world problems.

Helen Kopnina's chapter addresses the role of European policies and interest groups in shaping public debate on health and environment. The role of the European Commission, national environmental ministries and agencies as well as medical and environmental organizations in financing research into health is brought into focus. Resulting policies are discussed in the contexts of globalization in which (European) businesses work together on "socially and environmentally responsible" projects. This chapter examines the role of European institutions as well as lobby and interest groups in shaping health and environmental policies. The chapter describes the evolution of health and environmental policies at the European level, as well as outlines the necessity of social scientific research into the national-level policies, their interaction with European policies and the variability of outcomes thereof. Finally, the author ponders the role that social scientists can play in studying the work of political and interest groups, by evaluating the effectiveness of European policy at the national grass-roots level, and by engaging in multi-disciplinary debate with other stakeholders, such as policymakers and health and environmental specialists.

Stephen Zavestoski's chapter begins by briefly reviewing the evidence pointing to the global spread of environmental hazards over the last half of the 20th century. The next section examines evidence that just as economic globalization was leading to the global spread of environmental hazards, a parallel process was occurring in which social movements to resist environmental health threats were becoming increasingly global. The chapter's primary focus is an examination of three types of obstacles that these movements face as they attempt to prevent future exposures to environmental health hazards and hold accountable those responsible for past exposures. Legal obstacles include the absence of legal precedent for prosecuting multinational corporations, the challenge of acquiring expertise to engage legal systems in countries hosting the corporations responsible for an exposure event, and the development of legal framework at the international level that can minimize or prevent future exposures. Political obstacles include weak regulatory regimes at the domestic level, the favoring of economic development over human development, and limited ability to mobilize constituents outside of affected areas. Economic obstacles range from the costs entailed in monitoring and documenting health impacts in a community to the lack of monetary resources needed to launch trans-national campaigns. The analysis focuses on how social movements use trans-national organizing as a means of overcoming these obstacles.

In the conclusion to this book, we look back at the contributions to this volume by focusing on the problem-framing and problem-solving perspectives. A main critique of most contributors on the current state of the art of environment and health science is that important elements are not taken into account, thus resulting in a limited view on environment and health issues. First, we focus on how contributors want to broaden and deepen the health and environment science by taking into account complementary aspects of the complexity of environment and health issues. In achieving the objective of enriching the view on environment and health through this complementary social scientific lens, several obstacles are noticed. These obstacles arise from theoretical and social interpretations of the concepts of health and environment as well as from socio-political and economic aspects. From a problem-solving point of view, we highlight how contributors seek to overcome these obstacles and to what extent theoretical insights can be applied in practice. Finally, we reflect on the role of social science in relation to health and environment from the point of view of the contributors to this volume, and we discuss future research perspectives.

REFERENCES

Berkman, L. F., T. Glass, Brisette, I., T. E. Seeman (2002). From social integration to health: Durkheim in the new millennium. *Social Science & Medicine*. Vol. 51, Issue 6, 15 September 2000, Pages 843-857.

Brown, P. (2007). *Toxic Exposures: Contested Illnesses and the Environmental Health Movement*, Columbia University Press. New York.

Brown P. (2002). Preface to Special Issue on Environmental Health in *The Annals of the American Academy*, 584:7-12.

Brown, P. Zavestoski, S. Luebke, T., Mandelbaum, J., McCormick, S. and Mayer, B. (2003). The Health Politics of Asthma: Environmental Justice and Collective Illness Experience in the United States. *Social Science and Medicine* 57:453-464. Reprinted in Where We Live, Work, and Play: A Critical Appraisal of the Environmental Justice Movement, David Pellow and Robert Brulle, (Eds. Cambridge: MIT Press.

Brundtland, G. (Ed.) (1987). *Our Common Future: The World Commission on Environment and Development*, Oxford: Oxford University Press.

Catton, W.R., & Dunlap, R.E. (1978). Environmental Sociology: A New Paradigm. *The American Sociologist*, 13, 41-49.

Carolan, M. S. (2005). Society, Biology, and Ecology: Bringing Nature Back into Sociology's Disciplinary Narrative through Critical Realism, *Organization and Environment*, 18, 393-421.

Carrol, A. (1999). Corporate Social Responsibility *Business & Society*, Vol. 38, No. 3, 268-295 http://bas.sagepub.com/cgi/content/abstract/38/3/268.

Charity Navigator http://www.charitynavigator.org/index.cfm?bay=search.categories&categoryid=4.

Demers, M. (2005). *Walk For Your Life: Restoring Neighborhood Walkways to Enhance Community Life, Improve Street Safety, and Reduce Obesity*, by Vital Health Publishing, 2005.

Dietrich, W. (1992). *The Final Forest: The Battle for the Last Great Trees of the Pacific Northwest*. Penguin Books, New York.

Drexlar, M. (2008). *The People's Epidemiologists. Sibling specialists attack the political and economic causes of health disparities*. http://harvardmagazine.com/2006/03/the-peoples-epidemiologi.html.

Dunlap, R. E., F. H. Buttel, P. Dickens and A. Gijswijt (2002). *Sociological Theory and the Environment: Classical Foundations, Contemporary Insights*. Rowman & Littlefield Publishers.

Dunlap, R.E. (2002). Environmental sociology: A personal perspective on its first quarter century. *Organization & Environment*, 15(1), 10-29.

Durkheim, É. (1893) 1984. *The Division of Labour in Society*. Basingstoke: Macmillan.

Durkheim, É. (1950). *The rules of sociological method*. New York: Free Press. (Originally published 1895).

Durkheim, É. (1964). *The division of labor in society*. New York: Free Press. (Originally published 1893).

Durkheim, É. (1972). *Selected Writings*, edited by A. Giddens, Cambridge: Cambridge University Press.

European Automobile Manufacturers Association http://www.acea.be/

Forman, M. and M.S. Jørgensen (2001). The social shaping of the participation of employees in environmental work within enterprises—experiences from a Danish context, *Technology Analysis and Management*, Vol. 13, No.1, pp. 71-90.

Forman, M.; Jørgensen, M.S. (2004). Organising Environmental Supply Chain Management In: *Greener Management International*, Issue 45, Spring 2004, pp. 43-62.

Future Pundit (2004). *Plants Will Grow More Rapidly With Higher Carbon Dioxide* (February 17) http://www.futurepundit.com/archives/001938.html .

Gutberlet, J. (2008). *Recovering Resources—Recycling Citizenship*. Ashgate, Abingdom.

Harper, C. (2004). *Environment and Society: Human Perspectives on Environmental Issues*. Upper Saddle River, New Jersey: Pearson Education, Inc.

Hawken, P., Lovens, A. and Lovens, L. H. (1999). *Natural Capitalism: Creating a new Industrial Revolution*. Little, Brown and Company.

Helman, C. G. (2007). *Culture, Health and Illness*. Hodder Arnold, London.

International Organization of Motor Vehicle Manufacturers http://www.oica.net/

International Classification of Diseases (ICD) World Health Organization http://www.who.int/chp/about/integrated_cd/en/index.html.

Kirton, J. J. (2009). *Global Health*. Ashgate, Abingdon.

Kolk, A. and Pinske, J. (2008). *International Business and Global Climate Change*. Routledge.

Kroll-Smith, S., P. Brown, and V. J. Gunter (Eds.) (2000). *Illness and the Environment: A Reader in Contested Illness*, New York University Press.

Krueger R. J. and Gibbs D. (Eds.) (2007). *The Sustainable Development Paradox: Urban Political Economy in the United States and Europe*, The Guilford Press.

Latouche S. (2003). *Sustainable Development as a Paradox*, presentation at the Religion, Science & The Environment June 2003 Baltic Sea symposium, http://www.rsesymposia.org/files.php?catid=76&pcatid=46.

Latouche S. (2003). Would the West actually be happier with less? The world downscaled, In: *Le Monde diplomatique*, December 2003.

Levin, B. W. and C. H. Browner (2005). The social production of health: Critical contributions from evolutionary, biological, and cultural anthropology. *Social Science & Medicine*, Vol. 61, Issue 4, August, Pages 745-750.

McDonough, W. and M. Braungart (2002). *Cradle to Cradle. Remaking the way we make things*. North Point Press, New York.

McElroy, A. and P. K. Townsend (2003). *Medical Anthropology in Ecological Perspective*. Westview Press.

Mesch, D. J., P. M. Rooney, K. S. Steinberg, and B. Denton (2006). The Effects of Race, Gender, and Marital Status on Giving and Volunteering in Indiana *Nonprofit and Voluntary Sector Quarterly,* December 2006; vol. 35: pp. 565 - 587.

Moss, P. and Teghtsoonian, K. (2008). *Contesting Illness: Processes and Practices*. University of Toronto Press Inc., Toronto.

Murphy, R. (2002). The Internalization of Autonomous Nature into Society. *Sociological Review*, 50, 313-333.

Randles, S. and Mander S. (2009). Aviation, consumption and the climate change debate: *Are you going to tell me off for flying* 21, 93-113. Tyndall Center for Climate Change Research. http://www.tyndall.ac.uk/publications/pub_list_2009.shtml.

Rosa, E. A. and Richer, L. (2008). Durkheim on Environment. *Organization & Environment*, vol. 21, No. 2, 182-187.

Sahsuvaroglu, T., Jerrett, M., Sears, M. et al. (2009). Spatial Analysis of Air Pollution and Childhood Asthma in Hamilton: Canada: Comparing exposure methods in sensitive subgroups. In *Environmental Health*. April 8:14. pp1-13. http://www.ehjournal.net/content/8/1/14.

Shriver, T. and S. Cable (2008). The Institutional Context of Gulf War Illness Claims: A Commentary on Cohn, Dyson and Wessely. *Social Science and Medicine* 67(11): 1650-1653.

Speth, J. G. (2008). *The Bridge at the Edge of the World. Capitalism, the Environment, and Crossing from Crisis to Sustainability*. Yale University Press.

Trewavas, A. J. (2001). The population/biodiversity paradox. Agricultural efficiency to save wilderness. *Plant Physiology* 125, 174–179.

Trewavas, A. J. (2002). Malthus foiled again and again, in *Nature*, August 418, 668-670.

Verbrugge, L. M. and A. M. Jettie (1994). The Disablement Process. *Social Science Medical*, vol. 38, No. 1. pp. l-14.

Webb, N. J. and Abzug, R. (2008). Do Occupational Group Members Vary in Volunteering Activity? *Nonprofit and Voluntary Sector Quarterly,* Dec 2008; vol. 37: pp. 689 - 708.

World Health Organization (WHO) (2006). *Preventing disease through healthy environments: Towards an estimate of the environmental burden of disease*.

World Health Organization (WHO)(1995, 2004, 2005, 2006, 2007, 2008, 2009). *The World Health Report* http://www.who.int/whr/en/index.html.

Wilhelm, (2002) *The Distribution of Giving in Six Surveys*. The Working Paper, Indiana University //www.philanthropy.iupui.edu/Research/WorkingPapers/SurveyComparisons-Wilhelm.pdf.

PART I. PROBLEM-FRAMING AND DEBATES

Picture 4. Windmills.

Picture 5. Shore.

In: Health and Environment: Social Science Perspectives ISBN 978-1-60876-216-3
Editors: Helen Kopnina and Hans Keune

Chapter 1

SOCIAL SCIENTIFIC ENGAGEMENT WITH HEALTH AND ENVIRONMENT DEBATES: CLASSICAL ROOTS AND CONTEMPORARY SUB DISCIPLINES

Helen Kopnina and Hans Keune

ABSTRACT

In this chapter, we first sketch a historic picture by introducing health and environment from the perspective of some classical scholars of social science: Malthus, Marx, Durkheim and Weber. Main issues being raised in this respect are the relation between the natural and the social, human world on the one hand and the epistemological debate on objectivity and truth on the other. After the historical roots, we introduce modern developments in health and environment in order to sketch the background for modern social science's involvement in the field. Departing from the industrial revolution as a major turning point, we focus on issues of globalization, population growth and other developments in health and environment conditions. Finally, we introduce modern social science's involvement in health and environment. Subsequently the sub disciplines of environmental sociology, ecology of health, ecological sociology, environmental health and medical anthropology will be briefly introduced.

CLASSICAL THEORIES AND CONTEMPORARY INTERPRETATIONS

In order to comprehend the link between health and environment within social sciences, we will address various social scientific disciplines. The roots of these disciplines lie in classical philosophy, medicine, economics, political science and natural science. We shall start this chapter by briefly addressing the perspectives of some "classical" (mainly sociological) social scientists; Malthus, Marx, Durkheim and Weber on the relationship between the natural and social worlds. The grand questions occupying these classical scholars were partly related to the arguable uniqueness of human nature and human society and the place of humanity within the natural world.

Thomas Malthus

For Reverend Thomas Malthus (1766-1834), the natural world is related to the "natural state" or "natural law" governing human populations. Humans are seen as part of nature since it "is the constant tendency in all animated life to increase beyond the nourishment prepared for it" (Malthus, 1798: 5). In the following centuries, it was evolutionism that brought the distinction between mind and body into question: if man is considered a person for social purposes, he remains an organism from a biological point of view. Looking back once again, one sees Malthus as the source of the view of nature that led to Social Darwinism —the social struggle for existence, the survival of the fittest (Young, 1994).

In his *An Essay on the Principle of Population*, Malthus postulated that the growth of the human population will be inevitably bridled by poverty, war and disease. According to Malthus, subsistence severely limits population, and the increase in subsistence causes expansion in population. Population pressures stimulate increases in productivity and development of technology, which, in turn, stimulate further population-growth. Since this productivity cannot keep up with the potential of population growth for long due to limited natural resources, population requires strong checks to keep it in line with the carrying-capacity. Social and cultural adaptations (division of labour, child-caring practices, etc.) determine the expansion or contraction of population and production. Natural causes (such as accidents and old age), misery (war, pestilence, plague, and, above all, famine), vice (infanticide, murder) or contraception keep population in check. This implies that a segment of every human population is relegated to poverty and sickness ("the whole train of common diseases and epidemics," [Malthus, 1798: 14]) at one time (Trewavas, 2001). We may thus note that within Malthusian thought, human health was actually curtailed by "natural laws," and while these laws do not directly relate to "nature," modern terminology might translate them as "environmental determinants."

Reflective of the classical debates on the nature of "facts" and "knowledge," the majority of scholars within contemporary social science disciplines still position themselves within "realistic" (sometimes referred to also as "positivist") or "constructionist" (sometimes referred to as "interpretive") traditions. While constructionists concern themselves with social construction of environment or health, realists accentuate bio-ecological and physical aspects affecting both natural and human well-being. While realists may see humans as part of nature or as being "animals" themselves (recall the famous Freudian dictum "Das Mensch ist ein Tier"), constructionists focus on the uniqueness of human beings. From a realist perspective, both human health and environment (in a sense of "nature") can be defined in objective, biological terms. From a constructionist perspective, it is hardly possible to designate an "objective" (independent of human persecution of it) way of viewing something as large as society and/or nature. "Eccentric" position of humans implies that we both are the bodies and are in possession of our bodies. Before we address (post)modern conceptions of health and environment, it is worthwhile to address the classical schools of thought by trying to determine how these divergent perspectives were formed.

The relationship between scarce resources and a continuously increasing population turns the arguments in favor of population control into "self-evident truths," which can only be rejected by the unthinking and the dogmatic. From a Marxist viewpoint, such "self-evident truths" are but reifications of concrete historical, social, political, and economic relations that should be taken into account if the population issue is to be at all understood. Just as in the 18th century when the English ruling classes fought the impact of the French Revolution with military and ideological weapons among which Malthus' *Essay on Population* was perhaps the most important, today, the ruling classes are bringing back the Malthusian argument in an effort to increase their control over the growing number of the dispossessed. Like Malthus, contemporary socio-economic theorists view excessive population rather than social institutions and social relations as the main source and barrier to the solution of social problems. It is, therefore, the purpose of this paper to present a Marxist critique and a Marxist alternative to the Malthusian and Neo-Malthusian approaches to the study of population (Gimenez, 1973).

Malthus focused his analysis of capital accumulation and its effects upon specific sectors of production instead of looking at the relationship between total social capital and the total labor force (Gimenez, 1973). Contemporary scholars criticized Malthus for underestimating the technical developments and scientific innovation that allowed for much greater ability to foster and produce resources. The Malthusian idea that population growth is exponential while the growth of the food supply was arithmetical was proven wrong by many consequent scholars. Trewavas (2002) points out that since 1800, global food production has generally kept pace with the population growth, but increasing population did call for new ways "to increase yields while preserving natural habitats and biodiversity." Although the world population at the time of the French revolution was about one billion, it doubled by the middle of the nineteenth century. Concerns grew that a growing world population would once again be plunged into Malthusian disaster. Yet, new engineering technology allowed more advanced techniques of soil fertilization, substituting guano (seabird droppings) and potassium nitrate (saltpetre) for more durable fertilizers (Trewavas, 2002: 668).

Bowden, one of the contributors of this volume, describes conflicting views on scarcity and human capacity as a division between economic and biological conceptions. "The economic optimists argue that social arrangements—typically free markets, science and liberal democracy—provide the institutional arrangements that allow humans to solve any problem they confront. Economic markets raise capital and provide incentives for entrepreneurs to solve problems; science provides knowledge about the functioning of the natural world, knowledge that can be applied to the problem; and democracy allows participation by wider segments of the public, thus enhancing the probability that a solution will be found. In contrast, the neo-Malthusians...hold that the demand for resources, stemming from population growth and the increasing standard of living, grows exponentially while the ability to provide them grows arithmetically. As a result, there exist strict biophysical limits on resource supply—the carrying capacity of an ecosystem or the biosphere" (Bowden, this volume).

Homer-Dixon (1999) recognized that economic optimists and neo-Malthusians offered partial, or perhaps better put, conflicting "truths" in recognizing the deficiencies and complementarities of both theories. Homer-Dixon reasoned that through the application of ingenuity, humans can substitute for resources that are becoming physically exhausted, yet on the other hand, ingenuity cannot solve all problems. Proponent of "Natural Capitalism" Paul Hawken reflects:

> It is not the supplies of oil or copper that are beginning to limit our development but life itself. Today, our continuing progress is restricted not by the number of fishing boats but by the decreasing numbers of fish; not by the power of pumps but by the depletion of aquifers; not by the number of chainsaws but by the disappearance of primary forests. While living systems are the source of such desired materials as wood, fish, or food, of utmost importance are the *services* that they offer, services that are far more critical to human prosperity than are nonrenewable resources. A forest provides not only the resource of wood but also the services of water storage and flood management. A healthy environment automatically supplies not only clean air and water, rainfall, ocean productivity, fertile soil, and watershed resilience but also such less-appreciated functions as waste processing (both natural and industrial), buffering against the extremes of weather, and regeneration of the atmosphere (Hawken et al., 1999:2).

The real problem is not the biological limits of carrying capacity but, rather, the "ingenuity gap" that emerges when the requirement for ingenuity brought about by problems of greater and greater complexity cannot be met by the level of ingenuity a society is capable of supplying. Efforts at solving climate change could exemplify the "ingenuity gap." Optimists argue that new technologies will be able to reverse the global warming by cutting the rate at which fossil fuels are turned into carbon dioxide. Geo-engineering projects intended to reduce negative effects of climate change on health and environment were recently evaluated by a group of researchers from Britain's National Oceanography Center (Pollard, 2009) and from the Tyndall Center for Climate Change Research (Randles and Mander, 2009); they published very skeptical reports on the effectiveness and likely negative side-effects of most currently popular geo-engineering projects. The sobering conclusions of these reports suggest that there are no easy, cost-free and risk-free solutions to the climate problems. Further discussion of the "ingenuity gap" can be found in Bowden's chapter in this volume.

Yet another issue emerging from the tension between economic optimists and neo-Malthusians, is the belief, on the one hand, that health and environmental problems can be solved by human ingenuity, and that the problems affecting humanity and nature are much greater than our capacity to address them, on the other hand. The in-between view may be proposed in which not all problems can be solved by the same formula, and that, in fact, some solutions to the problems (AIDS in developing countries) may actually exacerbate other problems (overpopulation and environmental degradation).

Malthusian ideas influenced the work of the English naturalist Darwin (1809-1882) and Marx (1818-1883). Darwinists embraced some of the Malthusian ideas, particularly those related to natural law governing both human and animal population (Desmond and Moore, 1991). Marx and neo-Marxists, however, were more skeptical. In his footnote on Malthus in Capital, Marx reflects that the "great sensation" [*An Essay on the Principle of Population*] caused was due solely to party interest. The French Revolution had found passionate defenders in the United Kingdom; the "'principle of population,' slowly worked-out in the eighteenth century, and then, in the midst of a great social crisis, proclaimed with drums and trumpets as the infallible antidote to the teachings of Condorcet…was greeted with jubilance by the English oligarchy as the great destroyer of all hankerings after human development."

Karl Marx

Marx and Engels saw in Malthus' principle of population another instance of the way in which bourgeois economists manipulate social relations. According to Gimenez' (1973) analyses, Marx's answer to Malthus' principle of population is the principle of the reserve army of labor or relative surplus population that he elaborates in the course of his analysis of the general law of capital accumulation (Marx, 1970:612-712). Marx's analysis shows that whenever the reserve army of labor is relatively depleted and the level of wages tends to rise reducing the rate of surplus value, the capitalist class will adopt measures such as technological improvements and foreign investments (Gimenez, 1997).

According to Foster's analysis, imposition of the treadmill on English workers symbolized for Marx the tendency of the capitalist mode of production to degrade the work and, hence, the worker in mind and body (Foster, 2005:9). Post-industrialist writers extended the treadmill metaphor to include the "treadmill of accumulation" (Foster, 2005) and identified capitalism as the key to understanding the increased use of natural environment and resources. Schnaiberg (1980), for example, developed a conflict theory on human-environment interaction, arguing that modern capitalism driven by profitability obeys the imperative to grow and attract investments to survive in a competitive market. Starkloff (1995) argues that this "ecological" view of nature as a victim of human activity is opposed to Marxian and generally "modern" thinking. According to Starkloff, Marx sees nature as a human adversary, as discontinuous with culture and an opponent of laboring classes (Starkloff, 1995:99).

Marx analyses poverty and disease in the context of capitalism and class struggle resulting in innovation, rather than governed by natural laws. While at the ethical level, Malthus was criticized for lacking a caring attitude towards the situation of the poor, Marx showed greater moral concern about the well-being of fellow humans and saw their relationship with nature as an ambiguous one. For Marx, there is a duality in human and nature relationships: "man and his work on one side, nature and matter on the other" (Marx, 1973: 198). On the one hand, there is interdependency; on the other hand, dependency, which humans throughout history try to overcome (Starkoff, 1995:100). In Paris Manuscripts, Marx describes human beings as part of nature:

> Nature is man's inorganic body—nature, that is, insofar as it is not itself human body. Man lives on nature—meaning that nature is his body, which means he must remain in continuous interchange if he is not to die. That man's physical and spiritual life is linked to nature means simply that nature is linked to itself, for a man is part of nature (Economic and Philosophic Manuscripts of 1844, Collected Works Vol. 3:276).

In Marxist terms, hunting and gathering societies are ruled by the blind forces of nature, while industrial societies are characterized by their "rationality regulating their interchange with nature" by "bringing it under their common control" (Marx, 1978:441). Marx sees class repression as reduction of uniquely human capacities to those of animals:

> We arrive at the result that man (the worker) feels himself to be freely active only in his animal functions—eating, drinking and procreating, or at most also in his personal adornment—while in his human functions he is reduced to an animal. Eating, drinking and

procreating are of course also genuine human functions. But abstractly considered, apart from the environment of human activities, and turned into final and sole ends, they are animal functions (Marx, p. 125).

Marx classified tools as "means of nature" which are "means of power" that control nature (Marx, 1973:192). The transformative capacity of the bodies are shared by all nature's creatures, but the human use of tools as extension of their bodies make humans unique in nature (Ibid, 194).

Durkheim and Weber

Järvikoski (1996) compares Marx's and Durkheim's (1858-1917) perspectives on nature and society. He argues that the view of nature as a social category, found in Marx as well as Durkheim, is extremely important in environmental sociology. For Marx, man's relation with the natural world is not theoretical but practical (Järvikoski, 1996). Like all creatures in nature, man is bound by his physiological needs; and unlike other creatures, man possesses reason and consciousness. Human activity also transforms nature, and at the same time, human beings transform themselves. Marx sees "Man, a part and product of nature, makes nature a part of himself; it is at once the subject-matter of his activity and a prolongation of his body" (Kolakowski, 1978:137).

Rosa and Richter (2008) argue that it might profit scholars to re-visit Durkheim, not as the party guilty for dismissing the environment, but as a foundation for understanding the dynamics between human and environmental systems. For Durkheim, Rosa and Richter argue, environment is a precondition of social life:

Man depends upon only three kinds of environment: the organism, the external world and society. If we set aside chance variations due to the combinations of heredity… the organism is not modified spontaneously; it must be constrained to do so by some external cause. As for the physical world, from the very dawn of history, this has remained appreciably unchanged, if at least we take no account of innovations of social origin. (Note: transformation of the soil, the waterways, by the skills of farmers, engineers, etc.) (Durkheim, [1893] 1984:285-286).

Similarly to Durkheim, Weber (1864-1920) saw human agents as most significant in influencing the transformation of nature. The "disenchantment of the world" caused by the industrial revolution, led to the loss of both cultural and natural diversity. Weber differentiated human actions into traditional, affectional, value-rational and instrumental (Rhoads, 1991). These actions were often seen as divorced from nature:

> The trope of nature may seem a distinctly unpromising approach to a thinker so firmly focused on the specifically cultural—and, indeed, political—dimensions of human life. Even in those works where Weber's attention is turned to the pre-modern world, he is more concerned to emphasize the ways in which men shape and are shaped by their societies than their proximity to the organic (Thomas, 2006).

As the previously discussed scholars, Weber believed in primacy of human reason (over nature) having developed an elaborate theory of rationality guiding human action.

With regard to the qualification of scientific knowledge, Durkheim and Weber had some difference of opinion (Smelser, 1976), relevant to any social scientific study of complex issues such as environment and health. Durkheim may be labeled as a positivist with a great belief in the possibilities of an objective social science comparable to its counterpart in the natural sciences: the truth is out there, we have to objectively describe and understand it. Weber has a more subjectivist approach: the vast and unlimited complexity of reality prevents us from objectively describing and knowing reality. We have to more or less choose our reality by deciding which aspects of complex reality to take on board in our research and how to interpret the data. These choices cannot be based on any objective criteria, since there are none. Whereas Durkheim focuses on social facts, Weber stresses the importance of value judgments as part of the scientific endeavor (Weber, 1904).

Changes in the World since the "Classicists"

Although we could consider more episodes from the history of civilization development, we chose to focus upon the Industrial Revolution as a turning point in social scientists' perceptions of health and environment. We are aware, however, that when we speak of "pre-industrial," or "industrializing," or "traditional," or "developing" or "Third World" societies, these could hardly be generalized in opposition to industrial countries. For one, countries, societies and cultures are in a continuous state of flux due to unpredictable dynamics of globalization. The relationship of "pre-industrial" or "industrial" societies with health and environment issues are too complex to consider in this volume in more than superficial detail. Also, we realize that the effect of industrialization on different nation-states and its effect on health and environment is far from uniform. For the purpose of limiting this volume to contemporary debates, we shall consciously simplify and shorten the summary of industrialization to condense the most significant ideas related to today's concerns.

Since the time of the Industrial Revolution, the relationship between human welfare and environment has become increasingly complicated. With the emergence of a capitalist system in Western Europe, both "nature" and humans became largely seen and used as resources facilitating industrialization. As Marx witnessed, initial stages of industrialization often involved increased use of both human labour and natural resources. Early industrialists imposed hard labor and long hours, which caused a drop in health and an increase in mortality rates.

Increasing concern with the industrial workers' welfare in the later decades of industrialization in Western Europe reflected both the ideology of enlightenment and the economic necessity to take better care of the workers. Roberts (1999), a sociologist specializing in the history of leisure, describes the growth of concern for welfare, leisure and inevitably health of workers during the last decades of the industrial revolution. While agrarian life was primarily guided by "natural rhythms," the working life during the age of increasing manufacturing demands was almost "around the clock." Historians debate the human costs of industrial revolution. While in the long term it improved the lives of most capitalist countries' citizens, its costs in the short term were great. Martin Malia, a historian, describes the new industrial towns as having primitive housing and appalling sanitary conditions. In addition "the factory worker has lost the independence of the artisan or the peasant freeholder and became a mere cog in a vast machine controlled by capital" (Malia, 1996:43).

In the mid-nineteenth century, standard hours of work and greater social provisions were introduced due to a number of factors. Legislators began restricting the hours of work initially for vulnerable groups such as women and children and in hazardous industries such as coal mining. Some of these positive changes were initiated by the growing trade unions through adoption of "enlightened" or "progressive" approaches by the employees. As the industries became more capital intensive, considerations appeared that unhappy and sick workers were less efficient than the healthy and happy ones.

Industrialization gave rise to other institutions that further helped to maintain the equitable standard of living for even the lower strata of society. General well-being was assisted by better housing, water distribution and sanitation, waste collection and medical expertise available to almost anyone. Using leisure and other welfare provisions to motivate the workers was now the preferred "carrot" over the "stick," and the living standards became relatively higher for almost anyone in Western Europe and other industrial countries (Roberts, 1999:29).

In America in the 1910s, Henry Ford pioneered mass production and assembly lines. Consumer items were now available to and affordable for almost anybody. The working conditions of the factory workers and their welfare have largely improved, raising minimum wage to five dollars per hour and the daily working hours decreasing from nine to eight hours. Living standards and health started to be seen more as entitlements at the start of the twentieth century, and employers became partially responsible for their employees' welfare.

In Russia, industrial developments were propelled by political and economic changes after the Revolution in the second decade of the twentieth century. Industrial change was less gradual and more costly in terms of human sacrifice than in Western Europe, America and parts of Asia. Agricultural reform in Russia during the period of the first Five-Year Plan in 1927, led to huge shortages of food, rationing and the necessity to import grain (Malia, 1996: 150). Collectivization and industrialization during the Stalin era led to unprecedented human and environmental losses.

After the First World War, Communism, with its distinct brand of rapid industrialization, started to spread into countries of Eastern Europe, Asia, South America and Africa, This rapid industrialization, while brutal in its initial stages, was accompanied by the growth of the welfare state. Despite the terrible atrocities committed by Stalin in Russia, Mao in China and Pol Pot in Cambodia, these countries managed to reach the level of welfare (for those who survived the purges), which included universal medical coverage for the communist countries' citizens.

The second half of the twentieth century was marked by rising consumerism, preoccupation with (maintaining) high living standards, leisure and "wellness" industry in both capitalist and communist countries:

Leisure is not a cure-all, but once people have access to the best medical attention, and when they have achieved benign working and living conditions, their lifestyles are likely to become major discriminators of health status. The pursuit of health has become a prominent lifestyle goal among sections of the upper-middle classes (Roberts, 2002:12).

There was a significant advance of medical science in the second part of the twentieth century. Increasing spread of technological and medical innovation has not been distributed equally, rather privileging industrialized Western and partially Communist countries over developing countries. The rising standards of living in the Western countries (although not altogether equally distributed) were not reflected in the situation of the post-colonial and developing countries. While the spread of global goods and services, as well as the rise of consumerism, did affect most of the developing countries, they were often relegated to the status of production countries where cheap labour and natural resources could be found and to the "periphery" where goods were "dropped" and waste dumped.

However, along with the argued growth of the income gap between the developed and developing countries, some positive trends may be noted. Non-industrial or "developing" countries have profited somewhat from the medical and technological advancements that globalization gradually downed upon them. The "Green Revolution," in the post-World War II era, allowed Mexico, and later India and other developing countries, to develop wheat and other agrarian produce in such quantities as not only to feed their own populations but also to allow increasing volumes of export.

There has been significant decline of geographical areas characterized by "pristine" flora and fauna in the past hundred years in both developed and developing countries.

The industrial revolution that gave rise to modern capitalism greatly expanded the possibilities for the material development of humankind. It continues to do so today, but at a severe price. Since the mid-eighteenth century, more of nature has been destroyed than in all prior history. While industrial systems have reached pinnacles of success, able to muster and accumulate human-made capital on vast levels, *natural capital*, on which civilization depends to create economic prosperity, is rapidly declining, and the rate of loss is increasing proportionate to gains in material well-being. *Natural capital* includes all the familiar resources used by humankind: water, minerals, oil, trees, fish, soil, air, etc. But it also encompasses living systems, which include grasslands, savannas, wetlands, estuaries, oceans, coral reefs, riparian corridors, tundras, and rainforests. These are deteriorating worldwide at unprecedented rates. Within these ecological communities are the fungi, ponds, mammals, humus, amphibians, bacteria, trees, flagellates, insects, songbirds, ferns, starfish, and flowers that make life possible and worth living on this planet (Hawken, 1999:2).

While industrialized countries allowed large corporations and private parties to compete for the use of natural resources, Communist countries suffered the "tragedy of the commons" (the use and abuse of land and resources that "belonged to everybody and nobody").

This trend was accompanied by the introduction of numerous technological risks (Beck 1992, Giddens, 1990), the unhealthy impact of the use of numerous new chemicals (McCally, 2002) and Western lifestyle. These risks were particularly apparent in industrializing and developing countries where the volume of waste and chemical products, in combination with poor sanitation, have led to wide-spread pollution and incidence of disease.

Health and Environment in the Context of Globalization

Globalization is often expressed in metaphors: "Runaway world" or "time-space distanciation" or the "intersection of presence and absence" (Giddens, 1991), "time-space compression" (Harvey, 1989), "the age of connexity" (Mulgan, 1998:19), or the "global ecumene" (Hannerz, 2002). Anna Tsing, an anthropologist with a special interest in globalization, notes that social scientists' perceptions of globalization have been quite remote from that of other scientists as well as from activists (such as environmentalists and anti-globalists):

> Social scientists and historians have been rather disruptive of this global domain [of agency for global environmental science], although not always self-consciously, when their descriptions establish the incompatibility of various socially defined spatial scales and historical periods, as nature is made and remade in diverse forms that evade simple conflations...Through this distinctive antiglobalism, it can perhaps offer possibilities for nonglobalist global analysis in a different scholarly conversation, in which we might begin to get around blinding endorsements of futurism, conflation, and circulation (Tsing, 2002:481)

David Held suggests that globalization implies a speeding up of the flows of capital, people, goods, images and ideas across the world; intensification of the links, modes of interaction and flows that interconnect the world; as well as stretching of social, political, and economic practices across borders (Held, 2004). Held distinguishes between different schools of thought about globalization: the transformationalist, the international, the positivist and negativist. The first two schools of thought considered not as much the effects or consequences of globalization (perceiving them as neither good nor bad), but rather the question of origin and meaning and complexity of globalization. The latter two issued value judgments about the impacts of globalization on both society and environment.

For transformationalists, globalization, in the sense of "expansion of contact and trade" is not a new phenomenon. While transformationalist thinkers acknowledge that the speed with which technologies, communication, material goods, and cultural commodities spread across the globe has indeed intensified. The essential commodities or ideas transferred themselves—be it through trade, through the spread of political or religious ideologies, through war or "peaceful coercion"—are hardly new. The consequences of globalization are so complex and unpredictable due to multiplicity of "butterfly effects" affecting economic, social, political and cultural systems that we may hardly speak of "good" and "bad" effects. The questions of health and environment will need to be addressed at the very specific level in order to be able to say anything meaningful about them. Sweeping generalizations that we are all guilty can hardly account for the actual variability of both health and environment issues. The detailed accounts of health and environmental-focused studies by the contributors to this volume do provide us with greater depth and understanding of the "grass-roots" processes that lie at the core of far more complex global phenomena.

Another school of globalization theorists, the internationalists, believe that the nation-state is still much more important than all global, trans-national, international and supranational connections and interdependencies. The state and the local governments did not diminish but in fact were strengthened by the expansion (or the threat thereof) of global capital and (Western) ideologies. In the case of health, we may think of an example of how the national governments regulate the market of pharmaceutical industry, the movement of health workers, and expert exchanges; as well as the degree to which the health-care systems and institutions are organized within the national borders. In the case of environment, we may conceive of the role the national governments play in regulation of import and export of natural resources or raw materials, environmental protection laws, and many more resource- and policy-related issues.

The scope of this book does not allow us to reflect upon the extremely complex process of globalization in greater detail, nor do we want to indulge in superficial sketching of the processes affecting economic, political, social and cultural systems worldwide. For the purpose of this volume, we need to remain focused on the subjects of health and environment, while acknowledging the fact that the global context within which these are found needs to be explored in much greater detail in the future publications. We need to note, however, that since the subjects of health and environment are far from "neutral" within the social and political discourse, it is hard to avoid thinking of processes involving unequal distribution of "medical care" or "environmental degradation" as objective and value-free categories. Social scientists may conceive of many possible theoretical ways of addressing globalization. We have to realize that considering both positive and negative effects of globalization is value laden and not an objective or neutral endeavor. We thus self-consciously return to our brief and subjective summary of health and environment in the global world of the late twentieth and twenty-first centuries.

From the third quarter of the twentieth century, medical and social aid spread to the former colonies, located in Africa, South America and Asia, where both agricultural and industrial development, orchestrated from the "developed" world, took massive proportions. The "hard" advantages of capitalist growth were accompanied by neo-enlightenment (otherwise termed liberal, democratic, or humanitarian) sentiments in the increasingly prosperous West. In the Western or "developed" countries, concerns with the health of the less fortunate, either in their own society or elsewhere, has been an accompanying trend. Current public debates often center around the question of growing economic inequalities between the "West and the rest" and the media reflects (and in part, creates) concern for the "victims of development" (Collier, 2007). Concerns about the welfare of not only Western working population but also poverty, child labor, child mortality and disease in the Third World occur simultaneously with the growth of living standards and consumerism in the "West." Optimistic observers note that standards of living were rising and people, on average, were getting healthier.

Reduction of poverty, better health and education, and environmental regulation have been outlined as important targets within the UN's Millennium Development Goals (MDGs) for the 21st century.

Paul Collier (2007) emphasizes the fact that the MDGs are outdated since they focus on the world of the rich one billion facing the supposed world of the poor five billion. According to Collier, the more realistic situation is that almost 80% of the population lives, indeed, in developing countries that are, however, rapidly changing and evolving at an "amazing speed," while some countries are "falling behind and falling apart." While these countries exist in the twenty-first century, their reality is the fourteenth century: civil war, plague and ignorance (Collier, 2007:1). It is the five billion that are "developing" and one billion that are falling behind, with health problems as the most pressing of all.

Counteracting the optimism of early developers and industrialists, as well as humanitarians believing in positive trends towards increasingly egalitarian distribution of wealth and health care, skeptical observers have described the postcolonial period as "post-modern imperialism." In the Easterly's (2006) book *The White Man's Burden: Why the West's Efforts to Aid the Rest Have Done So Much Ill and So Little Good* the reverse side of "our" interference with "their" lives is aptly described.

Jeffrey Sachs's book *The End of Poverty: Economic Possibilities for Our Time* begins with the description of "medical ward" in Malawi, where people are brought to die of AIDS. The plight of Malawians to secure donors' funding for anti-AIDS drug treatment was termed a "perfect storm" by Carol Bellamy of UNICEF, a storm that "brings together a climatic disaster, impoverishment, the AIDS pandemic, and the long-standing burdens of malaria, schistosomiasis, and other diseases" (Sachs, 2005:10). According to Sachs, the world community has so far displayed a "fair bit of hand wringing and even some high-minded rhetoric, but precious little action" (Ibid, 2005: 10). In the case of developing countries, inequalities in distribution of medical labour and availability of medical treatment may be linked to political and economic reasons underlying actions of, for example, global pharmaceutical industry.

While health concerns are probably as old as humanity itself, environmental concerns seem relatively new, at least in the western society. The environmental movement may, in part, be inspired by idealism of early romantic and transcendentalist philosophers and poets, but is also derived from practical realization of fragility and scarcity of "nature" (in Emerson's definition of that which has been untouched by man). The "human plight" and its connection with the "environmental plight" in the post-industrial world is a growing concern. Most of the present-day environmental issues are framed in relation to human activities and—inevitably—health. The effects of rapid and uneven industrialization brought forth not only economic inequalities, but inequalities in geography of production and pollution affecting both human health and environment.

The use, abuse and overuse of natural resources triggered by globalization of production, has lead to increasing deforestation, erosion, as well as production of (toxic) waste, particularly affecting air, water and soil of poor countries. According to United Nations Environment Programme and the US Environmental Protection Agency, each and every second, 310 Kg of toxic chemicals are released into our air, land, and water by industrial facilities around the world. This amounts to approximately 10 million tons (over 21 billion pounds) of toxic chemicals released into our environment by industries each year (http://www.unep.org/; http://www.epa.gov/). Of these, over 2 million tons (over 4.5 billion pounds) per year are recognized carcinogens, at about 65 Kg each second (Occupational Safety and Health Administration (OSHA) Chemicals).

Some of these processes, such as deforestation, are irreversible. Simultaneously, industrial pollutants and destruction of biodiversity (and ethnic medical knowledge) have direct effects on human health. It may be argued that effects of Industrial Revolution were hard felt by both humans and non-human.

McDonough and Braungart (2002), the authors of the book *Cradle to Cradle* pioneered the principle of redesigning products by putting bio-effectiveness into practice. They provide a provocative summary of the negative outcomes of the Industrial Revolution by creating a virtual "assignment of designing the Industrial Revolution" with respect to negative consequences:

> The assignment would have to read something like this: Design a system of production that puts billions of pounds of toxic compounds into the air, water, and soil every year; produces some materials that are so dangerous they will require constant vigilance by future generations; results in gigantic amounts of waste...; creates prosperity by digging up or cutting

> down natural resources and then burying or burning them; erodes the diversity of species and cultural practices... (McDonough and Braungart, 2002: 18).

Alternatively, we can imagine a different world in which reversal of damage is a distinct possibility:

> Imagine, for a moment, a world where cities have become peaceful and serene because cars and buses are whisper quiet, vehicles exhaust only water vapor, and parks and greenways have replaced unneeded urban freeways. OPEC has ceased to function because the price of oil has fallen to five dollars a barrel, but there are few buyers for it because cheaper and better ways now exist to get the services people once turned to oil to provide. Living standards for all people have dramatically improved, particularly for the poor and those in developing countries...Houses, even low-income housing units, can pay part of their mortgage costs by the energy they *produce*; there are few if any active landfills; worldwide forest cover is increasing; dams are being dismantled; atmospheric C02 levels are decreasing for the first time in two hundred years; and effluent water leaving factories is cleaner than the water coming into them. Industrialized countries have reduced resource use by 80 percent while improving the quality of life...Is this the vision of a utopia? In fact, the changes described here could come about in the decades to come as the result of economic and technological trends already in place (Hawken et al., 1999: 1).

Proponents of globalization, industrialization and development argue that it is precisely through globalization and technological development that both health and environmental problems could be tackled. Technological advances anno 2009 have enabled increasingly successful treatment of formerly terminal diseases as well as offered a scale of possibilities concerning targeting (if not reversing) environmental challenges such as climate change. "Green energy" (bio-fuels and other alternatives) is one example of such desired solutions that are supposed to eliminate both toxic waste affecting human health and dependency on fossil fuels. But the reconciliation between neo-Malthusians and optimistic industrialists is not in sight.

Population Growth

An important topic of public and policy debates concerning health and environment is that of the population growth. Advances in medicine and technology, aided by global spread of agricultural and industrial technology, resulted in unprecedented population growth. According to the U.S. Census Bureau, the population grew by 79.4 million in 2007 (http://www.census.gov/ipc/www/idb/worldpop.html). As of April 2009, there are 43,678,000 births registered per year, as opposed to about 18,980,000 deaths per year (United Nations, Statistics Division; UNESCO Institute of Statistics).

The highest growth rates are registered in the Middle East, South Asia, Southeast Asia, Latin America, and primarily in Sub-Saharan Africa. According to projections by the U.S. Census Bureau, the annual world population growth will peak in 2011 at 80.9 million. By 2050, the global population is forecast to reach 9 billion. Population growth is still accompanied by poverty, low life expectancy and wide spread of disease, particularly in sub-

Saharan Africa (United Nations, Statistics Division; Economic Commission for Africa; Economic Commission for Asia and the Pacific, Statistics Division).

Skeptical observers have noted that The UN's Millennium Development Goals (MDGs) would be impossible to meet due to the increases in population growth. "No country has ever raised itself out of poverty without stabilizing population growth," said The All Party Parliamentary Group on Population, Development and Reproductive Health vice-chairman, Richard Ottaway, at a seminar on population issues in December 2006. "And the MDGs are going to be difficult or impossible to attain without a leveling out of population growth in developing countries" (Black, 2006).

Population growth and the negative effects thereof for both people (due to the supposed increase in the incidence of Malthusian "disasters" of disease, wars and famine) and nature (environmental degradation) have been popular topics of public debates and policy focus throughout the late ninety sixties and seventies. Up until the present time, however, only China has been relatively successful in instituting a viable state-controlled system of population growth control (the one-child-policy).

Simultaneously with concerns about the population growth, the critics of restrictions on population control raised their voices. Liberal Western critics pointed out that globalization has affected different parts of the world and people unequally, creating a huge underclass of impoverished "losers." Acknowledging the impact of colonialism and almost equally disastrous affects of post-colonial development on many countries in Africa, Asia and South America, "colonial guilt" spread among the liberal thinkers. Another trend responsible for increased consciousness about the living conditions of the developing countries is due to the growing individualism and the rise of the value of human life in the Western countries. Humanity came to be seen as "one," and the guilt about the poor and the diseased spilled over to the large segments of wealthy liberal western citizens.

Accompanying the trend was the unprecedented growth of migration from the South to the North and from the East to the West. While this migration has been increasingly more restricted in the recent decades (Kopnina, 2007), humanitarian aid to the developing countries has been on the steady rise. The question of population control has been practically forgotten in favor of humanitarian concerns. Malthusian concerns became suspect as morally unacceptable as Western liberals increasingly owned the "problems" of the poor.

We may recall the discussion of the "charity paradox" in the Introduction. Recent attention of the humanitarian charity Oxfam Novib towards climate change illustrates the desire to reconcile human and nature interests. Their recent publications call attention to the fact that poverty increases people's exposure, and climate change increases the risks; people living in poverty and poor communities are most vulnerable. This is due to a number of factors, such as the fact that most of these people live in overcrowded slum areas in unplanned, temporary settlements erected on unsuitable land with poor sanitation; and they have limited access to clean water, poor diet, and inadequate health-care provision, all of which undermine their resistance to infectious diseases. As a result, between 1990 and 1998, 94 per cent of the world's 568 major natural disasters, and more than 97 per cent of all natural disaster-related deaths, were in developing countries (Oxfam, 2009). Oxfam proposed a number of action areas to combat this problem, ensuring "that increasing numbers of people living in poverty who are harmed by climate-related disasters have equitable and efficient access to relief." Secondly, additional funding is required to protect the lives and livelihoods of the poor from "climate-related shocks and stresses, and to proactively adapt their livelihood strategies to the changing environmental conditions global warming brings." Finally, Oxfam calls for "tackling root causes of poverty," by supporting calls for "strong political action to reduce greenhouse gas emissions in order to prevent dangerous climate change" (http://www.oxfam.org.uk/resources/issues/climatechange/introduction.html). Remarkably, all these actions need to come from the West, and the effects of population growth are not addressed.

Thus, some fifty years after the heated concerns and debates about the population growth (and two billion people more inhabiting the earth), public and policy concerns have largely shifted to (un)equal distribution of wealth throughout the world and the "fight against poverty and disease" in the developing countries. One of the focal points of this "fight" is AIDS; we may remember from the discussion of the "charity paradox," states, ethical and social sentiments play a large role in how "the problem of AIDS" in the developing countries is conceived. Public and expert opinions differ as to the measures that need to be taken in order to combat apparent inequalities of access to medical technologies for treatment of AIDS.

Some observers have noted that while provision of AIDS-treatment medication saves millions of lives, these medications do not provide a long-term solution to address the causes of the disease and may, in fact, exacerbate it. Despite the unequal access to medicine in countries of Sub-Saharan Africa, their population has been on a steady increase. The use of birth control, particularly condoms, has not increased since wide-spread AIDS awareness campaigns. Cultural practices that favor having many children, as well as relative early introduction into liberal sexual relations without the use of contraceptives, factually meant that the birth rate and associated incidence of AIDS infections has not decreased. While medical and humanitarian NGO's blame local culture for blocking Western efforts to combat effects of HIV virus (such as traditional medicines and folk beliefs in the origin of HIV), they rarely focus on cultural practices involved in spread of AIDS.

Some observers point out that there is little attention for the effects of population growth on the environment and, in turn, adverse effects of environmental degradation on human health. The spread of diseases due to rapid urbanization, resulting in poor air, water and soil quality, as well as lack of sufficient natural resources necessary for sustaining the population, are hardly addressed. Western efforts to "feed the poor" by supplying them with food may work similarly to providing them with AIDS treatment—basically, being a short-term solution to a much deeper, long-term problem. Western involvement with economic and social development of much of the Sub-Saharan Africa may in fact spell continuation of health and environmental grievances, some observers argue. Other observers argue that withdrawing Western influence from the developing countries now, besides being economically and socially unforgivable, would spell a greater disaster that continuing the not-all-too-successful efforts to address the issues by using superior Western technology and medicine.

Health Issues in the Recent Decades

The prevalence of certain chronic diseases, such as diabetes, asthma, and cancer have changed somewhat due to better diagnostic and treatment technologies on the one hand and environmental factors (such as pollution or changing lifestyles) on the other hand. The treatments for many infectious diseases such as measles and tuberculosis became more efficient, and their prevalence has radically declined, particularly in the Western world. Yet, due to the graying population and the increased welfare of Western society, other diseases, such as Alzheimer's, diabetes and obesity, have increased (WHO 2004).

In the non-Western world, the scale and distribution of diseases as well as the availability of medication or expertise for their treatments have changed significantly. Aside from "traditional" diseases, such as malaria, lung and skin diseases have increased due to higher pollution levels. According to a WHO report (2004), half of the population in Asia has no access to essential medicines. In developing countries, traditional medicine is still widely used but not greatly effective against the gravest diseases. Potrata's and Akobirshoeva's chapter in this volume contains an engaging account of the decline in the use of traditional medicine in Central Asia and its effect upon both health and environment.

Thus, while some diseases became relegated to the poorer parts of the world (such as cholera, tuberculosis, and, for a large part, AIDS), others became designated as the "rich countries' diseases" (such as obesity). However, "rich country" diseases can also be found elsewhere in the world as the welfare spreads (though quite unevenly) and globalization slowly erases the "first," "second" and "third" world distinctions (WHO 2006).

There has been a general shift across the world towards the greater use of Western medicine, while the prevalence of indigenous medical knowledge (except for larger and more prosperous countries such as China) has reduced. Generally, medical diagnostics and treatments for a number of infectious and chronic diseases have improved. Large migratory movement of qualified medical specialists and knowledge are recorded throughout the world. Highly trained medical specialists moved from Eastern to Western Europe, as well as from Asia to North America and Australia (Kopnina, 2005; Kopnina, 2007). Due to the intricate forces of globalization, great inequalities and simultaneously great opportunities in the treatment of disease in world populations have occurred. "Medical tourism" (travel of rich tourists to poorer countries for cheaper operations) have become commonplace. The outsourcing of record keeping and the remote transcription of doctors' notes and x-ray analysis are becoming common (The Economist, 2008: 66).

Another recent trend is migration of medical personnel from the developing countries into the richer parts of the world. While prior to the 1970s, the main movement of health workers was directed from the developed to the developing world, this trend is now largely reversed due to increase flow of labour in the medical health market, especially affecting Sub-Saharan Africa. This trend is also due to the global decline in public investment in health and the easing of migration restrictions within OECD countries for high-skilled and medical personnel

A skeptical group of observers blames the West for the neo-colonial arrogance and the "we know better" attitude, which respects little of local knowledge and traditions. In this view, Western hegemony imposes its own version of "equality," propelled by underlying economic reasoning implying the need to "harmonize global markets" on the one hand, and the need to sate the post-colonial guilt by "doing good deeds." At best, some argue, these efforts are the result of ignorance, at worst, the result of cunning selfishness in the world driven by desire for profit. Third World poverty, particularly the sad disintegration of sub-Saharan Africa into the chaos of Malthusian disaster, is an embarrassment to those who have made development, industrialism, and, in fact, capitalism sound like the Holy Grail.

Environmental Issues in the Recent Decades

The statistics vary greatly on the amount of pristine nature areas that have disappeared in the past century across the globe. While in Europe, most of the pristine nature areas have ceased to exist during or shortly after the industrial revolution, and the developing countries have lost most of their wilderness during the past few decades. Deforestation, erosion, water, soil and air pollution as well as unprecedented rapid loss of biodiversity have doubled in the past decade. Areas (illegally) logged or cleared for farming encompass the majority of countries in Africa, Asia, South America and the Pacific. Costa Rica alone has lost some 70% of her original forest in the past three decades. Even areas supposedly protected by the governments, such as those along the Amazon basin, become developed and "cleared" for faming or industrial purposes (United Nations Industrial Development Organization (UNIDO)). The emerging economies of India, Brazil and China are rapidly turning their nature into "natural resources" and speeding this process up for the poorer countries (such as Chinese businessmen investing in industrial development of Africa). Anno 2009, the rate of deforestation continues in the tropics at around one acre per second (United Nations Framework Convention on Climate Change (UNFCC)).

In Western Europe, pristine nature areas already experienced deforestation hundreds years ago. It is likely that the woods whose paths classical thinkers used to wander in the past centuries while pondering their theories no longer exist or have been turned into well-groomed urban parks or designated recreational areas. In Eastern Europe, particularly those countries that recently entered the EU, large-scale developments, such as road building, and rapid industrialization and urbanization of what used to be "wild" areas, are on the way. In countries such as Bulgaria and Romania, the borders of national parks are often being violated and sold to rich private owners for the purposes of "development." The Baltic states develop their coastal regions, the speed disproportionate to tourist demand.

The causes of this unprecedented demise are well-known but still widely disputed. Just as in the case of the uneven distribution of the world poverty and disease, globalization and particularly the forces of advanced post-industrial capitalism are blamed for, some would say, catastrophic disappearances of pristine nature, biodiversity and extinction of many flora and fauna species. Other observers (both within scientific and public camps) are less concerned and see the purported "catastrophe" rather in terms of (economic) opportunity and the sign of great progress.

Technological risks like nuclear accidents and industrial catastrophes have become an integral part of our modern society (Beck, 1992; Giddens, 1990). The impacts of risks from modern technologies seem to have become much more extensive. Tens of thousands of chemical substances are on the market. In recent years, the international level experts increasingly support the conclusion that environmental problems have serious consequences for public health (McCally, 2002; Harremoës et al., 2002). The number of environmentally related cancers is rising, an increasing number of people suffer from respiratory problems, fertility problems are on the rise, and so on. The (causal) relation between environment and health is very complex though and to a large extent unknown. For a number of individual toxic substances, the health effects from high doses are well known. Unknown are the effects of small doses of different substances over a longer period. Also unknown are the combined effects of different substances. There are clues though about DNA-damage, hormone disruptions and loss of sperm quality.

While often recognizing the interrelatedness and interdependency of human and natural worlds, debates are often characterized by decades of favoritism of one side or the other. The population growth concerns were great in the nineteen sixties and seventies; poverty and disease control in the nineteen eighties and nineties; and the current arguments in favor of reversing the climate change are popular anno 2009. These concerns are certainly fueled by economic and political forces underlying public debates.

From Classical Scholars to Modern Academic Subdisciplines

It would be perhaps wrong to identify "change" in contemporary theory. Rather, we note a continuous pendulum-like shift in focus from "human" to "ecological" concerns that has occurred even before the writing of Malthus, Marx and other "classicists." At the level of scientific or expert discourse (often represented through academic publications), as well as public or lay discourse (often represented through the media), we may identify a few schools of thought. Contemporary debates still reflect upon the difficulty of relating humans to either the "natural" or "social" sphere. In relation to health, for example, Ananth (2008) argues that a naturalistic concept of health, drawing on evolutionary considerations associated with biological function, homeostasis, and species-design, is defensible without jettisoning norms in their entirety. Other health specialists, notably medical anthropologists, accentuate distinctiveness of culture and subjective perceptions of health and illness. Those concerned with the environment, still tend to address it either as dependent or intertwined with external and independent of human concerns.

Within social science, attention has shifted away from the grand theories of human nature, society and culture and toward more empirical, concrete examples of how global and local (and indeed, "globalocal") relationships shape the world. Medical and natural sciences moved towards greater integration of other disciplines' insights (like those of psychological anthropology and cultural psychology) on the one hand, to deeper specializations within own field (such as human genetics or molecular biology) on the other hand. Holistic science of Victorian gentlemen has made way for highly specialized and fragmented knowledge generated from many centers.

Considering that a number of sub disciplines are involved in the study of health and environment, we need to point out that methodologies involved in such research vary greatly. Social scientists engaged in health and/or environmental research employ methodologies ranging from qualitative (for example, employed by medical anthropologists and environmental sociologists), quantitative (clinical psychologists, health economists, medical sociologists) and experimental (including different sub disciplines) (Green and Thorogood, 2006; Mehta and Ouellet, 1995).

The scientific disciplines currently involved in the study of health and environment range from immuno-biology to medical sociology (Skair, 2005) to the specific discipline of medical ecology. The disciplines diverged into distinct strands combining elements from both natural sciences and humanities and leading to more eclectic discussions. Regrettably, little dialogue and real "interdisciplinary cooperation" can be observed between them. Hereby we present a cursory exposé of some social scientific perspectives on environment and/or health, further expanded upon by the contributors to our book, without claiming to be exhaustive. We shall now briefly discuss some of the subdisciplines represented by the contributors to this volume.

Environmental Sociology

The classical theories were complemented by modern sociologists attempting to encompass contemporary environmental issues (Catton et al, 1978; Dunlap, 2002; Martell, 1994). The works of Catton and Dunlap challenged the anthropocentrism as well as the Human Exemptionalist Paradigm dominating mainstream sociology (Catton and Dunlap, 1978). The Human Examptionalist Paradigm theory traced from the second half of the twentieth century claims that humans are a unique superior species exempt from environmental forces. Within this paradigm, human dominance over the environment was justified by the uniqueness of culture, which is far more adaptable than biological traits.

Environmental sociology is a branch of sociology informed by an understanding of ecological limits (Mehta and Ouellet, 1995). Catton and Dunlap's aim was to bring to the larger discipline of sociology a recognition of its anthropocentric worldview and to incorporate environmental issues into sociology, identifying the specific theories "that had led sociology down a path of environmental blindness" (Rosa and Richter, 2008:182). Catton and Dunlap (1978) developed a new paradigm termed New Ecological Paradigm, which recognized the innovative capacity of humans, but stressed that humans were still ecologically interdependent as other species. Just as Malthus has predicted, there is a finite level of natural resources and waste repositories. The unbridled human population growth and the need for still higher living standards as well as the limitation of the natural environment do cause both natural and social grievances, the authors argue. The examples of "climate crisis" and "financial crisis" debates anno 2009, fit well with the neo-Malthusian description. The dictum of "risk society" has gained its prominence. "It is a society increasingly preoccupied with the future (and also with safety), which generates the notion of risk" (Giddens 1999: 3) Risk can be defined in the risk society as a systematic way of dealing with hazards and insecurities induced and introduced by modernization itself (Beck 1992: 21).

Guy and Shove (2001) call for examination of the environmental "problems" [*sic.* quotation marks] in the context of sociological theory in order to create a "distinctive strain of environmental sociology." The authors argue that sociological theory can make a significant contribution to environmental debate and policy. Yet, sociologists have often missed the opportunity to engage with societal debates, including pragmatic concerns about health and environment. This can be partially explained by the sociologists' engagement with their own theoretical concerns, and partially by the fact that sociologists have been drawn into limited roles of considering "human dimension" of environmental concerns (Guy and Shove, 2001:2).

A societal-environmental dialectic was proposed by Allan Schnaiberg (1975; 1980) transfiguring environmental sociology. Schnaiberg developed ideas of economic synthesis and managed scarcity, stating that the desire for economic expansion will prevail over ecological concerns. The logic of advanced capitalism will stimulate policymakers to stimulate economically advantageous policies at the expense of environment. The managed scarcity synthesis refers to the fact that governments will attempt to control only the most dire of environmental problems to prevent health and economic disasters. These policies may be termed "environmental," while the driving factor is economic damage caused by environmental degradation. As we, the twenty-first century Westerners currently experience, there is a widespread concern about renewable resources, particularly energy. According to Schnaiberg, production and consumption methods would adhere to sustainability regulations until the economic need diminishes. This does not necessarily mean that "environmental" policy efforts will be abandoned, but that the most powerful economic and political forces will preserve the status quo and bolster their dominance by either continuing with these policies or by falling into a stalemate. Schnaiberg also hypothesized that tumultuous environmental or social events may result in the redistribution of economic and political resources (Schnaiberg, 1980).

We may remember the discussion of Marxian treadmill from part one. In 1980, Schnaiberg expanded the treadmill of production metaphor and developed a conflict theory on human-environment interaction in which he argued that corporations are driven by higher profitability and thereby must continue to grow and attract investments to survive in a competitive market. This identifies the imperative for continued economic growth levels that, once achieved, accelerate the need for future growth. This growth in production requires a corresponding growth in consumption. The process refers to the sustainability paradox; economic growth is perceived to be socially desirable but environmental degradation is a consequence that, in turn, disrupts long-run economic expansion.

Ecology of Health

Ecology of health deals with the impact that the structure of institutions has on how patients are treated and on how the interpersonal relation between clinicians and patients unfolds. While studies of recreational drug users indicate that drugs have a range of physical and psychological effects, the interpretation of those effects depends on the user's prior knowledge or that of the social peer group (such as other drug users) (Becker, 1973; Lindesmith, 1968). In other words, the drug-taking activities are structured by the social context in which the drug user is told by the drug-using peers what physical or psychological sensations to expect and how to recognize their effects (Zinburg, 1984). In the case of prescription drugs, it is the prescribing doctor and not the peers who supplies drug information to the user and acts as the authority figure on the effects (or side effects) of the drugs (Becker, 1973; Haafkens, 1997).

Anthropologists, such as Evans-Pritchard, have taken Weber's view of rationality to explain the native beliefs about the causes of disease. While the native Azande attributed many ailments to the workings of the witchcraft, their supposedly "magical" thinking was based on the rationality of cause and effect and observable (although, according to Evans-Pritchard, empirically doubtful) phenomena. While Evans-Pritchard himself confesses to seeing something light (which Azande would classify as witchcraft) flying from hut to hut, his Western rationality does not allow him to accept native explanation as "true" or" valid." Yet, he credits the natives with rationality, enabling them to explain the cause of illness (Evans-Pritchard, 1974).

Ecological Sociology

Ecological sociology is related to environmental sociology and yet is argued to be different from it. Bowden, a contributor to this volume, argues for the need of distinction between the disciplines of environmental and ecological sociology. Bowden reflects that the examination of social class distinctions in public opinion toward the environment merely transferred the gaze of traditional sociological theory to a new substantive area—attitudes about the natural environment. In contrast, Bowden argues, Catton and Dunlap (1978) argued for the inclusion of both environmental and social variables as cause and/or effect. The studies of, for example, the impact of social class on environmental degradation or of energy shortages on society, are viewed as "real" environmental sociology. Examining the work of Catton and Dunlap, Bowden emphasizes two distinct definitions of environmental sociology, which differ from what he terms "ecological sociology": one methodological and the other paradigmatic. Bowden draws a distinction between environmental sociology (defined methodologically) and ecological sociology (defined paradigmatically).

Environmental Health

The field of environmental health, or environment and health, seems to some extent rather young when compared to the distinct fields related merely to the environment on the one hand, or to human health on the other hand. The relative "youth" of this field may come as a surprise to many as the combined attention for environment and health today seems so obvious. Part of the explanation in the Western world may be that until recently, the field of public health "flourished" in the modern era without explicit attention for environment-related issues. In the Western world, many environmental issues were dealt with rather well in earlier days, of which the modern sewer system, proper waste disposal in general and care for healthy drinking water conditions are prominent examples. With the rise, though, of the Risk Society (Beck 1992, Giddens, 1990), public health again, like in the old days, had a protective duty with regard to health risks from the environment. Building on the experiences with high-dose risks for employees working with dangerous substances in occupational health, attention shifted to the issue of low-dose risks caused by a variety of substances in the environment, or perhaps better put, in our society as a whole.

In recent years, a variety of publications and scientific journals emerged with some focus on and reference to environment and health. We can distinguish among views on these developments both from the natural sciences and from the social sciences. We will introduce them to illustrate the complexity and the richness of this field. Moreover, the need for interdisciplinary cooperation in this field is not only mirrored in the combination of domains (environment and health), but also in the need to tackle not only natural scientific complexity, but also social complexity.

From a social scientific perspective, Brown (2002) defines environmental health broadly as "the totality of unhealthy living and working conditions," and more specifically as "health effects caused by toxic substances in people's immediate or proximate surroundings." Main features mentioned by Brown when describing the "environmental health-scene" are complexity of the relation between environment and health and newness and controversy. Brown thus not only focuses on the knowledge side of the environmental health issues but also on the actors (social movements, scientists, government officials, industry, to name a few) and diverging views and interests. Brown broadens a merely scientific view (what seems to be the problem and how can it be solved) to politics: different problem-framings, different interests and different problem-solving strategies. Who decides based on whose knowledge thus becomes a relevant question.

From a natural science perspective, Phillippe Grandjean (2005, 2008) is rather critical of environmental health-science: due to complexities, traditional science will not be fit to tackle environmental health-problems, especially with regard to policymaking. Standard scientific approaches do not fit issues in focus and method (too much focus on simplified models and effects of single hazards one by one) and in interpretation (too strict methodological standards). Thus, they do not supply policymakers with policy-relevant knowledge. Grandjean stresses the need for a different scientific approach and application in policy practice of a precautionary approach.

David Briggs (2008) combines the scientific with the political in his plea for more integrated methods of assessment. Key challenges do not only relate to the content of analysis, environment and health problems, but also to the involvement of relevant (actor) perspectives. With regard to complexity, Briggs (like Grandjean) criticizes traditional forms of assessment and proposes to focus on a real-world perspective in which the issue of problem-framing becomes of main importance. The involvement of relevant actors, according to Briggs, need not be limited to scientists alone: also the involvement of stakeholders is important, at an early stage. With regard to policymaking, like Grandjean, Briggs chooses to pursue a precautionary approach. Real life state of affairs according to Briggs is that, despite numerous pleas for and ambitious ideals with respect to environment and health, the application of integrated approaches in research practice is still in its infancy.

Medical Anthropology

Medical Anthropology often places emphasis on relativity of medical knowledge and diversity of medical traditions, beliefs and practices. Medical anthropology emphasizes that societal practices play an important role in creation of medical knowledge. "Medical knowledge is not purely scientific but shapes and is shaped by the society in which it develops" (White, 2002:i). Medical anthropologists alert us to the fact that certain political and economic incentives that effect the actions of Western pharmaceutical industry often underplay and even negate the local medical expertise and traditional knowledge.

A BBC report about the "white priestess of black magic" describes Susanne Wegner, an Autralian-born "frail 94-year-old artist, with one seeing eye" as "bent double by age, the high-priestess of Nigeria's Yoruba spirit-world." In the town of Osogbo, she is in charge of the sacred grove, a "place where spirits of the river and trees are said to live." Since Wegner's arrival, traditional religion gave way to Christianity. Missionaries branded the river god Orisha worship as "black magic," involving human sacrifice. When Wegner arrived with her husband in the nineteen fifties, she met Adjagemo, a high-priest of creator-god Obatala who "led her into the spirit world" and while she did not speak Yoruba, and he did not speak English, their "only intercourse was the language of the trees."

Wenger's ideas about the preservation of the forest have become central to the survival of the traditional beliefs...Mr. Adigun Davies, a former curator with the government museums directorate who first met Mrs Wenger in 1989, says the battle to save the grove was a "tug of war." He recalls her lying down in the path of a bulldozer brought by a man who bought the grove from a relative of a traditional leader and who wanted to build a house on the land.

"It's a disgrace to the Yoruba that the person who came to save our culture was a European," he says. "But Susanne's knowledge of the behaviour and character of all the deities means she has actually become Yoruba now." (Walker, 2008, http://news.bbc.co.uk/go/em/fr/-/2/hi/africa/7595841.stm)

While "constructionist" view of disease and treatment involves subjective interpretation and experience thereof, reasoning about the disease may be seen as universally rational:

> Magic, for example, has been just as systematically "rationalized" as physics. The earliest intentionally rational therapy involved the almost complete rejection of the cure of empirical symptoms by empirically tested herbs and potions in favor of the exorcism of (what was thought to be) the "real" (magical, demonic) source of the ailment. Formally, it had exactly the same highly rational structure as many of the most important development in modern therapy (Weber, 1949:37).

REFERENCES

Ananth, M. (2008). *In Defense of the Evolutionary Concept of Health*. Ashgate, Abingdom.

Beck, U. (1992). *Risk Society: Towards a new modernity*, London, Sage, (1986).

Becker, H. S. (1973). Consciousness, power and drug effects, *Society*. 10/May-June: 67-76.

Black, R. (2006). *Birth rate harms poverty goals* http://news.bbc.co.uk/1/hi/sci/tech/6219922.stm

Bowden, G. *From Environmental to Ecological Sociology* www.tasa.org.au/conferencepapers04/docs/RURAL/BOWDEN_a.pdf

Briggs D. J. (2008). A framework for integrated environmental health impact assessment of systemic risks. In: *Environmental Health* 2008, 7:61 doi:10.1186/1476-069X-7-61.

Brown P. (2002). Preface to Special Issue on Environmental Health., in: *The Annals of the American Academy*, 584:7-12.

Brundtland, G. (Ed.). (1987). *Our Common Future: The World Commission on Environment and Development*, Oxford: Oxford University Press.

Catton, W.R., & Dunlap, R.E. (1978). Environmental Sociology: A New Paradigm. *The American Sociologist*, 13, 41-49.

Collier, P. (2007). *The Bottom Billion: Why the poorest countries are failing and what can be done about it.* Oxford University press, Oxford.

Desmond, A. and Moore, J. (1991). *Darwin*. London: Michael Joseph, the Penguin Group.

Dunlap, R. E., F. H. Buttel, P. Dickens and A. Gijswijt (2002). *Sociological Theory and the Environment: Classical Foundations, Contemporary Insights*. Rowman & Littlefield Publishers.

Dunlap, R.E. (2002). Environmental sociology: A personal perspective on its first quarter century. *Organization & Environment*, 15(1), 10-29.

Durkheim, É. (1893) 1984. *The Division of Labour in Society*. Basingstoke: Macmillan.

Durkheim, É. (1950). *The rules of sociological method*. New York: Free Press. (Originally published 1895).

Durkheim, É. (1964). *The division of labor in society*. New York: Free Press. (Originally published 1893).

Durkheim, É. (1972). *Selected Writings*, edited by A. Giddens, Cambridge: Cambridge University Press.

Easterly, W. (2006). *The White Man's Burden: Why the West's Efforts to Aid the Rest Have Done So Much Ill and So Little Good*. Penguin Press, New York.

The Economist (2008). *Briefing Globalisation and health care. Operating profit*. August 16, pp. 66-68.

Evans-Pritchard, E. E. (1973). *Witchcraft. Oracles, and Magic Among the Azande*. Clarendon Paperbacks. Oxford.

Foster, B. F. (2005). The Treadmill of Accumulation: Schnaiberg's Environment and Marxian Political Economy *Organization Environment* ; Vol. 18/1; 7-18.

Giddens, A. (1990). *Consequences of Modernity*. Cambridge: Polity Press.

Giddens, A. (1991). *Modernity and Self-Identity*. Cambridge: Polity Press.

Giddens, A. (1999). Risk and Responsibility *Modern Law Review* 62(1): 1-10.

Gimenez, M. E. (1973). The population issue: Marx vs. Malthus. Revised version of a paper presented at the Pacific Sociological Association Meeting in Honolulu, April 1971. Published in *Den Ny Verden* (Journal of the Institute for Development Research).

Green, J. and Thorogood, N. (2006). *Qualitative Methods for Health Research*. SAGE Publications. London.

Grandjean P. (2005). Non-precautionary aspects of toxicology. In: *Toxicology Applied Pharmacology* 2005; 207(2 Suppl): 652-7.

Grandjean, P. (2008). Seven Deadly Sins of Environmental Epidemiology and the Virtues of Precaution. In: Epidemiology & Society, *Epidemiology*. 19(1):158-162, January 2008.

Guy, S. and E. Shove (2001). *The Sociology of Energy, Buildings and the Environment: Constructing Knowledge, Designing Practice*. Routledge Research Global Environmental Change Series, 5.

Haafkens, J. A. (1997). *Rituals of Silence. Long-term tranquilizer use by women in the Netherlands. A Social Case Study*. Academic Dissertation. Vrouwenstudies Geneeskunde, AMC J2-215, Amsterdam. *Handbook of Environmental Sociology* (2002) Greenwood Press.

Hannerz, U. (2002). Notes on the Global Ecumene. In *The Anthropology of Globalization: A Reader*. Ed. by J. X. Inda and R. Rosaldo. Blackwell Publishers, Malden and Oxford.

Harremoës P., Gee D., MacGarvin M., Stirling A., Keys J., Wynne B., Guedes Vaz S. editors, (2002). *The precautionary principle in the 20th century. Late lessons from early warnings*, Earthscan Publications Ltd, London England.

Harvey, D. (1989). *The Condition of Postmodernity*. Cambridge, MA. Blackwell Publishers.

Hawken, P., Lovens, A. and Lovens, L. H. (1999). *Natural Capitalism: Creating a new Industrial Revolution*. Little, Brown and Company.

Held, D.(2004). *A Globalizing World? Culture, Economics and Politics*. Routledge, London and New York.

Homer-Dixon, T. (1999). *Environment, scarcity and violence*. Princeton: Princeton University Press.

Järvikoski, T. (1996). The Relation of Nature and Society in Marx and Durkheim *Acta Sociologica*, Vol. 39, No. 1, 73-86. http://asj.sagepub.com/cgi/content/abstract/39/1/73

Keune, H., Morrens, B. Loots, I. (2008). Risk communication and human biomonitoring: which practical lessons from the Belgian experience are of use for the EU perspective? *Environmental Health*, **7**(Suppl 1):S11 http://www.ehjournal.net/content/7/S1/S11

Kolakowski, L. (1978). *Main currents of Marxism. Its Rise, Growth and Dissolution*. Vol. 1. Oxford: Clarendon Press.

Kopnina, H. (2005). *East to West Migration: Russian Migrants in Western Europe*. Ashgate, Aldershot.

Kopnina (2007). *Migration and Tourism: Creation of New Social Classes*. Cognizant, New York.

Malia, M. (1996). *The Soviet Tragedy. A History of Socialism in Russia, 1917-1991*. The Free Press, New York.

Malthus, T. (1798 1st edition, plus excerpts 1803 2nd edition). *An Essay on the Principle of Population. J.Johnson, London.*

Martell, L. (1994). *Ecology and Society: an Introduction*. Polity Press.

Marx, K. (1973). *Capital*. Vol. 1. New York: Vintage.

Marx, K. (1974). *Early writings*. New York: Vintage.

Marx, K. (1978). *Capital*. Vol. 3. New York: Vintage.

McCally M. (Ed.) (2002). *Life Support, The Environment and Human Health*, The MIT Press, Cambridge, Massachusetts, London England.

McDonough, W. and M. Braungart (2002). *Cradle to Cradle. Remaking the way we make things*. North Point Press, New York.

Mehta, M. and Ouellet, E. (1995). *Environmental Sociology: Theory and Practice*, Toronto: Captus Press.

Mulgan, G. (1998). *Connexity: Responsibility. Freedom, Business and Power in the new Century*. London, Vintage.

Occupational Safety and Health Administration (OSHA) *Chemicals* http://www.epa.gov/triexplorer/list-chemical-osha.htm

Oxfam Novib (2009). http://www.oxfam.org.uk/resources/issues/climatechange/introduction.html

Pollard, R. T. et al. (2009). Southern Ocean deep-water carbon export enhanced by natural iron fertilization. In Nature 457, pp. 577-580, 29 January.

Rhoads, J. K. (1991). *Critical Issues in Social Theory*, Penn State Press.

Roberts, K. (1999). *Leisure in Contemporary Society*, CABI Publishing, New York.

Rosa, E.A., York, R.F., & Dietz, T. (2004). Tracking the anthropogenic drivers of ecological impacts. *AMBIO: A Journal of the Human Environment, XXXIII*, 509-512.

Sachs, J. (2005). *The End of Poverty: Economic Possibilities for Our Time*. The Penguin Press, New York.

Sakwa, R. and A. Stevens (2006). *Contemporary Europe*. Palgrave Foundations

Smelser, N J (1976). *Comparative Methods in the Social Sciences*, Prentice Hall, Englewood Cliffs.

Schnaiberg, A. (1980). *The environment: From surplus to scarcity*. New York: Oxford University Press. 2003 http://media.northwestern.edu/sociology/schnaiberg

Schnaiberg, A., & Gould, K. A. (1994). Environment and society: The enduring conflict.

Sklair, L. (1995). *Sociology of the Global System*. London and Baltimore, MD: Prentice Hall.

Speth, J. G. (2008). *The Bridge at the Edge of the World. Capitalism, the Environment, and Crossing from Crisis to Sustainability*. Yale University Press.

Starkloff, R. (1995). Nature: Adversary or Victim? The Problem of Control in the Relationship between Nature and Society. In Mehta, M. and Ouellet, E. (1995) *Environmental Sociology: Theory and Practice*, Toronto: Captus Press.

Thomas, P. (2006). Being Max Weber. *New Left Review*, 41. September-October. http://www.newleftreview.org/?view=2641

Toxicology, National Research Council, *Waste Incineration and Public Health*, by National Academies Press.

Tsing, A. (2002). The Global Situation. In *The Anthropology of Globalization: A Reader*. Ed. by J. X. Inda and R. Rosaldo. Blackwell Publishers, Malden and Oxford.

United Nations. Statistics Division (2009). http://unstats.un.org/unsd/default.htm

United Nations. Economical and Social Commission for Asia and the Pacific (2009). http://unescap.org/stat/

United Nations Environment Programme http://www.unep.org/

United Nations Population Fund (2009). http://www.unfpa.org/worldwide/

United Nations Framework Convention on Climate Change (2009). http://unfccc.int/2860.php

UNESCO Institute for Statistics http://www.uis.unesco.org/ev.php?ID=2867_201&ID2=DO_TOPIC

US Environmental Protection Agency http://www.epa.gov/

Walker, A. (2008). The White Priestess of "Black Magic." BBC News, September 10, 2008. http://news.bbc.co.uk/go/em/fr/-/2/hi/africa/7595841.stm

Wallace, W. L. (1990). Rationality, Human Nature, and Society in Weber's Theory. *Theory and Society*, vol. 19, No. 2 (April), pp. 199-223.

Weber M. (1904). Objectivity in Social Science and Social Policy, in: *Archiv für Sozialwissenschaft und Sozialpolitik*, translated in English in: Weber M (1949) The Methodology of the Social Sciences, translated and edited by Edward A. Shils and Henry A. Finch, Glencoe: The Free Press.

Weber, M. (1949). *The Methodology of the Social Sciences*. Glencoe, Free Press.

White, K. (2002). *An Introduction to the Sociology of Health and Illness*. SAGE, London.

World Health Organization (WHO)(1995, 2004, 2005, 2006, 2007, 2008, 2009). *The World Health Report* http://www.who.int/whr/en/index.html

World Health Organization (WHO) (2006). Preventing disease through healthy environments: Towards an estimate of the environmental burden of disease.

Wynne, R. (2008). *Employment guidance services for people with disabilities/ Chronic illness and job retention* http://www.introdm.be/idm/idm01.nsf/52b2da8b666e069080256aaa002ab228/995022d8285bee6fc125727300383d77/$FILE/Presentation_Wynne.pdf

Young, R. M. (1994). Malthus and the Evolutionists: The Common Context of Biological and Social Theory. In *Darwin's Metaphor: Nature's Place in Victorian Culture*. Cambridge University Press, Cambridge.

Zinberg, N. E. (1984). *Drug, set, and setting. The basis for controlled intoxicant use*. New Haven: Yale University Press.

In: Health and Environment: Social Science Perspectives ISBN 978-1-60876-216-3
Editors: Helen Kopnina and Hans Keune

Chapter 2

WHY IS SCIENTIFIC INFORMATION ON SERIOUS ENVIRONMENT AND HEALTH PROBLEMS NOT EASILY RECOGNIZED IN OUR SOCIETY?

Hans Keune and Nik Van Larebeke

ABSTRACT

Why is scientific information on serious environment and health problems not easily recognized in our society? In a discussion between a natural and a social scientist, the complexity of this issue is addressed from a scientific, a social and a personal point of view. A part of the answer is to be found in the debate on scientific proof and its problems related to traditional statistical analysis. The Physics Chemical Hygiene is presented as an alternative approach. A part of the answer is to be found in the complicated relation of environment and health science and society. The problem of economic and social consequences of recognizing health problems caused by environmental hazards is discussed and illustrated by practical cases in Flanders. Also the issue of social choice related to environment and health is discussed: does society prefer wealth or health. Furthermore, some aspects of the social fabrication of environment and health are discussed. The issue of biased expertise is of concern here: scientists working secretly for industry and the growing influence of industry experts in official risk assessment. Finally, the difficulty of applying expert knowledge on environment and health to personal daily life is addressed.

INTRODUCTION

In this chapter, we discuss the complexity of translating scientific information on health problems caused by environmental pollution into adequate action, e.g., policymaking. We focus not only on characteristics of environment and health problems but also on the social practice in which these problems are being dealt with. Scientific knowledge about the impact of environmental pollution on problems like cancer, diabetes, osteoporosis, Alzheimer's disease, Parkinson's disease, renal failure and cardiovascular disease (for more detail see *Current environment and health problems*) perhaps may seem self-evident as a basis for policy action. Nevertheless, in real life this is not so obvious. We address the often-heard complaint: why does this knowledge not automatically lead to problem-solving actions? This is addressed through a self-report of a discussion between a natural scientist and a social scientist; in fact the author of this chapter. Professor Nik Van Larebeke[1] (NVL) is spokesman of the Flemish Centre of expertise for Environment and Health (CEH)[2]. Hans Keune[3] (HK) is a political scientist and, as such, is working for the same CEH (http://www.milieu-en-gezondheid.be/English/index.html). We think such discussion report to be relevant for several reasons.

The combination of disciplinary expertise (natural and social science) is potentially fruitful to a broader understanding than merely a specialist view. Typically, specialists focus on only one part of a complex field like environment and health. An often-underestimated element of interdisciplinary exchange or collaboration, especially concerning rather different disciplines like social and natural science, is the challenge to understand each other. This discussion is an example of this.

Next to understanding, the interdisciplinary co-production of understanding in a complementary manner, resulting in more than simply adding different forms of expertise, poses quite a challenge.

Furthermore, we combine the rich reservoirs of experience in the field of environment and health of two experts involved in this matter for some time. We hope to open up a potency that may not be predictable in outcome, but may present an abundance of referential material for in-depth discussion. Referential material is not limited to science, but also policymaking and social debate on environment and health.

And, last but not least, referential material of being not only an expert at work, but also an ordinary citizen ...We do not aim to produce scientific evidence or present unambiguous answers to complicated issues but wish to reflect on current practice from our point of view and experience. The result will contribute to wider critical debate on these important issues.

The discussion and reporting approach is inspired by qualitative research traditions in narrative research (Mishler, 1991, 1995; Riley and Hawe, 2005) and (exemplary) action research (Boog et al., 2008; Coenen, 1987). Several elements are discussed that are of value here. The narrative type of qualitative methodology gives voice to both professional and personal views and may provide in-depth insight into the complexity of content and social aspects of the issue. It opens up our view on the dynamics of science and policy practice, which normally is not discussed in scientific publications on issues like environment and health. The use of "key informants" is a relevant approach for such research and could be extended to a wider variance of actors. This contribution is only part of the whole picture of

---

[1] He is a medical doctor (University of Ghent, 1972) who worked on the mechanisms of cancer induction and on the prevention of cancer since 1970, and who, since 1994, is mainly active in the human biomonitoring.

[2] The Centre of expertise for Environment and Health was funded and steered by the Flemish government. In this Centre, environmental health experts from all Flemish universities and two research institutes jointly investigate the complex relationship between environment and health. In addition, a social scientific expert unit is part of this Centre. After a first programme period, the Centre was granted continuation for the next five years (2007-2011), again including the social scientific unit.

[3] His work on environment & health is mainly related to issues like risk (perception, communication and assessment), knowledge (production, integration, interpretation and application), complexity, inter- and trans -

actors (potentially) relevant and involved in this debate. Moreover, in terms of exemplary action research, we see this as an example of explorative brainstorming that is part of the first (thematic) phase of action research: considering an issue, how can we characterize this issue, who are important actors in the field and how can we deal with it? In order to stir up discussion about the difficulties of translating scientific information into adequate action, we hope this will be a valuable contribution to the wider debate and an invitation to further deliberation.

We did not have a discussion agenda but agreed to the title of the discussion and to its importance. This unstructured format is valuable in order not to restrict the natural flow of discussion, as it was not predictable how this might become most fruitful at the start. The discussion of more than two hours was recorded and transcribed by HK. In this chapter, we use parts of the discussion to clarify the explanations for our starting point: *Why is scientific information on serious environment and health problems not easily recognized in our society?* Whenever we think it appropriate, we add background information relevant for and complementary to our discussion. Both authors are responsible for this discussion report.

In the experience of HK, whenever he tries to explain the work he is doing, people respond like this: "So, if these problems are measured and scientifically qualified as in need of policy action, why, then, does this not automatically happen? Why make such a fuzz of complex debates about these issues, and what seems a lot of lost time between determining the problems and sufficient policy action?" HK then answers: "in real life practice, it is not as easy as it seems..." and then loses his audience in talking about natural scientific complexity, social complexity, and related issues (see chapter Keune et al. in this volume).

Why is it so complicated, and why is this hard to explain? Time for a thorough discussion with a natural scientist (NVL); who cannot only shed light from his experience in the natural science side, but also from his vast experience in addressing these issues to different audiences including policymakers, scientific experts, and the general public. In this discussion report, we pinpoint who is "speaking" by using the abbreviations of the names.

Part one focuses on environment and health science. We introduce current problems based on scientific knowledge. We then examine the issue of scientific proof and its problems related to traditional statistical analysis. Finally, we introduce Physics Chemical Hygiene as an alternative approach.

In part two, we focus on the complicated relationship between environment and health science and society. We discuss the problem of economic and social consequences of recognizing health problems caused by environmental hazards. We present a Flemish example in which mysterious health problems may have environmental causes but where scientific research seems complicated both from a scientific and policy perspective. We also present a case in which problem knowledge does not easily lead to policy action.

Finally, we pose the problem of social choice related to environment and health: wealth or health. Part three examines the social fabrication of environment and health, including the issue of biased expertise. We illustrate this with the example of scientists working secretly for industry, and the growing influence of industry experts in official risk assessment. We reflect on how complicated it is to apply expert knowledge on environment and health to personal daily life.

---

disciplinary (boundary work science—government—society and participation), action research, group assessment/decision support methods.

1. Environment and Health Science

Current Environment and Health Problems

There is substantial natural scientific knowledge on the health effects of environmental pollution. Epidemiological studies increasingly associate environmental pollution with cancer. Even though, due to complexity,[4] epidemiological methods have a limited sensitivity in relation to the effect of particular environmental factors, an extensive series of epidemiological observations indicates that pollutants can be an important determinant of cancer in humans (Cislaghi & Nimis, 1997; Clapp et al., 2005; 2006; 2007; van Larebeke et al., 2005; Belpomme et al., 2007; De Bont & van Larebeke, 2007; Sasco, 2008). Cancer incidence in industrial nations increased throughout the 20th century, and this rise was not confined to types of cancer associated with the use of tobacco (Adami et al., 1993). The rise of environmental pollution in developing countries has been followed in the last decades by an increase in the incidence of cancer (Sasco, 2008). Occupational epidemiological studies have shown a large number of agents associated with an increased risk of cancer in humans (Clapp et al., 2005; 2006; 2007). An impressive study in Northern Italy demonstrates a striking association between biologically active air pollution measured through a decrease in lichen diversity and lung cancer mortality in men (Cislaghi & Nimis, 1997). For more technical details on the complexity of cancer in relation to environmental problems, see Appendix at the end of this chapter.

All this evidence points towards long-term effects of pollution, and the phenomenon is not exclusive to adults. Studies also indicate increased sensitivity to toxic agents early in life, *in utero* and during the postnatal period. Studies show that prenatal or very early life exposures have much more influence on later health than previously thought (Gluckman et al., 2007; vom Saal, 2007). It is evident that during early development, there are windows during which differentiation of cells and tissues is extremely sensitive to disruption by toxic influences. These developmental processes rest upon gradients of naturally occurring molecules, on cell-cell contacts and on contacts between cells and the matrix that surrounds them. It is easy to understand that exogenous molecules can disturb these delicate interactions. Prenatally and in infancy, there is also an increased sensitivity to mutagenic agents (Perera et al., 2002). Experiments on animals and biomonitoring—and to a lesser extent, epidemiological data on humans—indicate a series of diseases of civilisation may be determined by early life exposures to pollution. These include reproductive/endocrine disorders (breast and prostate cancer, endometriosis, polycystic ovary syndrome, infertility, diabetes and the metabolic syndrome, changes in sexual maturation, obesity) (Prins et al., 2007; Prins et al., 2008; Soto et al., 2008; Soffriti, et al. 2008; Hanson & Gluckman, 2008), brain and nervous system disorders (Alzheimer's disease, Parkinson's disease, Attention Deficit and Hyperactivity Disorder) (Barlow et al., 2007; Ross et al. 2007), cardiovascular and pulmonary diseases (atherosclerosis, asthma, chronic obstructive pulmonary disease, hypertension) (Hanson & Gluckman, 2008; Walter et al., 2000), and immune and autoimmune disorders (systemic or tissue-specific autoimmune disease, immunosuppression).

---

[4] The multitude of agents, the multicausal nature of carcinogenesis and the frequent occurrence of negative confounding (a control population is frequently exposed to other agents having a similar effect)

Internal exposure to cadmium is associated with an increased risk of renal damage and in particular with tubular dysfunction (Staessen et al., 1994). Several studies showed an increase in PCB levels in the blood of diabetes patients with a significant dose-response relationship with risk of diabetes (Lee et al. 2007a). Seveso inhabitants exposed to dioxins during the environmental disaster in 1976, showed a higher prevalence of diabetes than unexposed inhabitants. Organochlorine pesticides, too, were linked to diabetes. Another example of an association with increased levels of persistent organic pollutants is the metabolic syndrome (Lee et al., 2007a), which is characterised by an increase in arterial blood pressure and in serum levels of glucose and triglycerides, and by a low High Density Lipoprotein concentration and too much fat around the waist. Also Alzheimer's disease (Ferreira et al., 2008) has been associated with exposure to pesticides in several studies and is also found to be associated with exposure to aluminium.

Another example of a disease sensitive to several agents is Parkinson's disease (Brown et al., 2005), which has been associated with exposure to pesticides, heavy metals and solvents. Osteoporosis and lowered bone density (Alfven et al., 2002; Alvaer et al., 2007) are found to be associated with exposure to cadmium and lead. More intense air pollution, including fine particles and nitrogen dioxide, was also negatively associated with bone density in a study on elderly men aged 75 to 76. Chronic Obstructive Pulmonary Diseases (Walter et al., 2000) have been associated with a diversity of sources: smoking, passive smoking, fine particles, heavy metal fumes, and environmental air pollution. Fine particles, NO2, SO2 and ozone could all contribute to this effect of environmental air pollution. Finally, consider cardiovascular disease (Miller et al., 2007), for which substantial data indicate an important role of environmental air pollution as a determinant. In a recent American study on 65,893 postmenopausal women without previous cardiovascular pathology in 36 cities in the USA, an increase in $PM_{2,5}$ with 10 $\mu g/m^3$ was associated with an increase of 24% in the risk of a cardiovascular incident and with an increase of 76% in cardiovascular mortality.

So with such an abundance of evidence, why is scientific information on serious environment and health problems not easily recognized? We start with the issue of scientific proof of a relation between environment and health. Even though the above exposé of scientific studies underlining this relation seems convincing, the proof is a topic of intense and complicated debate. We will start by introducing a case in court where a worker tries to prove that his health problems are caused by his work.

Scientific Proof

NVL illustrates his view with a court case of a worker trying to demonstrate that his work had a negative impact on his health. As a medical expert, NVL was asked to state whether he thought this case of cancer was occupationally induced, due to the fact that the patient had been working with toxic substances. "It's scientifically absurd to ask for proof that a certain substance is the cause for cancer in a human being, when you already know that this proof can never be given" (van Larebeke, 2004). "Apart from absurd, it's also unfair," NVL continues. The scientific proof of a cancer induced by a substance, e.g., a colouring matter, cannot be proven in an individual case, even when it is known that the substance is carcinogenic by nature. What complicates research on this type of question is that it would be unethical to perform experiments in which a group of people is exposed to such substance without exposure to likewise substances.

Can we compare this to the case of smoking, HK asks? There is enough support for the idea that smoking is bad for health, but how this works is not fully clear. Gee (2008) points out how understanding of how the negative health effect of smoking works is very limited. But, nevertheless, it is generally concluded that severe health risks are involved. To reach this conclusion took several decades:

> For example, some 4,000 chemical substances have been identified in tobacco smoke, of which 167 are currently classified as toxic. However, the disease process that leads to cancer or heart disease in some smokers, but not in others, is still largely unknown after more than 40 years of research. Despite the absence of knowledge about the specifics of the disease process within mixtures, it has still been possible to prevent some harm, albeit belatedly, by reducing exposures to the whole mixture, such as tobacco and combustion smoke, welding and rubber fumes, and fine particles of air pollution. The practical difficulties of studying and understanding complex multicausal biological processes have meant that the attraction of a monocausal approach remains strong, despite the need to be more realistic about biological processes. (...) Reductionism, and the metaphor of the body as a machine, are powerful paradigms that continue to support the idea of linear relationships between specific causes, long after knowledge about irreducible uncertainties, emergent properties and non-linear dynamics have became available. (...) It took more than 40 years of much scientific endeavour and debate between the 1940s and the 1980s, before what was known about smoking and lung cancer was applied to protect public health, following sustained opposition from economic and political interests. In this case, the opportunity for precautionary action on a likely hazard in the 1960s was lost: by the 1990s, only the prevention of known harm was possible.

NVL explains that the court case is completely different for several reasons. Firstly, it is inarguably proven that smoking induces lung cancer and several other diseases. Second, it is an exceptional case: tobacco contains strong carcinogens, and we can easily distinguish between smokers and non-smokers, which makes it easier to investigate (with passive smoking this is not the case). This brings NVL to another point: unethical scientists who deliberately confound evidence to prevent proof of health effects for the sake of corporate interests—there are many well-known accounts of scientists, including within the tobacco companies, who are paid to do so. We revisit this idea in *Biased expertise*. Another example of relatively easy scientific proof is asbestos (see e.g., Gee and Greenberg, 2002): here, the carcinogen results in a tumour that normally rarely occurs otherwise.

HK poses the question of what scientific proof actually is? HK refers to his experience in the framework of the CEH, where a lack of significant statistical evidence seems to lead to the conclusion of policymakers to do nothing (see chapter Keune et al. in this volume). One could assume it is best to leave it like this scientifically, until better data are available, but one might also consider looking at the issue entirely differently. NVL agrees. Indeed, HK continues, the issue of limited data may be unsolvable, and the yardstick of significant statistical correlation makes it almost impossible to produce proof. Such paralysis by traditional analysis is well documented, as we will discuss in the next section.

Paralysis by Traditional Analysis

The traditional scientific approach of dealing with environment and health complexity is increasingly under attack (Harremoës et al., 2002; Grandjean, 2005, 2008; Gee, 2008; Van der Sluijs et al., 2005; chapter Keune et al. in this volume). Questions are raised with regard to both the choice of research questions and the testing of those questions. We may typify this as a "the glass is half full or half empty" discussion: do we depart from the idea that risks are an exception until it is proven otherwise, or do we depart from the idea that risks may occur until we have enough clues that there is no cause for concern? And which demands do we put to proof when testing those hypotheses? Can we investigate and prove complex environmental problems with the traditional reductionist laboratory approach?

An important point of departure for traditional research on environmental issues is the (often not communicated) idea that there is a problem only when proven so. This, contrary to the medical ideal that the patients' health comes first: "better safe than sorry" (Harremoës et al., 2002). Technically, this means that in mainstream environment and health research, care for good (statistical) interpretation of research results mainly focuses on preventing false positives (Type I errors), results that by chance or due to technical errors wrongly show a correlation. The prevention of false negatives (Type II errors), the occurrence by chance or technical errors of wrongly concluding that there is no problem gets much less attention. Thus, the chance that real problems remain undetected is bigger than the chance of detecting problems that do not exist.

Another restriction of the mainstream traditional scientific approach is the application of research standards that are suitable for less complex research, in which data collection and analysis are less problematic and the issue itself far less complex. Environmental research not only is complex, it also is expensive and, when it also incorporates research into health effects, it often concerns only a small number of people. Moreover, apart from occupational exposure, it typically involves low-dose exposure, which causes problems in traditional statistical analysis. It is often erroneously concluded that when these strict statistical tests are not passed, there is no problem.

If we also consider that the burden of proof is with those who examine the possible occurrence and prevention of problems and not with those who introduce potential problems, then from a precautionary perspective, this becomes a political issue in which scientific choices are meaningful. The European Union Registration, Evaluation, Authorisation and Restriction of Chemical substances (REACH; http://echa.europa.eu) law, though weakened in its final version, is meant to take a different approach to protection of human health. The REACH Regulation gives greater responsibility to industry to manage the risks from chemicals and to provide safety information. Manufacturers and importers are required to gather information on the properties of their substances, which will allow their safe handling, and to register the information in a central database run by the European Chemicals Agency (ECHA) in Helsinki. The Agency acts as the central point in the REACH system, managing the system databases, coordinating in-depth evaluation of suspicious chemicals, and running a public access database for consumers and professionals. The Regulation also calls for the progressive substitution of the most dangerous chemicals when suitable alternatives are identified (http://ec.europa.eu/environment/chemicals/reach/reach_intro.htm).

If a pragmatic balance between scientific complexity and the urgency of decisions is important, then according to Van der Sluijs et al. (2005), we must look at a more qualitative approach. Such thinking, according to Van der Sluijs et al., is more and more recognized with regard to chemicals. It comes down to no longer trying to quantify (potential) risks, because this is too complex, but rather qualitatively characterising potential hazards as a basis for assessment. In the case of chemicals, we can better consider intrinsic characteristics (is it carcinogenetic or mutagenic, or toxicity with regard to fertility) than the often-misleading quantitative dose–response relations or exposure models.

So, what can we consider as evidence? Are statistical correlations proof, or do we need understanding of how things work? Or, do we consider some kind of agreement on the assessment of pieces of information as proof, like a grand jury in judicial cases determines whether enough evidence exists? NVL presents a rather pessimistic picture: "With regard to statistics, reality is even darker than what you sketch. What a lot of researchers or institutions are trying to sell is the need for all statistical associations to be corrected for multiple testing. But then it really becomes crazy..." One example is the Bonferroni correction (see Grandjean, 2008). Multiple testing can lead to a problem: when 100 statistical tests are performed, five of them are expected to be significant ($p<0.05$) by chance. However, the number of tests performed does not, in itself, have an influence on the p-value for a particular association, such as between the internal exposure to a pollutant and mutation frequency in blood cells, which might be associated with a p-value=0. 04. Correction for multiple testing implies that the observed association is no longer considered to be significant. If a second test is performed, even if unrelated to the previous one, to study the association between the blood level of another pollutant and height, correction for multiple testing would imply that the association between the first pollutant and mutations in blood cells would no longer be significant. So correction for multiple testing is inappropriate.

Solving the problem that statistical correction for multiple testing tries to address in an inappropriate way depends on repeating the observation. The great epidemiologist Kenneth Rothman (1986) wrote, in his standard work *Modern Epidemiology*: "If no adjustments should be made to p-values when they are reported individually in separate publications, it follows that the process of lumping the results together in one place should not affect the results themselves, regardless of when and how the lumping is done. Therefore, no adjustments for multiple comparisons should be made even if a large number of comparisons are reported at one time, provided that it is clear how many comparisons have been made and that all 'negative' (that is, 'non-significant') results have been reported along with the 'positive' or 'significant' results."

HK wonders if "too strict" statistical testing thus hides a lot of valuable information. "One may in fact wonder whether the often mentioned problem of lack of data in fact is a problem of statistics." NVL agrees that correction for multiple testing discards valuable information. In the analysis within the CEH, NVL never accepted correction for multiple testing. He illustrates the position by an example. When in the CEH, ten different geographical areas are compared and nine areas show similar characteristics and only one clearly differs from the other nine, this will not show in the results if correction for multiple testing is performed. In that case, the statistical conclusion would be that no significant differences exist. HK: "Whereas those exceptions probably are most interesting?" Indeed, NVL responds. HK points at another issue of disappearing differences: the same happens with the differences between individual human beings participating in the human bio-monitoring research.[5] Differences between individuals may disappear in the scientific analysis on a group level. But, of course, we have, here, the problem of low numbers, which gives us low statistical power. According to NVL this is not necessarily so, because the individual differences may be rather big. Also, one never knows for sure whether chance plays a role. Therefore, it is necessary that data resulting from biomonitoring or epidemiological studies are confirmed by additional observations. "But is this practically feasible?" HK wonders. According to NVL, this is mainly a matter of political will. But, we then, of course, have the problem of a time lag, HK points out: if you do not measure again quickly, things may have changed drastically. According to NVL, this depends on what you investigate. "Many environmental health problems are slow phenomena." NVL gives the example of one of the first findings of human bio-monitoring research in Flanders:[6] the fertility problems in the rural area of Peer (Dhooge et al., 2001). To the surprise of all involved, this so-called control study area that was supposed to be "more healthy and clean" compared to another study area, a well-known environment and health-risk area next to a waste incinerator, contained problems that were even worse than in the control area, such as diminished sperm quality. NVL: "In fact, we discovered an unknown problem. And, in fact, it's really hard to believe that we would not try to confirm these results, so that inhabitants have some certainty about their health condition. But apparently, certainty is not wanted it seems."

NVL returns to smoking evidence: "By means of DNA-adducts, we can demonstrate which substances (at minium) are involved in causing lung cancer, e.g., benzoapyrene. But, this does not mean that benzoapyrene is the only agent involved in the induction of lung cancer through smoking. There are 50 other[s]..." He illustrates his point with regard to the occupational cancer court case. The fact that this worker developed cancer at that time, according to NVL, was due to the work he was doing. This does not necessarily mean that he would not have developed this type of cancer several years later, had he not done this kind of job. Several other factors play a role, and that the occupational factor may only be "The drop that floods the bucket."[7] If, for example, you have six genes and you develop cancer when in one cell all six become mutated, if one of them will not mutate spontaneously, you will not develop cancer. If this sixth gene becomes mutated due to an external agent, you will develop cancer. If this external agent is introduced due to occupation, this cancer is occupational, even though five of the six mutations are not occupation-related. "So, the difference in sensitivity between someone who develops a cancer and someone who doesn't may be limited, but in fact is the difference between cancer and no cancer." HK adds that this also is due to

[5] Human bio-monitoring is the measurement of pollutants and health effects in human beings, using biomarkers. Exposure markers measure the amount of chemicals present in tissue fluids, such as blood. Markers for effect measure the possible biological and health impacts of these exposures. Apart from the environment, other factors may influence these markers, such as lifestyle, health status, working conditions, and food intake. If we want to investigate the contribution of environmental exposure to the biomarkers, we need to correct for the contribution of other factors, the confounding factors and co-variables. Many of these co-variables are quantified through questionnaires. In the 2001 to 2006 project of the CEH, chlorinated hydrocarbons such as PCBs, dioxin-like compounds, persistent pesticides such as HCB, and pp'-DDE (a key metabolite of pp'-DDT), heavy metals (such as lead and cadmium) and metabolites of PAHs and benzene were measured. To determine the possible biological and health impacts of these pollutants, biomarkers of effect, such as cardiovascular markers, markers of DNA damage and of cancer risks were used. Markers of respiratory problems (such as asthma) were collected via an extensive questionnaire.

[6] This was part of a study before the CEH was established. This study had the intention of investigating whether human bio-monitoring was feasible in Flanders.

[7] Dutch proverb, meaning something like: the last (tiny) thing that makes you snap. English equivalent: The straw that broke the camel's back.

increased health, because in earlier points of human history, people did not grow old enough to develop cancer. NVL agrees they would not have developed occupational cancer.

According to NVL, if we cannot do experiments on human beings and we don't have enough observations, the questions should be: how can we mechanistically understand the issue, and can we assess the probability of carcinogenetic or other health effects? Or, can we estimate the unlikelihood of these effects? The question of can we conclude with absolute certainty whether a substance has carcinogenetic effects on human beings is scientifically uninteresting according to NVL. HK asks, considering the complexity and research limitations, is it absurd to stick to a positivistic approach? NVL confirms: we should use a probabilistic approach.

We may conclude that the traditional "truth"-seeking approach of science may be questioned when it concerns a complex issue like environment and health. Now, we consider an alternative approach: Physical Chemical Hygiene.

Physical Chemical Hygiene

NVL criticizes an evidence-based approach to environment and health issues and promotes a knowledge-based approach. This is different from his work as a general practitioner, where he focuses on evidence-based interventions. He discusses the example of medicine. If one has a medicine of which not all side effects are known because there is lack of experience in practical use, but you know that it will help tackle a serious health problem, he will use it. HK: "Extinguishing fire may have negative side effects, but stopping fire is of priority importance." NVL illustrates this with the example of AIDS medicines that are prescribed, even though not all side effects are known, since stopping this devastating disease is most important.

But, HK responds, environment and health issues are completely different. "In environment and health issues, the potential problems (of e.g., chemicals) are administered. And generally, it's taken for granted that they cause no health problems, until proven so." NVL: "In my field of expertise (HK: environment and health), we can, in fact, conclude that evidence is often irrelevant because producing it is not feasible." HK asks if the alternative would be to stop everything until it's proven that there is no health risk. Is this feasible? NVL: "We have to leave the idea of zero risk. And we have to start thinking in mechanistic and probabilistic terms." He introduces the concept of Physics Chemical Hygiene (PCH). We know that mutations (change in genetic material, e.g., DNA damage in a cell) are dangerous. We know that the relation between mutations and cancer is exponential. And we know that there are many, many mutagens, the physical or chemical agents that change the genetic material of an organism. These are facts. There are also many hormone-disrupting substances, including PCBs and dioxins, which influence sexual maturation, cancer risk, neurological development, and so on. Now, according to the concept of PCH, the exposure to mutagen and hormone disruptive substances should be limited as much as possible, without exactly trying to find out in detail whether a known mutagen is in fact harmful for human beings.

When we return to the example of the apple growth promoter discussed later in *Biased Expertise*, it is not of priority importance to know exactly whether the natural or the industrial growth promoter is more harmful; this would cost a fortune to discover. Exposure to both should be limited as much as possible, because enough is known about how both substances are dangerous. HK responds: "Thus, knowledge development still is important, but, action is required much earlier when some level of knowledge is established." NVL confirms. According to the concept of PCH, we take action with regard to many agents. But "we impose no extreme measures on one agent. The chance that you take adequate action by focusing extremely on one agent, but not to a hundred other, is small. You try to take action on a broad range, and thus accept the risk of taking only limited action on one agent that really represents a catastrophe. But, the chance of catching fish is much greater." So, HK continues, we take action on the basis of state-of-the-art knowledge, and, in time, this may be evaluated, taking into account future knowledge. NVL agrees and stresses we should not think in absolute terms. But, HK continues, how can we assess any state-of-the-art of knowledge to be a relevant basis for action? NVL responds this is a social issue. The yardstick we have to use is hazard assessment, not risk assessment. In risk assessment, they aim too much at very specific knowledge, which is hard to come by. HK: "They are captured in complexity." NVL confirms: "This keeps us from taking action. Society's capacities should not all go to one specific case, leaving a hundred others unattended." Hazard assessment is aimed at taking action once we know that something is wrong, without knowing exactly to what extent damage is done.

The *Scientifics* of environment and health partly explains the frustration of knowledge being translated into action. We now turn to the *Politics* of environment and health to look for more clues. We start with the political sensitivity of recognition of health issues caused by environmental factors.

2. Science and Society

Trial and Retribution

When posing a scientific question (of which NVL claims no answer can be found) is to be considered absurd, is it then merely irrelevant to science, HK asks. Because one can almost occupy oneself without limits with complicated issues, as in math where some problems have been studied for ages before they reach final conclusion, like Fermat's last theorem (Aczel, 1996), or indeed smoking for health issues (see Box 2). Another example is the continuous scientific research on health risks related to lead, notwithstanding that knowledge on health risks related to lead is substantial, which is not true for many other substances that are rarely studied (Grandjean, 2008). Or, do you mean that when such issues are of societal relevance and are a current problem, one has to use a different approach? NVL replies by stating that when it is scientifically an uninteresting question, it is ethically irresponsible to wait for absolute answers. Especially if one knows that science will not be able to produce such answers now, and probably never will.

HK continues: and we were just discussing one substance, we did not even touch upon the tenfold complexity of a combination of substances. NVL: "I think it to be ethically irresponsible to use the absence of proof that a substance is carcinogenetic for human beings in order to hide the fact that a substance in itself is carcinogenetic and to pretend that this is of no societal relevance. Of course it's proven carcinogenetic for rats or mousse or guinea pigs, thus we have a colossal issue here." NVL continues by pointing to the fact that millions of people suffer occupational cancer, whereas only a very small portion is recognized as such. "Even epidemiologists paid by industry state that all cancers that are not lifestyle related, are due to occupational cancer; environmentally induced cancer they do not believe in." According to a report of the International Agency for Research on Cancer (IARC), published under the form of an article by Boffetta et al. (2009), less than 1% of cancers in French men are due to pollutants, but occupational exposures were considered to cause 3.7% of cancers in French men. However, very few of these occupationally caused cancers are recognized as such. HK refers to the title of this chapter. NVL: "So, if you know that occupational exposure is one of the triggers for developing cancer, and some percentage (perhaps even 10%) of cancers is occupationally induced and should be recognized like that, then we do have a budget problem." HK refers to the Lekkerkerk case in the Netherlands: Lekkerkerk is a village in the Netherlands where in the early 1980s, it became public that a newly built residential area was located on heavily polluted soil. This contributed to the instigation of a law with regard to soil sanitation. HK: "A Dutch environmental health scientist once told me that the commotion at that time made government promise to solve all such issues. When they found out that there appeared to be many of such cases, their enthusiasm slacked considerably." It is impossible to retroactively solve all problems. We have to find another solution. Otherwise, all those companies will go bankrupt. NVL agrees: "I do not consider myself to be competent for defining a final solution. But, I do think that denying these problems is not a good solution." Perhaps recognition of all occupational cancers is not bearable for our society, but this does not exclude us from the responsibility of taking preventive measures. HK: "In fact, we deny the problem and thus will never solve it."

The way we choose to deal with knowledge about complex problems is not without societal consequences. It implies that problem knowledge may cause problems from a policymaking perspective. In the next section, we present a Flemish case illustrating the relationship between scientific research and policymaking.

To Investigate or Not to Investigate: The Mysterious Case of Dendermonde

An interesting example of research on a possible environment-related cancer incidence where the scientific question is legitimate is the complicated case of Dendermonde, a district in Flanders. Nevertheless, NVL does not expect clear-cut answers.

As of 1990, in the district of Dendermonde (together with another district called Aalst) official statistics showed the mortality rate to be higher than in other districts. Until recently, there was no plausible explanation. The question is whether environmental problems have something to do with it.

Dendermonde was one of the top-priority cases that resulted from selecting hot spots in Flanders for human bio-monitoring research measuring pollutants and health effects in human beings. In the CEH, we experimented with the input of a diversity of actors with regard to selecting hot spots: between 2009 and 2011, a limited number of local environment and health issues will be investigated by human bio-monitoring. Important procedural questions are: "On which issues to focus our research effort?" and "Which problems are most urgent and in need of human bio-monitoring?" We designed a procedure to make an inventory of possibly relevant issues and to select a limited number for research. We opened up discussion and invited other scientists, policy representatives and stakeholders to propose research options. This diversity of actors was also involved in the selection process. The case of Dendermonde ended up as the second of five most priority hotspots in Flanders.

Preliminary statistical analysis of cancer incidence in the Dendermonde district, performed by scientists from the CEH, did not result in undisputable conclusions, mainly due to the inadequate quality of registration data. Nevertheless, according to the analysts, there are clues of a higher cancer incidence in this region for several types of cancer, with sex-specific differences in different municipalities in the district region.

This analysis caused some discussion within the CEH and with policy representatives. A major concern was a fundamental difference of opinion between the analysts (with NVL as principal investigator) and other scientists, as well as policy representatives, with regard to use of statistical techniques. The decision not to use a type of statistical correction technique, commonly employed in health research, to avoid the influence of possible analytical distortions due to chance or data limitations, was heavily criticized. The response was that such strict statistical corrections would reduce the probability of falsely concluding that there is a problem, but increase the likelihood of falsely concluding that there is no problem. It was already known that the probability of increased cancer incidence in Dendermonde was greater than the probability of the absence of such an increase. There is also the matter of public accessibility of the data of the cancer incidence registry. The data used in the analysis were not publicly accessible, but were provided to the researchers for another project in the framework of the CEH. Informal contacts with people in charge of the cancer registry learned that more recent data on cancer incidence promise to be of much better quality than the ones used in the current analysis. These new data also are not public though, thus approval is needed for further analysis. And, this approval, it seems, is not straightforward, either.

When the hurdle of approval for the use of better data on cancer incidence is taken, NVL expects that clear and significant differences will be evident. But, he also expects that no explanation for these differences with other districts will be found. The intention of further field research by means of human bio-monitoring is welcomed by NVL. Nevertheless, he has limited expectations of such research being able to explain the higher cancer incidence: he estimates the chance that we will not find clues at 50%. He points out that an easy response might be not to do this type of research, but, one may wonder, which are better and reasonable alternatives? NVL responds: “Of course, if we do not deal with this issue, we know for sure that the problem will remain the same. If, though, we investigate it, we have a 50% chance that we will find an explanation, but also a 50% chance of having to tell the people living there: we have investigated it, but, unfortunately, we do not understand it.”

HK continues by saying that where occupational cancer recognition of an individual may lead to huge financial consequences, it’s not even clear what exactly the problem is. The only thing that is clear is that there is a problem. According to NVL a minister may appear to be taking the problem seriously by doing something, but on the other hand, may also choose not to come up with bad news by ignoring the problem and favouring clear success stories. The prize one pays for the latter strategy is that trust in politics drops. “I give lectures throughout the country, and what I hear everywhere is that the general public does not have any confidence in what politicians say about these problems, because they act as if there are no problems.” HK continues by saying that with policy representatives, this is mirrored by the fact that their own awareness-raising campaigns are not taken seriously by the public. But when he proposes it might be interesting to investigate the reasons, they do not consider this worthwhile. When developing an awareness-raising campaign, they fear potential panic amongst the public should messages clearly state the severity of problems. This results in bleak messages like "perhaps you might consider doing a little less of this or that...." HK: "But, if people are not alarmed about these issues, they will not take warning messages very seriously." NVL agrees.

The implications for policymaking of not knowing the details of a problem may thus cause concern and intensify debate. Knowing about a problem, as discussed earlier in this chapter, may also cause conflicting views. We discuss another Flemish example.

To Know and Not to Act

We discuss a Flemish case of knowing but not acting: chicken eggs. According to NVL, research from scientists and governmental experts has shown with total certainty there is a problem of polluted eggs from chickens kept on polluted soil in private backyards. The contaminated eggs from those chickens create a health risk for the consumers. Still, nothing happens. The question is why? NVL: "I also think it's a pity, I also would like to believe that the chickens we keep are healthier than industry chickens. But when we scientifically know this is not necessarily the case, why do we not act on it?" In this case, NVL does not think financial implications are the driving force. How can we explain this? NVL thinks politicians do want to dispute the common sense of people. People think they have control over what their chickens eat, but they don't really. "On the one hand, they do not want oppose the intuitive notion of control of the general public. On the other hand, they do not want to confront people with the fact that the pollution is everywhere."

According to HK, government experts say there is no certain proof; they do not want to cause panic, and there seems to be no alternative. The problem is not limited to chicken, but this is also true for home-grown vegetables. People want to look after their own food, and you have to discourage them; that's difficult to sell. NVL responds it's not a general rule that all soil is polluted. One could investigate the soil in different areas. Is this feasible for governments, HK asks, both financially and logistically? According to NVL this is possible, since Belgium has many biomedical labs. A lot of money is wasted on irrelevant measurements. This issue is more important, and the outcome will be stable because persistent substances are involved. HK: do you mean that some clinical research should be replaced by research nearer to sources and causes of harm to health? NVL confirms. At the same time though, HK continues, we can say that the people's gardens are not the direct source because many of the pollutants are not produced there, but elsewhere. Finding out exactly where sources are is very complicated. According to NVL a lot of such sources are well known, and their emissions should be filtered. A decade ago, the emission of dioxins by Flemish waste incinerators was diminished by a factor of hundred to thousand by implementing adequate filters, which shows that important progress can be made. Another issue is people burning their waste in their backyards.

The political sensitivity of environment and health issues partly can be explained by the crude choice between wealth and health. This frustrates the alternative approach of physical chemical hygiene.

Mercedes or Health

According to NVL, the physical chemical hygiene approach is not indicated when sufficient evidence is available and allows priorities to be defined, e.g., in relation to cigarettes, asbestos, cars, trucks and combustion motors. We need no more profound knowledge about the devastating effect from current traffic on the environment. Here, the rationale of PCH is not relevant; here, we need something much more powerful: political will. "Society has to choose between Mercedes (as a symbol of sophisticated powerful cars) and health." HK: "Whenever I explain about my work, about the discovery of problems, the existence of some amount of knowledge about these problems, internationally respected scientists warning about the severity of problems, people ask: why then nothing seems to happen?" NVL: "A German politician once said publicly that the majority of scientists are afraid to make statements that are not in accordance with orthodoxy. I have a lot of experience with that; frequently, I make statements that are not mainstream."

HK wonders, whether this is not common for scientists who are at the end of their career? "I can imagine that someone like Philippe Grandjean, who nowadays is not only well known but also a firm criticaster of uncritical mainstream science, twenty years ago might not have been so critical himself." NVL agrees: "I have always been critical, but I would not recommend this to anybody; I did not have much of a career for that reason." He quotes a friend scientist, currently head of the department of experimental oncology to which NVL's research unit belongs, who told him: "Scientists should do science and not philosophy." NVL: "Considering that I am about to retire, and that both the dean of the faculty and my friend are interested in prolonging the activity of my research unit, the Study Centre for Carcinogenesis and Primary Prevention of Cancer, in the future, I have to remind them that my successor should do good research in carcinogenesis in the first place, but that he/she should also pay proper attention to societal issues related to cancer prevention. Otherwise, the work will lose its relevance."

NVL: "We need a very transparent approach. We have to leave the tendency of pretending that everything is all right, that there are no problems. We have to face the problems. At the same time, we must give up on the target of zero risk, which, of course, is not feasible. Aiming for zero risk will increase the tendency to be silent about problems." HK: "Indeed, in our experience in the CEH, we noticed that several people involved in both environmental health science and policymaking more or less tend to panic with regard to problems and information and communication about problems. The standard reflex seems choosing a perspective of diminishing or being silent about problems instead of a perspective that highlights problems. They prefer to give a clear and reassuring message, which is not realistic." NVL: "We have to stop misinforming the public." An example of disinformation would be if there has been an explosion, the advice is to close all doors and windows, but not to worry as there is no danger to public health. We should inform the public realistically as much as possible and use probabilistic terms in describing situations. This will cause concern amongst the general public, but we have to get used to reality as it is. "In all kind of texts, you read about 'the right for health.' In biology, though, justice is irrelevant. And demanding right to a healthy environment is the same as demanding zero risk. A fully healthy environment is not feasible. This would e.g., mean that air should be fully free of polycyclic aromatic hydrocarbons. This would prohibit any combustion process. So, we do not have a right for a clean environment, but, we have the right to an environment that is as clean as possible. And, it could well be that the way we perceive this nowadays is totally different from how it will be perceived in 50 years from now."

3. The Social Construction of Environment and Health

Knowledge—or lack of knowledge—of complex environmental health issues may be difficult for society and policymaking to handle. The understanding may not only be hampered by the elasticity of knowledge on the subject, but also clouded by conflicting interests involved in the expertise dealing with it. The elasticity of knowledge gives way an opportunity for an unscientific manner: the problem of biased expertise.

Biased Expertise

NVL has two answers for why there seems to be so much controversy among scientists. One is that some original mastermind follows non-mainstream routes of thinking and focuses on a small number of the numerous characteristics or mechanisms of a system. For this reason, he will attribute a higher importance to his specific expertise, which results in a selective view. The other reason is the trend of academic research becoming less and less independent. Contacts with industry have become much more intense, and there are hardly any scientists to be found who are fully independent. "Even I have extra earnings from working as an industry expert." The same goes for Hardell (see below), one of the scientists known for unmasking the secret ties between industry and science. NVL: "We are all in it. But, you can hardly say that scientists like Hardell are dishonest about it and try to deny the existence of problems. Most scientists, though, deny problems related to the industry they work for."

NVL refers to several accounts of dubious relations between scientists and industry when it comes to assessment of health risks in relation to products or industrial activities. One example concerns the risk posed to the American population by a naturally occurring toxin (aflatoxin) and an industrial apple growth promoter (Alar^{R}). A well-known establishment scientist stated that the risk posed by consumption of peanut butter (which can be contaminated by aflatoxin) was a hundred times higher than the risk posed by consumption of apple juice (which can be contaminated by Alar^{R}) (Ames & Gold, 1989). An expert working for the U.S. national institute for occupational safety and health (NIOSH) calculated, using a more sophisticated approach, that these risks were more or less equal (Finkel, 1995). The 95% confidence interval of the ratio of the risks associated with these two exposures was calculated as 100 and 1/100. In this situation, it was possible for Ames and his co-worker to choose the different parameters involved in this comparison so that the natural aflatoxin appeared to pose a hundred times greater risk than the artificial apple growth promoter, while different parameters could as well have been chosen to result in a hundred times lower risk.. Thus, even though this scientist cannot be accused of drawing an unscientific or illegitimate conclusion, it's only part of the picture, and thus arbitrary. HK mentions the example of a scientist who at a 2008 WHO/EU-conference on health and environment (http://www.euro.who.int/symposium2008) stated how he had been approached by an oil company: for $10,000 USD he was asked to undersign a statement in favour the industry's view on climate change. There are many such examples. We will discuss examples from science-industry relations and one example of governmental institution-industry relations.

Smokescreen Evidence: Science and Industry

Hardell et al. (2006) introduce us to the mechanics of industry involvement in cancer research, as one example of the demystification of independent scientific judgments of which David Michaels' (2008) *Doubt is Their Product: How Industry's Assault on Science Threatens Your Health* is another. Hardell et al. (2006) refer to several cases in which scientists secretly paid by industry but who were not transparent about it took a stance against scientific results critical of activities or products from the industry financing them. A Swedish professor was recently unmasked as working for Philip Morris for 30 years (Diethelm et al., 2005). Even after first evidence of the relationship was published in 2002, the professor still denied any ties. Hardell et al. (2006) present another Swedish case of a consultancy firm acting as an intermediary between chemical industry and scientists paid by this firm. Another example is a UK professor with recently uncovered ties to asbestos manufacturers and Monsanto.

Next to secret ties between industry and experts, Hardell et al. also present staggering figures about the amount of industry-funded scientific involvement: "A recent study found that one-third of all original research articles published in the *New England Journal of Medicine* and the *Journal of the American Medical Association* were funded by for-profit health care companies [Friedman and Richter, 2004]. Furthermore, one in four original research articles published in these journals had one or more authors with corporate financial relationships and conflicts of interest. The authors with conflicts of interest were two times more likely to report results supporting their sponsor's products [Friedman and Richter, 2004]. For obvious reasons, these numbers are biased. Only those with known conflicts of interests are recognized. Those with hidden ties are not found in the correct column."

Industry-paid scientists fire their secret arms at negative outcomes for industry by presenting their results that show opposite findings as "independent," or try to cast doubt by criticizing the quality of the research that conveys serious risks. Hardell et al. (2006) present examples of such statements: "lacking academic rigor without any regard for research method" and one directly aimed at Hardell himself "Your Review of Hardell's work, with the additional evidence obtained directly from him at interview, shows that many of his published statements were exaggerated or not supportable and that there were many opportunities for bias to have been introduced in the collection of his data. His conclusions cannot be sustained and, in my opinion, his work should no longer be cited as scientific evidence."

Some defence lines published by those accused raise doubts to almost all claims of independence. All involved have some interest in such issues, they argue. This, of course, is correct. Scientists are also entrepreneurs who are dependent on funding for the research that makes their research careers float. In fact, they even have an interest in the existence of problems, one might cynically state.

Monograph Vol. 15 - 22 1977-1980	Public health	Industry	?	Observers		
				PH	I	?
Number (total)	93	11	17	13	18	6
Percentages	77%	9%	14%	35%	49%	16%
Chairs	8	0	0			
Vice-Chairs	6	2	0			
Monograph Vol. 62-79 1995-2001	**Public health**	**Industry**	**?**	**Observers**		
				PH	**I**	**?**
Number (total)	76	83	99	6	35	10
Percentage	29%	32%	38%	12%	69%	20%
Chairs	5	6	1			
Vice-Chairs	7	2	0			

Figure 1.

Whilst Hardell at el. (2006) and others stress the importance of rules with regard to industry involvement, the problem is even more fundamental. Whom can we trust? Is transparency the solution? Perhaps official government bodies like the IARC, part of the World Health Organisation, are a better "environment" for more independent knowledge assessment than the difficult environment of international science. Unfortunately these institutions also appear not to be immune to similar developments.

The Politics of Expert Judgement: Industry Involvement in the IARC

James Huff (2002, 2007) shows how government bodies involved in environment and health research and policy have become increasingly influenced by industry interest and expertise. Huff (2007) discusses several cases of industry involvement with the U.S. National Institute of Environmental Health Sciences (NIEHS) and the International Agency for Research on Cancer. We will focus here on the International Agency for Research on Cancer (IARC; http://www.iarc.fr/). The IARC is part of the World Health Organisation, and its mission is to coordinate and conduct research on the causes of human cancer, the mechanisms of carcinogenesis, and to develop scientific strategies for cancer control. The Agency is involved in both epidemiological and laboratory research and disseminates scientific information through publications, meetings, courses, and fellowships. An important part of the work of the IARC is the Monograph Series. The IARC Monographs identify environmental factors that can increase the risk of human cancer. These include chemicals, complex mixtures, occupational exposures, physical and biological agents, and lifestyle factors. National health agencies use this information as scientific support for their actions to prevent exposure to potential carcinogens. Interdisciplinary working groups of expert scientists review the published studies and evaluate the weight of the evidence that an agent can increase the risk of cancer (http://monographs.iarc.fr/).

Huff (2002), a former Chief of the Unit responsible for the IARC Monographs, wrote about the unprecedented and growing industry influence on the Monographs. In the case of chemical exposures, this resulted in a lower risk evaluation for chemicals. This has serious implications for workers' and public health. According to Huff, the downgrading of chemicals by the IARC is done on the basis of little in-depth scientific evidence. Interestingly, the evaluations of carcinogenicity of chemicals have been influenced by "mechanistic" considerations. This disputes the mechanisms of carcinogenicity in animals to be similarly operative in humans. The increase of downgrading chemicals runs parallel to the increase of the percentage of industry experts in the expert groups assessing the health risks of chemicals (see Figure 1).

Huff J. (2002); Alignment according to Huff

This uncovers the issue of trust: who is to be trusted, and how can we enhance trust and knowledge, assessments and action? Can we trust ourselves? If knowledge about these issues in society does not lead to adequate actions, perhaps this is different in the private life of environmental health experts?

I Rest my Case (at Home)

NVL illustrates how this is not necessarily the case when he gives the example of his wife burning candles in the house. Another domestic issue relates to the occupation of his wife; being a veterinarian, she needs to stock a number of big sacks with special food items in the garage. To facilitate access to the garage, she took away a door between the staircase and the garage, increasing the contamination of the rest of the house by emissions from the car. These examples in NVL's private life are cases of indoor pollution of which he, as an expert, says are a threat to health. At home, though, he cannot easily convince his wife to change her way of living in relation to cosiness and comfort for the sake of reducing health risks (which probably will only materialize in a distant future). She says she cannot live comfortably when she has to adapt her life to the standards NVL promotes in his professional life. Thus, she more or less denies the problem.

Opinions and standards can change over time. The first time NVL left a restaurant when someone started smoking, this caused some discussion with his wife. After some time though, she came to agree with him. NVL explains: "People have to let go of habits they are used to, things they feel psychologically attached to, and this contradicts a purely rational approach based on natural scientific insight." HK gives another example. When we enter a room that smells fresh due to air purifiers or due to the smell of cleaning chemicals, many people associate this with "clean" and "healthy" conditions, whereas current knowledge on indoor air pollution advises us to be alarmed about such smells. We, thus, need to adapt to different triggers, which is difficult. NVL gives a personal example of how adaptation (of taste) takes time. When he was young, he felt attracted to a smoky atmosphere: "I have never smoked, but a smoky atmosphere, like in a bar, had a positive connotation to me. Nowadays, I cannot relax in a smoky atmosphere; I have to think about benzopyrene every second…."

Why, when it comes to unhealthy food and domestic habits like air purifiers and toys for kids, do novelties seem to be introduced much more rapidly in daily life, and become habitual changes quickly? This is another method by which habits are changed without people being warned of the health risks. This probably has to do with marketing, NVL responds. A personal dilemma in relation to expert advice is the advice of NVL's cardiologist to do daily exercise. When the available information on air quality shows levels of fine dust are relatively high, NVL will not go running, because the negative impact will outweigh the positive effect. So you have to put some extra effort in looking for the data, and ask yourself: will I go running or not? Finally, NVL puts these issues in perspective by noting that where we deal with environment and health, similar concerns maybe raised with respect to climate change, where behavioural change also is complicated to accomplish. "We have to realize that with climate change, we are really heading for disaster." HK points at the responses he sometimes gets from people who have no academic background with regard to trust in science and policy. With regard to science, they point out that for every point of view "a" scientist can be found. They hardly pay attention to such messages. With regard to politicians, they have hardly any trust left. This is sustained, of course, when problems are treated so convulsively and without transparency. HK: "It could be very informative when academics would spend more time with people with now academic or expert background."

Conclusion

Why is scientific information on serious environment and health problems not widely appreciated in our society? Posing such a question is easier than answering it, let alone changing the issue in a problem-solving perspective. Both the complexity of the natural scientific aspects of environment and health problems and the social complexity play a role and interact. The natural scientific complexities are difficult the deal with due to both the intrinsic properties of the phenomena under study and the scientific approach itself. The latter is not only hampered by practical and ethical complications, but also by choice: analysing and interpreting "truth" in a traditional statistical fashion not only seems irrelevant here, it also is unwise from a public health perspective.

The alternative may be found in changing from an evidence-based to a knowledge-based approach in which qualitative assessment becomes more important. This is not without social and political consequences, which may in themselves hamper adequate action. The social choice between wealth and health is one thread in the social fabric. It also appears in expert assessment in which conflicting interests play a role. Finally, personal daily life of the environment and health expert shows how practical change will take time and will only follow knowledge in mysterious ways.

Acknowledgments

The studies of the Flemish Center of Expertise on Environment and Health were commissioned, financed and steered by the Ministry of the Flemish Community (Department of Economics, Science and Innovation; Flemish Agency for Care and Health; and Department of Environment, Nature and Energy). We thank the colleagues from the Centre of Expertise for Environment and Health for their cooperation: Prof Willy Baeyens, Prof Greet Schoeters, Prof Ilse Loots, Prof Stefaan De Henauw, Prof Geert Molenberghs, Bert Morrens, Maaike Bilau, Isabelle Sioen, Liesbeth Bruckers, Gudrun Koppen, Elly Denhond, Dr. Vera Nelen, Els Van de Mieroop, Greet Van Kersschaver and Karen Goeyens.

We very much thank Brooke Magnanti (United Bristol Healthcare Trust), Adri Smaling (University for Humanistics Utrecht), Lou Keune (University of Tilburg) and Helen Kopnina (University of Amsterdam) for commenting on the text.

APPENDIX: THE COMPLEXITY OF ENVIRONMENTALLY INDUCED CANCER

Basic Insights Concerning Carcinogenesis

Malignant tumoral transformation is based on accumulation of somatic mutations in several genes in the same cell. In addition, changes in the expression of genes, whether a particular gene is translated into proteins and to which extent this happens, also play an important role (Alberts et al., 2002).The formation of a cancer cell is thus a rare event, resting on a process that starts at conception, when the accumulation of mutations begins.

Most cancers probably arise from stem cells, as other cell types are often discarded before they could give rise to cancer. Agents that directly (by interacting with DNA) or indirectly (by disturbing DNA replication, DNA repair or the cell-division process) increase the number of mutations in cells are able to cause cancer. They are called genotoxic carcinogens in this text. Agents that induce changes in gene expression favouring carcinogenesis have been designated as epigenetic carcinogens, non-genotoxic carcinogens or tumour promoters. Each of these terms has been used in a different context and stresses a different characteristic of carcinogenesis. These classifications are somewhat artificial, as, for instance, some agents have as well genotoxic as tumour-promoting properties. In this text, we use the term non-genotoxic carcinogen. Receptor-binding substances probably are the most important non-genotoxic carcinogens. Receptor-binding substances are able to bind with certain specificity to cellular receptors that are complex proteins, which are active in transmitting signals to other parts of the cell, including to the cell nucleus, and can act like hormones.

Implications of the Fact that Carcinogenesis Rests on Mutations in Several Genes

As mutations accumulate in several genes in the same cell, the likelihood that this happens is given by the multiplication of the likelihood of the separate events, which means that the risk of cancer increases exponentially with the frequency of mutations (Alberts et al., 2002). So a small increase in the mutation rate will cause an important increase in cancer risk.

Picture 6. Complexity opinions.

Exogenous vs. Endogenous DNA Damage and Mutations

Endogenous DNA damage is caused by reactive oxygen substances, oxidized lipids, endogenous oestrogens and endogenous alkylating agents resulting from normal metabolism. Alkylating agents are reactive substances that can attach alkyl groups (such as the CH3 methyl group) to other molecules.

Endogenous DNA alkylation reactions play an important role regulation of gene expression. Exogenous DNA damage is, in quantitative terms, almost negligible compared to endogenous damage (De Bont & van Larebeke, 2004), but a small increase in total DNA damage due to some exogenous agents can significantly increase mutation rate.

Different types of DNA damage are repaired by different DNA repair mechanisms (Wood, 1996). Most forms of endogenous DNA damage are repaired swiftly and correctly without causing mutations by the Direct Repair and Base Excision Repair modes of DNA repair. These are relatively simple and are probably permanently activated actions, as leaving frequently occurring damage unrepaired would be incompatible with survival of cells and tissues. Also, mutations arising from errors in DNA replication occur frequently and are subject to error-free repair by the mismatch repair mechanism.

However, some types of exogenous agents cause damage that is less common and is repaired by other DNA mechanisms that require more time and/or are more error-prone. This is the case of polycyclic aromatic hydrocarbons and aflatoxins, inducing bulky DNA adducts, which are repaired by the Nucleotide Excision Repair, which is more complex and slower than the Direct repair or Base Excision Repair pathways. Of particular importance are the DNA double strand breaks, which occur rarely under physiological conditions, but which are induced efficiently by exogenous agents such as ionising radiation and some chemical substances. These double-strand breaks are repaired by special DNA repair mechanisms, which frequently lead to mutations (Vamvakas et al., 1997).

Genotoxic Agents Can Be Relatively More Efficient at Low Doses

At any one time, living organisms tend to invest only in the expression of those proteins they need. This is may be the case in relation to DNA repair pathways. Evidence indicates that some repair mechanisms are not fully induced in the absence of significant exposure to some exogenous agents. Very low dose or low intensity exposures to agents such as ionising radiation touch the cells off-guard, in the absence of full induction of the relevant DNA repair pathway. Such exposures can induce mutations with a relatively high efficiency in terms of induced mutation frequency per unit of dose. So very low dose (environmental) exposures to mutagenic agents probably can be more effective in inducing mutations than would be extrapolated from experimental (high dose) or occupational studies.

Subgroups with Higher Susceptibility

For many cancers, data show that people with genetic characteristics predisposing them to cancer in general or to certain types of cancer, or increasing their susceptibility to exogenous substances, have a higher incidence of cancer. Ideally, norms concerning exposure should provide protection for these susceptible persons. Interestingly, the difference between susceptible subgroups and the general population is most pronounced at low levels of exposure to exogenous agents. That these susceptible people are often observed to suffer from increased cancer risk points to the importance of the corresponding exogenous agents as causes of the corresponding cancers. In the Flemish population, persons with a high susceptibility to exogenous agents, deduced carrying more at-risk alleles concerning phase II biotransformation enzymes, die in middle age more frequently than the rest of the population.

Non-genotoxic Processes and Agents in Carcinogenesis

Disturbance of expression of normal genes is also involved in carcinogenesis. An increase in entropy, both in terms of DNA sequence and in terms of gene expression, may increase the risk of cancer. Besides mutations in a series of genes, a certain "climate" of gene expression probably is a necessary condition for malignant transformation to occur. That may be one of the reasons why induction of cancer has a long latency period.

Receptor-binding substances, which act similarly to endogenous hormones, could be recognised as the most important non-genotoxic carcinogens. Evidence shows receptor-binding agents display dose-response relationships that obey the Michaelis-Menten equations, as expected on mechanistic grounds, and so are per unit of dose more active at low doses than at higher doses. In recent years, substantial evidence has accumulated showing that these agents affect gene expression or cell functions at concentrations more than a thousand times below toxicological no-effect levels. Also, it has been shown that xeno-hormones can induce or repress transcription of many genes that are turned on or off by the corresponding endogenous hormones. However, they also can induce or repress transcription of genes that are not regulated by the corresponding endogenous hormones.

REFERENCES

Aczel A. D. (1996). *Fermat's Last Theorem: Unlocking the Secret of an Ancient Mathematical Problem*, Basic Books UK.

Adami H.O., Bergstrom R., Sparen P., Baron J, (1993). Increasing cancer risk in younger birth cohorts in Sweden, in: *Lancet*, 341(8848): 773-7.

Alberts, B., Johson A., Lewis, J., Raff, M., Roberts, K., Walter P. (2002). *Molecular biology of the cell.* Garland Publ., Inc.

Alfven, T., Jarup, L., & Elinder, C. G. (2002). Cadmium and lead in blood in relation to low bone mineral density and tubular proteinuria. In: *Environ. Health Perspect.* 110, 699-702.

Alvaer, K., Meyer, H. E., Falch, J. A., Nafstad, P., & Sogaard, A. J. (2007), Outdoor air pollution and bone mineral density in elderly men—the Oslo Health Study. In: *Osteoporos. Int.* 18, 1669-1674.

Barlow BK, Cory-Slechta DA, Richfield EK, Thiruchelvam M. (2007). The gestational environment and Parkinson's disease: evidence for neurodevelopmental origins of a neurodegenerative disorder, in: *Reprod Toxicol* 2007 Apr;23(3):457-70.

Belpomme, D., P. Irigaray, L. Hardell, R. Clapp, L. Montagnier, S. Epstein, and A. J. Sasco. (2007). The multitude and diversity of environmental carcinogens, in: *Environ. Res.* 105 (3):414-429, 2007.

Boffetta, P. M. Tubiana, C. Hill, M. Boniol, A. Aurengo, R. Masse, A. J. Valleron, R. Monier, The G. de, P. Boyle, and P. Autier. (2009). The causes of cancer in France, in: *Ann. Oncol.* 20 (3):550-555, 2009.

Boog B., Slagter M., Zeelen J. and Preece J. (Eds.) (2008). *Towards Quality Improvement of Action Research, Developing Ethics and Standards*, Sense Publishers, Rotterdam.

Brown, R. C., Lockwood, A. H., & Sonawane, B. R. (2005). Neurodegenerative diseases: an overview of environmental risk factors, in: *Environ. Health Perspect.* 113, 1250-1256.

Cislaghi C. and Nimis P. L. (1997), Lichens, air pollution and lung cancer 5. In: *Nature* 387 (6632):463-464, 1997.

Clapp RW, Howe GK and Jacobs M (2005). *Environmental and occupational causes of cancer. A review of recent Scientific Literature.* Published by the Lowell Center for Sustainable Production, University of Massachusetts, Lowell.

Clapp RW, Howe GK and Jacobs M (2006). Environmental and occupational causes of cancer re-visited. In: *J Public Health Policy* 27: 61-76, 2006.

Clapp RW, Jacobs M M and Loechler EL (2007). *Environmental and occupational causes of cancer. New evidence, 2005-2007.* Published by the Lowell Center for Sustainable Production, University of Massachusetts, Lowell.

Coenen, H. (1987). *Action Research as Exemplary Learning. A contribution to the foundation of the methodology of action research*, Groningen/ Utrecht, (in Dutch).

De Bont R., Van Larebeke N. (2007). Adducts, SCE's and Mutations following Benzo(a)pyrene Exposure: a Review of quantitative Data followed by some Considerations regarding Risk. In: *Chromosomal Alterations: Methods, Results and Importance in Human Health*, edited by Guenter Obe and Vijayalaxmi, Springer Verlag. ISBN 978-3-540-71413-2

De Bont R. and van Larebeke N. (2004). Endogenous DNA damage in humans: a review of quantitative data, in: *Mutagenesis* 19 (3):169-185, 2004.

Dhooge W., Stuyvaert S., Kaufman J. M., Koppe G., Nele V., Schoeters G., Van Larebeke N., and Comhaire F.(2001). Observations as to male fertility in the Flemish environment and health studies, in: *Folia Histochem. Cytobiol.* 39 Suppl 2:38-39, 2001.

Diethelm PA, Rielle JC, McKee M. (2005). The whole truth and nothing but the truth? The research that Philip Morris did not want you to see. *http://image.thelancet.com/extras/ 03art7306web.pdf* (Accessed April 12, 2009).

European Commission (2005). The attitudes of European citizens towards environment, Eurobarometer, http://europa.eu.int/comm/public_opinion/archives/ebs/ebs_217_en.pdf, (Accessed April 12, 2009).

Ferreira P.C., Kde A. Piai, A. M. Takayanagui, and S. I. Segura-Munoz (2008). Aluminum as a risk factor for Alzheimer's disease. In: *Rev.Lat.Am.Enfermagem.* 16 (1):151-157, 2008.

Friedman LS, Richter (Ed.) (2004). Relationship between conflicts of interest and research results. In: *J Gen Int Med* 19:51–56.

Gee D. (2008). Establishing Evidence for Early Action: the Prevention of Reproductive and Developmental Harm, In: *Basic & Clinical Pharmacology & Toxicology*, 102, 257–266, Doi: 10.1111/j.1742-7843.2008.00207.

Gee D. and Greenberg M. (2002). Asbestos: from "magic" to malevolent mineral, in: Harremoës P., Gee D., MacGarvin M., Stirling A., Keys J., Wynne B., Guedes Vaz S. (editors) (2002), *The precautionary principle in the 20th century, Late lessons from early warnings*, Earthscan Publications Ltd, London England, pg 52-63.

Gluckman PD, Hanson MA, Beedle AS (2007). Early life events and their consequences for later disease: a life history and evolutionary perspective, in: *Am J Human Biol.* 2007;19:1-19.

Grandjean P. (2005). Non-precautionary aspects of toxicology. In: *Toxicol Appl Pharmacol* 2005; 207(2 Suppl): 652-7.

Grandjean, P. (2008). Seven Deadly Sins of Environmental Epidemiology and the Virtues of Precaution. In: *Epidemiology en Society, Epidemiology.* 19(1):158-162, January 2008.

Hanson MA, Gluckman PD. (2008). Developmental origins of health and disease: new insights., in: *Basic Clin Pharmacol Toxicol* 2008 Feb;102(2):90-3.

Hardell L., Walker M. J., and Walhjalt B. (2007). RE: A rebuttal: Secret ties to industry and conflicting interests in cancer research, March 2007;50(3):227-233 In: *Am.J.Ind.Med.* 51 (9):717-718, 2008.

Hardell L., Walker M. J., Walhjalt B., Friedman L. S., and Richter E. D. (2007). Secret ties to industry and conflicting interests in cancer research, In: *Am.J.Ind.Med.* 50 (3):227-233, 2007.

Hardell L., Walker M. J., Walhjalt B., Friedman L. S., and Richter E. D. (2007). RE: a rebuttal: secret ties to industry and conflicting interests in cancer research. In: *Am.J.Ind.Med.* 50 (9):697-698, 2007.

Harremoës P., Gee D., MacGarvin M., Stirling A., Keys J., Wynne B., Guedes Vaz S. (Eds.) (2002). *The precautionary principle in the 20th century, Late lessons from early warnings*, Earthscan Publications Ltd, London England.

Huff J. (2002). IARC monographs, industry influence, and upgrading, downgrading, and under-grading chemicals: a personal point of view. International Agency for Research on Cancer, In: *Int.J.Occup.Environ.Health* 8 (3):249-270, 2002.

Huff J. (2007). Industry influence on occupational and environmental public health, In: *Int.J.Occup.Environ.Health* 13 (1):107-117, 2007.

Keune H., Loots I., Bruckers L., Bilau M., Koppen G., Van Larebeke N., Schoeters G., Nelen V., Baeyens W. (2008). Monitoring environment, health and perception: an experimental survey on health and environment in Flanders, Belgium, In: *International journal of global environmental issues*, Special issue on "Risk Perception and Social Trust," 8:1/2(2008), p. 90-111.

Lee DH, Lee IK, Porta M, Steffes M, Jacobs DR, Jr. (2007a). Relationship between serum concentrations of persistent organic pollutants and the prevalence of metabolic syndrome among non-diabetic adults: results from the National Health and Nutrition Examination Survey 1999-2002, in: *Diabetologia* 2007 Sep;50(9):1841-51.

Lee DH, Lee IK, Steffes M, Jacobs DR, Jr. (2007b). Extended analyses of the association between serum concentrations of persistent organic pollutants and diabetes. In: *Diabetes Care* 2007 Jun;30(6):1596-8.

Lucas RM, McMichael AJ (2005). *Association or causation: evaluating links between "environment and disease."* Bull WHO 2005, 83:792-795.

Michaels D. (2008). *Doubt is Their Product: How Industry's Assault on Science Threatens Your Health*, Oxford University Press.

Miller KA, Siscovick DS, Sheppard L, Shepherd K, Sullivan JH, Anderson GL, Kaufman JD. (2007). Long-term exposure to air pollution and incidence of cardiovascular events in women. In: *N Engl J Med* 2007 Feb 1;356(5):447-58.

Mishler, E. (1995). Models of narrative analysis: a typology, in: *Journal of Narrative and Life History*, 5, 87–123.

Mishler E. (1991). *Research Interviewing: Context and Narrative*. Harvard University Press, Cambridge, MA.

Perera F, Hemminki K, Jedrychowski W, Whyatt R, Campbell U, Hsu Y, et al. (2002). In utero DNA damage from environmental pollution is associated with somatic gene mutation in newborns, in: *Cancer Epidemiol Biomarkers Prev* 2002 Oct;11 (10 Pt 1):1134-7.

Prins GS, Birch L, Tang WY, Ho SM. (2007). Developmental estrogen exposures predispose to prostate carcinogenesis with aging, in: *Reprod Toxicol* 2007 Apr;23(3):374-82.

Prins GS, Tang WY, Belmonte J, Ho SM. (2008). Perinatal exposure to oestradiol and bisphenol A alters the prostate epigenome and increases susceptibility to carcinogenesis, in: *Basic Clin Pharmacol Toxicol* 2008 Feb;102(2):134-8.

Riley T. and Hawe P. (2005). Researching practice: the methodological case for narrative inquiry, in: *Health Education Research* 2005 20(2):226-236; doi:10.1093/her/cyg122

Ross MG, Desai M, Khorram O, McKnight RA, Lane RH, Torday J. (2007). Gestational programming of offspring obesity: a potential contributor to Alzheimer's disease, in: *Curr Alzheimer Res* 2007 Apr;4(2):213-7.

Rothman KJ. (1986). *Modern Epidemiology*. Boston: Little, Brown. & Company.

Sasco AJ (2007). Cancer and globalization. In: *Biomed Pharmacother*. 2008 Feb;62(2):110-21. Epub 2007 Dec 5.

Soffritti M, Belpoggi F, Minardi F, Maltoni C. Ramazzini (2002). Foundation cancer program: history and major projects, life-span carcinogenicity bioassay design, chemicals studied, and results, in: *Ann NY Acad Sci* 2002; 982:26-45.

Soto AM, Vandenberg LN, Maffini MV, Sonnenschein C. (2008). Does breast cancer start in the womb?, in: *Basic Clin Pharmacol Toxicol* 2008 Feb;102(2):125-33.

Staessen JA, Lauwerys RR, Ide G, Roels HA, Vyncke G, Amery A. (1994). Renal function and historical environmental cadmium pollution from zinc smelters, in: *Lancet* 1994 Jun 18;343(8912):1523-7.

Vamvakas S., E.H. Vock and W.K. Lutz. (1997). On the role of DNA Double-Strand Breaks in Toxicity and Carcinogenesis., in: *Crit. Rev. Toxicol.* 27: 155-174.

Van der Sluijs J.P., Kaiser M., Beder S., Hosle V., Kemelmajer A., Kinzig A. (2005). *The Precautionary Principle*, UNESCO COMEST Report.

Van Larebeke N. A., Birnbaum L. S., Boogaerts M.A., Bracke M., Davis D. L., Demarini D. M., Hooper K., Huff J., Kleinjans J. C., Legator M. S., Schoeters G., VähäkangasK. (2005). Unrecognized or Potential Risk Factors for Childhood Cancer. In: *Int J Occop Environ Health* 2005;11:199–201.

Van Larebeke N. (2004). *Health effects of environmental exposure to mutagen or hormone disruptive agents* (Gezondheidseffecten van blootstelling aan omgevingsconcentraties van mutagene of hormoonverstorende agentia), study commissioned by the Flemish Environmental Agency, MIRA, MIRA/2004/03, UGent (in Dutch).

vom Saal F.S. (2007). Could hormone residues be involved?, in: *Hum Reprod* 2007 Jun;22(6):1503-5.

Walter, R., Gottlieb, D. J., & O'Connor, G. T. (2000). Environmental and genetic risk factors and gene-environment interactions in the pathogenesis of chronic obstructive lung disease. In: *Environ. Health Perspect.* 108 Suppl 4, 733-742.

Watanabe H, Suzuki A, Kobayashi M, Lubahn DB, Handa H, Iguchi T. (2003). Toxicogenomic difference between diethylstilbestrol and 17 beta-estradiol in mouse testicular gene expression by neonatal exposure. In: *J Mol Endocrinol*. 2003;31:487-497.

Wood R.D. (1996). DNA repair in eukaryotes, in: *Annu. Rev. Biochem.* 65: 135-167.

In: Health and Environment: Social Science Perspectives ISBN 978-1-60876-216-3
Editors: Helen Kopnina and Hans Keune

Chapter 3

DEFINITIONAL STRUGGLES, ENVIRONMENTAL RISK ASSESSMENTS AND THE LEVEL OF TRUST: THE SUDBURY SOILS STUDY, 2001-2008

François Dépelteau

ABSTRACT

This chapter describes the role of political and commercial interests in the problem-solving strategies employed by local governments in cases related to environmental and health damage. Specifically, the controversial case of soil pollution in Sudbury (Canada) and different stakeholders involved in attempts to solve health and environmental damages in a situation of conflict of interest are discussed. In this case, the companies who are the main polluters pay the scientists investigating the case and have an implicit veto right on any scientific and non-scientific decisions. The author argues that this is problematic with regard to democratic values and social trust in risk management.

1. THE SUDBURY SOILS STUDY

Approximately 158,000 people live in the city of Greater Sudbury (Northern Ontario, Canada). The development of Sudbury has been deeply shaped by the ups and downs of two mining companies: Inco (today, Vale Inco) and Falconbridge Nickel Mines, Ltd. (today, Xstrata Nickel). By the end of the 1920s, Inco controlled 90% of the world nickel market. By the 1970s, Falconbridge had become the third-largest nickel producer in the Western part of the world. Mining industry denotes heavy pollution. From 1971 to 2000, the Ontario Ministry of the Environment (MOE) conducted tests on the level of soil contamination in the Sudbury area. The concentrations of cobalt, copper, nickel, and arsenic frequently exceeded MOE generic soil quality criteria. Due to these findings, in 2001, MOE recommended that a more detailed soil study be undertaken to fill data gaps and that a human health and ecological risk assessment be undertaken. Both Vale Inco and Xstrata Nickel "voluntarily accepted these recommendations" (SARA Group, 2008b: 1-2) and funded the risk assessment. After eight years, the total cost of the risk assessment totaled approximately 13 million Canadian dollars. Voluntary actions have been made on many occasions by companies since Canada adopted the approach of deregulation (Mcdonald, 2007). Voluntary actions by companies allow them to improve their green images and to have more control over the adoption of new norms, policies, regulations, etc. related to the environment. In Sudbury, Vale Inco and Xstrata Nickel became two key players in the risk assessment.

In 2003, the SARA Group was selected to undertake the risk assessment. The Sudbury Soils Study (SSS) was created, and it has to answer this question: "Do Sudbury soils containing metal and arsenic levels above the generic guidelines pose an unacceptable ecological or human health risk?" (SARA Group, 2008b: 1). One "Technical Committee" was formed to "oversee" the study, comprised of the two mining firms, MOE, the Sudbury & District Health Unit, the City of greater Sudbury, and the First Nations and Inuit Health Branch of Health Canada. Decisions have been taken on a consensual basis, which means that no decision could be made without the approval of Vale Inco and Xstrata Nickel—the ones who are responsible for most of the targeted pollution.

The SARA Group has always maintained that they use the "best science" one could find to conduct this type of risk assessment. According to the first volume of the report distributed by the SARA Group in May 2008, the risk assessment has been:

> (…) conducted in accordance with the available Canadian, U.S. EPA, and MOE framework and regulatory guidance for SSRAs (Site-specific Risk Assessment), with emphasis on those provided in the *Guideline for Use at Contaminated Sites in Ontario*' (SARA Group, 2008b: 1-2).

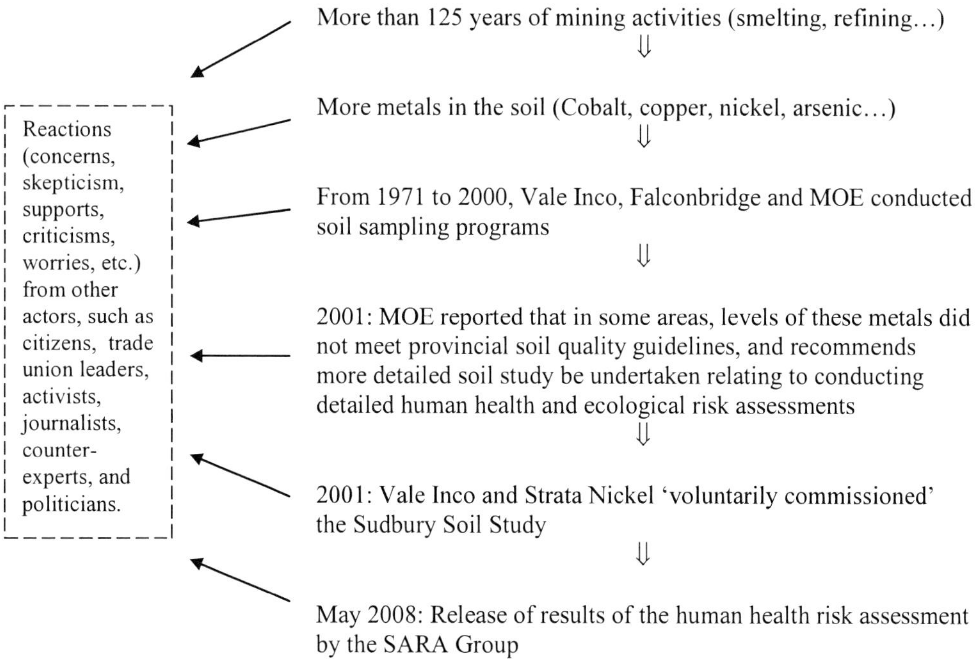

Figure 1.

Picture 7. Pollution.

The SARA Group's risk assessment results surprised many Sudbury residents who were accustomed to living in Sudbury's degraded environment. The study "predicted little risk of health effects on Sudbury area residents associated with metals in the environment" (SARA Group, 2008b: 33). It also concluded that "there were no unacceptable health risks predicted for exposure to four of the six Chemicals of Concerns studied: arsenic, copper, cobalt, and selenium." (Ibid). The risk "for typical exposures to lead" "are within acceptable benchmarks for protection of human health" (Ibid)—except in some areas of Sudbury where "levels of lead in some soil samples indicate a potential risk of health effects for young children" (Ibid). There is some risk "of respiratory inflammation from lifetime exposures (70 years) to airborne nickel in two areas" (Ibid), but "it is unlikely that any additional respiratory cancers will result from nickel" (Ibid). Finally, "anglers, hunters and First Nations people who may consume more local and wild game are at no greater risk of health effects due to metals in the environment than the general population" (Ibid).

Like many risk assessments in Canada and elsewhere, the SSS has raised concerns and strong criticism. By relying on the SSS case study, we want to propose some guiding principles and concepts to show how the level of public trust towards risk assessments is related to definitional struggles and democratic problems. In brief, we will explain that definitional struggles and the low level of public trust towards this risk assessment comes from concerns related to two important democratic issues: i) the central role played by the two mining companies in the realization of the risk assessment, which create a situation of conflict of interest, and ii) the lack of key resources available to citizens who cannot evaluate the scientific validity of the risk assessment by themselves due to their limited scientific skills and knowledge. In this context, these two democratic problems reinforce each other in terms of definitional struggles and lack of trust: since they cannot evaluate the risks by themselves, residents of Sudbury are asked to *believe* the conclusions of experts, which are not seen as being independent from the mining companies. This dynamic is also fueled by more than 125 years of polluting activities and previous studies and concerns about effects of metals on human health (see Figure 1).

The predictable result of all these dynamics is a low level of public trust, which could eventually jeopardize the risk assessment.

2. Definitional Struggles and Social Games

We are not proposing one general theory that is able to predict the evolution of public trust and risk assessments. We are using a toolbox of guiding principles and concepts that allows relevant and accurate descriptions of non-deterministic social patterns. Being inspired by the risk society approach (Beck, 1999, 1992; Strydom, 2002) and the relational approach of B. Latour (2005, 2004, 1987), we present risk assessments as "definitional struggles" involving relations between interdependent human actors—such as experts, managers of mining industries, public servants, activists, journalists, politicians, and citizens—and non-human actors such as lead, arsenic, cobalt, copper, and nickel.[1] In this respect, health and environmental issues do not exist outside of social relations between key actors. These issues are not "matters of fact" imposed by "Science" and they are not only subjective social constructions. These issues are "matters of concern" (Latour, 2004, 1987) linked to unequal power relations between human and non-human actors, where experts play a major role in terms of experimentation and legitimacy. In this sense, the credibility of experts is crucial, especially when the risks are invisible to the lay-people.

It is not possible to reduce these types of stories to the relations between experts and lay-people. The case of the SSS shows that the intensity and the form of definitional struggles depend on the strategies, emotions, mobilization of resources, reactions, etc. of activists, journalists, citizens, experts, public servants, politicians, and managers of private firms such as mining industries. In this sense, risks are more relational than "Scientific." Their perception and management are related to social relations rather than just cold and neutral technical choices and the discovery of facts. Risk assessments are social also because their evaluation is influenced by social factors. Risks always stay away from lay-people's sight due to being indirectly visible only through the mediation of experts' representations and discourses. Lay-people cannot compare experts' representations of risk with the invisible objective reality of these risks. Usually, they can only evaluate experts' representations by relying on their limited scientific skills and knowledge (if any), other criteria such as the

[1] Considering metals and other "things" as non-human actors comes from Latour's idea (2005, 2004) that things do have some "agency" (like human beings) as long as they leave some traces and when they are part of associations with other things and human beings. It is clearly the case with arsenic, nickel, lead, etc. in our story. By being associated with human beings and other things, they have contributed to the story (or the structuration) of the SSS. By being associated with other actors, their presence pushed for one risk assessment and still generate frustration and worries.

image, the prestige or the reputation of the experts, and their (perceived or real) connections to other actors and interests. In other words:

> When people assess risk, they are not only responding to the risks attached to events or technologies and the quantifiable harm they might cause; they also consider the behaviours and responsibilities of the relevant institutions. (Higgins and Natalier, 2004: 83)

Once again, in this type of definitional struggle and social game, trust is one of the main assets experts may rely on. Experts try to present themselves as the most trustworthy people by diminishing the legitimacy of the claims made by competing actors (counter-experts, activists, etc.). Usually, the goal of this game is to make sure that the people will trust your definition of the risk more than your competitor's definition. This is a process of "translation" where, for instance, experts (or counter-experts) work hard to build one network in which the representations of allies, sympathizers, citizens, etc. correspond to their own representation of the risk. This network-building process is founded on various forms of communication (public meetings, press conferences, publications, etc.) and the use of techniques (such as charts, graphs and peer reviews) (see Latour, 1987).

We cannot rely on the classical separation between the irrational public playing social games and the rational experts relying on cold methods and facts. Experts are playing social games partly because the "acceptance of any risk is more dependent on public confidence in risk management than on the quantitative estimates of risk" (Starr, 1985). By working with and for governments and corporations, risk assessors may also be connected to some economic and political interests. These interests have nothing to do with "pure" science. According to Slovic (2000), the following social dynamics explain how risk assessments have become so contentious in the last decades:

> (...) risk assessment and risk management (...) have become much more contentious. Polarized views, controversy and overt conflict have become pervasive. Frustrated scientists and industrialists castigate the public for behaviors they judge to be based on irrationality or ignorance. Members of the public feel similarly antagonistic toward industry and government. (Slovic, 2000: 316)

In reaction, governments and industries have made significant efforts to change relations between experts and lay-people by relying on "the young and still primitive field of risk communication in search of methods" (Slovic, 2000: 318). It appears that risk communication has been an ineffective method in dealing with trust issues. As P. Slovic noticed:

> (...) there is rather little evidence that risk communication has made any significant contribution to reducing the gap between technical risk assessments and public perceptions, or to facilitating decisions about nuclear waste or other major sources of risk conflict. The limited effectiveness of risk-communication efforts can be attributed to the lack of trust. (2000: 318)

It is not enough to just improve the communicative skills of risk assessors because the lack of public trust is related to democratic issues. We think that any efficient solution for this type of problem has to be related to these democratic problems. In this respect, social scientists can make a contribution by identifying these issues and proposing some solutions.

As social actors being involved in the structuration of the collective, social scientists can use their expertise and social opportunities, strategies, etc. (classrooms, congresses, publications, press conferences, etc.). Telling the story of the Sudbury Soils Study is one small example of the possible contributions that can be made by social scientists. The goal is not to close the movement of the structuration of the collective by revealing the truth. The goal is to contribute to this movement by: raising issues (such as one conflict of interest), asking different questions, looking for answers, proposing other concepts, talking about other actors (mothers, grandfathers, activists, etc.) and their concerns, understanding the choices of some methods or techniques used by experts, comparing different but similar stories, etc.

3. Democratic Issues and Trusting the System

There are many different reasons as to why the nature of risk assessments is contentious. One could argue that science of risk assessment is too young and fragile to offer one solid protection against public fears, concerns, and mistrust. The perception and management of risks is also quite often related to "an adversarial legal system that pits expert versus expert, contradicting each other's risk assessments and further destroying the public trust" (Slovic, 2000: 324). Furthermore, and maybe more importantly, the rise of participatory democracy and related organized challengers (environmental organizations, pressure groups, green parties, etc.), which are getting more efficient by using their own counter-experts and the media, can seriously challenge any risk assessments and their experts—even with limited scientific skills.[2]

Once again, we think there are also problems related to undemocratic characteristics of risk assessments. The SSS case study strongly suggests that the low level of public trust is partly related to the relations between the risk assessors and some economic and political interests. Conflicts of interest decrease the level of public trust. As Slovic stated: "scientific analysis of risks cannot allay our fears of low-probability catastrophes or delayed cancers unless we trust the system" (2000: 324). In our case, it seems that the apparent conflict of interest of two mining industries (Vale Inco and Xstrata Nickel) is seriously affecting the level of trust towards the results of the Sudbury Soils Study. The risk assessment was "voluntarily commissioned" by the two mining industries, who happened to be two significant polluters and key stakeholders of the risk assessment. Many activists, journalists, citizens, and even some politicians have a hard time believing that the risk assessment in Sudbury has been based on independent and "good" science.

In this respect, one of the most important challenges facing people living in risk societies is to democratize sciences and relations of definition of risks by clearly and fully disconnecting experts from any specific economic and political interest. Impressions—justified or not—that experts and their sciences are under the influence of governments and/or

[2] We also have to note that the emergence of the risk society can be associated to the diffusion of scientific skills and knowledge in the whole population. The importance of these risks and the new dependency and mistrust towards experts and sciences have encouraged many citizens, activists, etc. to develop some scientific skills and knowledge in their attempts to avoid the role of simple believers. In our case, many activists are learning about science by contesting the official risk assessment conducted by the SARA Group. However, this knowledge does not significantly reduce the gap between the experts and the others. The official experts usually maintain some evident advantages in terms of scientific knowledge and access to other key resources.

corporations can quickly destroy the public trust, especially when the risk-bearers do not have the capacity to evaluate the risk assessment by themselves. When the public trust is low in relation to one real or imagined conflict of interest, risk-bearers can be compared to blind and deaf people invited to cross a street by someone who can see and hear the cars but who is not trusted. This would not be a pleasant experience, especially if by deciding to live in one polluted area, you accept to be this blind and deaf risk-bearer for the rest of your life. Trust is always fragile; it is so difficult to build and so easy to destroy (see Slovic, 2000: 319).

The SSS has raised skepticism, criticisms, and debates since its release in 2001. Experts and managers of risk assessment have been often shocked by these reactions. Not surprisingly, they have usually associated criticism against their work to irrational factors and lack of information. This is one typical reaction from experts who oppose their "Science" to the ignorance and/or the lack of good faith of other human actors who contest their methods and conclusions (see Slovic, Flynn and Layman, 2000: 275).

If we accept Beck's thesis of the risk society, criticisms against experts and their works are not simply irrational. Industrial activities and sciences have produced new invisible risks such as global warming, nuclear power, the depletion of the ozone layer, and the contamination of soil, air and water. These new risks affect our health and environment at different levels, and we need to know how dangerous they are and what action should be undertaken to protect ourselves. In this world, the definition, management, and distribution of risks are becoming more and more crucial for all of us, even if we cannot see them by ourselves. Of course, this situation creates new debates and conflicts. As Beck wrote:

> (…) in advanced modernity, the social production of wealth is systematically accompanied by the social production of risks. Accordingly, the problems and conflicts relating to distribution in a society of scarcity overlap with the problems and conflicts that arise from the production, definition and distribution of techno-scientifically produced risks. (Beck, 1992: 19)

Experts play a crucial role in the definition and management of risks. They are usually the only ones who can represent these risks through their scientific procedures and techniques. Scientific explanations, diagnosis, data, etc. are the main sources of legitimacy in terms of perception and management of risks. This is why challengers or skeptics often rely on counter-experts to re-evaluate one pre-existing definition of risk.

Even if scientists and (counter)experts are still significant actors capable of making socially accepted knowledge claims, "knowledge is not restricted to science but includes everyday knowledge of lay-people, citizen groups, organizations, and societal institutions, often mediated by the media" (Zinn, 2008: 26). This decentralization of knowledge partly results from the democratization of higher education and the proliferation of knowledge through mass media and internet. Citizens, activists, trade unionists, etc. can now attempt to be quasi-experts and challenge the experts by making "scientific" statements. By being associated with non-human actors such as metals and cancers, citizens can also use their own empirical senses and common sense knowledge to raise questions and doubts about previous risk definitions. The days where experts could stop any discussion with their science are over. There are many reasons as to why lay-people do not blindly trust sciences and experts anymore. "Matters of facts" are replaced by "matters of concerns," and it is becoming more and more evident that "sciences in the making" or "in action" are part of complex associations involving various actors (Latour, 2005, 1987). This is partly because technoscience has lost its previous quasi automatic and guaranteed legitimacy by producing the new risks that people have to face today. Further, some lies and frauds have been made with the association of some experts. Another reason is because experts do not always agree—especially when it is about political issues, for example, the number of victims of the Chernobyl accident or the safety of nuclear power plants. Industrial modernity has been replaced by "reflexive modernity," where science is not simply associated with progress (Beck, 1992). Therefore, we assist with the proliferation of definitional struggles where sciences and experts play a central role, even if they are often contested through "sub-politics" (Beck, 1992).

Another reason for the proliferation of definitional struggles is the high complexity of many risks (Ruckelshaus, 1983). Most of the new risks are very difficult to define and manage because:

- They refer to various levels of risks (low, high, acceptable, not acceptable…) that are not "objective" and are, therefore, usually socially constructed and contested. The acceptability of one risk depends partly on subjective factors such as the culture and the interests of the actors. (Do we let children play in contaminated soil if we think that as long as they do not eat the soil it is relatively safe?) Levels of acceptability cannot be simply scientifically imposed by experts to parents, workers, etc. They are constantly negotiated through relations of definition—at least in a democratic society.
- Various risks interact with each other, and it becomes difficult to establish clear scientific knowledge based on relation of causality. If the rate of cancer is higher in one part of the world, is it caused by the contamination of the soil or by the interaction of other factors such as life habits? How can we prove that one specific cancer is caused by one particular factor? How do we calculate the cumulative effects on health by the accumulation and interaction of different substances without underestimating them? Do we have enough knowledge to make these calculations? How do we deal with contradictory scientific explanations? How do we deal with different norms adopted by different experts and governments? Etc.

If we think about the importance of many new invisible risks, the end of scientific certainty, the relative diffusion of knowledge, and the complexity of these risks, we can quite simply understand why risk assessments are contested and often lead to definitional struggles. The challenge is not to produce "perfect" risk assessments that do not create any criticism. Undemocratic risk assessments usually reject any criticism by not answering "annoying" (but relevant) questions, refusing to build bridges over the ravine separating the experts and non-experts, assuming that the role of others (residents, journalists, activists, etc.) is to simply believe the experts, or by simply physically or symbolically destroying any contester. The challenge of risk assessments is to produce democratic relations of definition where non-experts can express themselves and have good reason to trust the conclusions made by the experts.[3] Otherwise, we end up dealing with a "crisis of confidence" characterized by "a

[3] As Welch et al. (2005) noted, social trust is founded on "mutually shared expectation," the absence of scandals, the intensity of relations between the actors, the fact that the actors know the motivations of the others actors they trust, the idea that there is one common interest (like finding the truth about the nature of the risks), and a

profound breakdown of trust in the scientific, governmental and industrial managers" of risks (Slovic, Flynn and Layman, 2000: 281). The level of trust inspired by the experts is the result of social relations (Welch et al., 2005) between various actors such as experts, challengers, journalists, politicians and members from the public. Clearly, the democratic nature of risk assessments is a crucial issue.

4. The Social Dynamic of Risk Definitions

Risks are socially constructed and reconstructed through relations of definition involving various actors and their emotions, values, interests, etc.

Parents and grandparents, for instance, are worried about their children and grandchildren. They want to know if Sudbury is a clean place to raise their (grand)sons and (grand)daughters. They want to know how safe it is to let their children play in the park or in their backyard due to possible soil contamination. They want to know if they should grow tomatoes in their garden and let their children eat them. They want to know if local food at the downtown market is toxic or not. By raising these questions, they upset local producers who fear that they will not be able to sell their products. These concerns, worries and fears have already been expressed many times in relation to the SSS. Two brief examples:

- In March 2004, by noting that "Nickel giants *Inco* and *Falconbridge Ltd.* are paying for the study," the Sudbury Star (reproduced in www.minesandcommunities.org) reported that a "new Sudbury woman is puzzled as to why Sudbury's drinking water isn't tested for the presence of nickel, as it is for other minerals of concern such as copper and lead." The article also reported that "Reed said research she has obtained shows that Sudbury has the highest level of nickel in its drinking water in the world. The information also says that nickel can be a carcinogen or cancer-causing agent. Reed's information comes from the *Agency for Toxic Substances and Disease Registry*, an agency of the *U.S. Department of Health and Human Services*." According the journalist, these comments were made at the public session of the Sudbury Soils Study's technical committee at Tom Davies Square.
- In June 2008, Sudbury citizens expressed their concerns and fears in an article published by another regional newspaper, *Northern Life*. Kim Johnson is raising her three-year-old granddaughter in Copper Cliff—one of the most contaminated areas of Sudbury. She wondered if she could get her soil tested. "Clearly, there is a risk in this locale." "What about families that can't afford a soil test?" "How will you ever know if it's detrimental for your family?" Duane and Lisa Bradley also live in Copper Cliff. "I'm just worried about the children playing in the sand," said Duane. "Is it harmful to the children? I always take my daughter Briana to the park. She makes sandcastles. If she has a cut and gets sand in the cut, will that be a concern?" (Reproduced in *Republic of Mining*, 16th June 2008)

---

certain level of equality between the actors. In simple terms, arrogant, distant and corrupted experts have less chances to be trusted than honest experts who do their very best to insure a dialogue with the non-experts.

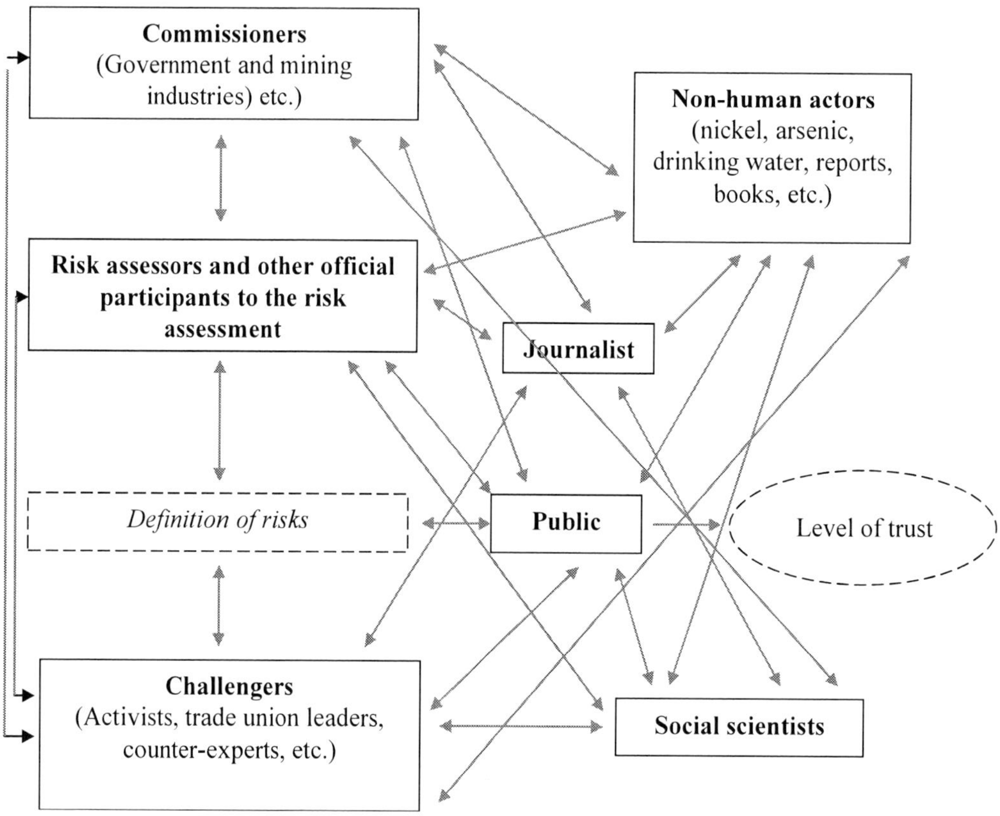

Figure 2. Simplified Representation of the Main Social Relations.

Many Sudburians are also concerned about the value of their properties. If it eventually turns out that their properties are significantly contaminated or perceived as such, and if nothing is done to fix this problem, the value of their homes could be reduced significantly.

The power of these emotional and cultural factors depends partly on how the SSS and its experts are perceived by the population. Do people trust the results or not? In May 2006, the local newspaper *Northern Life* revealed the result of its own poll about the study. Their question was clear and simple: "Do you feel assured by the results of the Sudbury Soil Study?" The result was: Yes, 31.13% and No, 68.87%. (*http://polls.ontarionorth.com/ xlalistprevious.asp?z=1*). The sample was not a representative one; however, it shows that there is some significant discomfort or mistrust regarding the SSS.

These signs of mistrust and/or skepticism mean that many individuals could be easily mobilized by challengers such as trade unions and environmental groups. These activists and organizations have been active since the beginning of the SSS. Let us give some examples:

> In October 2008, under the umbrella of a new group called the *Community Committee*, various actors denounced the SSS in clear and simple ways by referring to a counter-report prepared by a pollution watchdog, *Environmental Defense Canada* (See www. environmentaldefence.ca) and two counter-experts.
>
> John Fera (president of *Local 6500 Steelworkers*) stated that "The process has been dominated by the companies who are responsible for the mess." Brennain Lloyd (*Northwatch*) said: "We need the Ontario Government to help the public formulate a response to the Soil

> Study. The Ecological Risk Assessment has yet to be released, and we don't want another green-wash." Rick Grylls (president *Local 598 Mine Mill*) added: "It is not acceptable to exclude the extra health risks for the 25-35,000 current and past workers in Sudbury. The community has to decide the acceptable levels of risk and what should be done to deal with the problems." Monique Beaudoin (health promoter for the *Centre de santé communautaire de Sudbury*) also added: "the francophone community and the community in general have the right to information in their language and to the tools and resources that will allow them to participate effectively in the Ecological Risk Assessment. The environmental health of our community is at risk, and the public has a right to be fully informed and to determine the level of risk it is willing to live with."

If these activists are successful in building an efficient collective action based on skepticism, worries, and resentment, and manage to connect themselves to the emotions and values of many citizens along with being supported by journalists and, eventually, politicians, they can create a stronger local movement against the SSS. The emergence of this type of movement also depends more on the actions and reactions of other actors (including power relations) such as the politicians, media, local food producers, and experts of the SSS, rather than any "pure" facts that any expert could find and raise to stop the debate. Scientific assessments include much more than experts testing the soil, making predictions about the effects of contaminants on human beings, etc. Risk assessments are networks of social relations, including power relations.

Several politicians have reacted to the SSS by showing some willingness to eventually intervene in the process. The Green candidate Gordon Harris said during one electoral campaign that "voters have told him their big concern is whether the results of the Sudbury Soil Study are believable" (http://www.northernlife.ca/News/LocalNews/2008/07-24-08-politicians.asp?NLStory=07-24-08-politicians). More recently, the NDP MP France Gélinas raised questions on what was presented as "conflicting soils study results" by the *Sudbury Star*. She raised questions such as: "What will be done to clean-up affected areas and properties and what will be done for those whose health is at risk?" The Northern Development and Mines Minister Michael Gravelle answered: "It's something we take incredibly seriously and one, indeed, that we remain very strongly committed to." (http://www.sudburystar.com/ArticleDisplay.aspx?e=1287859) With newly elected left-wing politicians in Northern Ontario (including in Sudbury) and the relative emergence of the Green Party in Canada, it would be surprising if these politicians stayed passive if challengers of the SSS manage to get some success in terms of social mobilization and media coverage.

As far as the media is concerned, two temporary conclusions can be made. First, local media are highly interested in covering this story. It is not difficult for challengers to get media attention to spread some of their messages to the public. On the other hand, the members of the technical committee of the SSS have been reluctant to express themselves in the media. Second, there are signs showing that the local media is sympathetic to the activists and the concerns of some of their readers and fellow citizens. In one of its recent editorials, *The Sudbury Star* recommended that city council assure more leadership regarding the SSS and to democratize the process:

> So, given that two studies using the same data have produced such different conclusions, what are Sudburians to do now? The answer to that lies partly in another recommendation made in the Environmental Defense Canada report: let Sudburians decide. It's a job that should be led

> by city council, perhaps through a mayor's task force, to bring all the parties—ordinary citizens, scientists, labour and environment ministry staff, the mining companies and union leaders—together to decide what to do. The provincial ministries, by the way, should pay for the taskforce's work. (www.thesudburystar.com/ArticleDisplay.aspx?e=1267096)

The local public radio stations (CBON and CBC) distributed four reports mentioning most of the concerns regarding the conflict of interest, lack of answers to important questions, etc., raised by other actors including activists, former employees of the companies, and local doctors. In reaction, Vale Inco sent a letter to Radio-Canada/CBC saying that the SSS answered the questions, and that Radio-Canada/CBC was damaging themselves and the community by being alarmist. (See http://www.radio-canada.ca/regions/ontario/Radio/Sitems/Index.asp?pk_region=25&id=73&IDEmissionFR=356&IDCat=6&leMois1=2009/03&sub=%2E%2E%2F%2E%2E%2F)

In this example, the definitional struggle relates to who is allowed to define the risk and raise questions about its methodology. By making themselves known and raising questions, other actors refuse to see official experts as the only actors who should define the risks and the acceptable levels of risks. These discourses clearly reflect concerns and definitional struggles related to the democratic nature of the risk assessment.

As we mentioned before, the key role played by the two mining companies in the SSS fuels these democratic concerns. Vale Inco and Xstrata Nickel have financed the study and participated on the Technical Committee as full members with the right to vote (with one veto right since the decisional process has been a consensual one). We think adopting this structure was a mistake in terms of trust-building. It is, indeed, very difficult for activists and citizens to trust the soil study when two major players—the mining industries—are so deeply involved, creating a conflict of interest. It is very difficult to trust experts who are paid by the main polluters and who work with these powerful companies. By giving the two mining companies such a central role, other commissioners and experts took the risk of placing themselves in a vulnerable position. It was easy for activists, citizens, journalists, and some politicians to raise critical questions and to be worried.

It does not mean that we can accuse the experts of the SSS of doing a poor job or of lying to the population. They cannot be guilty by association. As one retired trade union leader (Homer Seguin) said in May 2003, in relation to this apparent conflict of interest: "What is gonna be the use, in the end, if the study is truthful and nobody believes it or pays attention?" (*The Sudbury Star,* May 14 2003, reproduced on www.minesandcommunities.org). These concerns were expressed once again on the first page of *The Sudbury Star* on March 3, 2009:

> As a result, activists are now contemplating the merits of launching an extensive testing program that would be independent of any influence from the mining companies.
>
> "It's a really good idea and we're looking into it," confirmed Kuyek.
>
> "Like 64 per cent [*sic*] of Sudburians, we don't really trust the findings of the Human Health Risk Assessment." (St. Pierre, D. 2009)

With this type of social dynamic, the actions and reactions of the experts of the SSS significantly influence citizens' evaluations of the risk assessment. Their actions and reactions can contribute to build positive or negative relations with activists and citizens, thereby affecting the level of intensity, type of the definitional struggle, and level of trust. There are examples showing the goodwill and open-minded attitude from some experts associated with the SSS. However, signs of impatience and possible frustration from other experts have been reported in the media:

> One process observer for the study declared in May 2008 (by noting that union representatives were present at the technical committee meetings): "If I had seen any arm twisting by the companies I would have gone straight to the press. I did not." (Reproduced on Northern Life.ca)
>
> In October 2008, the Community Committee on the Sudbury Soils Study criticized the SSS by relying partly on a report produced by Dr. Kapil Khatter (one counter-expert). The credibility and the good faith of Dr. Khatter were quickly questioned by the director of the SARA Group, who was defending the legitimacy of the SSS: "I am especially frustrated that Dr. Khatter did not contact us directly. Now, he is travelling outside of Canada, unreachable for three weeks. How convenient." "I have never heard of Dr. Khatter or his work. As a scientist, I am used to peer review, but we have not been given any facts from Dr. Khatter." (Northern Life 28th October 2008, reproduced on Republic of Mining's website)

These reactions might intimidate or infuriate some activists and citizens. This frustration is understandable due to experts to dealing with negative critiques coming from external actors with – sometimes – superficial knowledge about the SSS and its methods. However, it is quite normal and acceptable for definitions of risks to be contested. By relying on Latour's (2005) and Beck's works (1992) on the democratization of sciences, we can even say that good risk assessments involve lay-people and their questions, fears, emotions, etc. On one hand, the involvement of these actors encourages good science by multiplying perplexity and consultation. It supports reflexive modernity through sub-political spaces occupied by counter-experts, activists, citizens, politicians, moralists, etc. Intense definitional struggles might indicate that the risk assessment is conducted in a democratic society where the methodological decisions and conclusions are not controlled by powerful companies, governments, and "their" experts. It may also show some form of productive consciousness about the limits of sciences instead of believing in the virtue of Science. On the other hand, definitional struggles are not necessarily good ones in terms of trust and building acceptable results when the expert's facts and the public's concerns are disconnected due to processes of exclusion or personal attacks. In this sense, the real question is: What kind of relational definition do we need in order to produce democratic risk assessments to reach acceptable results? Generally speaking, we are saying that good risk assessments should be based on democratized science when competent experts are not co-opted or corrupted by any specific political and/or economic interests. Further, sound risk assessments should be based on productive relations between the official experts and various actors such as politicians, journalists, citizens, activists, and social scientists. These general and vague explanations do not resolve all the problems. In fact, these explanations create new moral and technical issues related to their implementation; however, they might contribute by improving the efficiency of risk assessment. These considerations lead us to a needed re-evaluation of the policy of deregulation, the role of companies, and the effects of some important inequalities.

5. Deregulation, Companies and Inequalities

Our study of the SSS reveals that one major task is to precisely define the democratic relations of definition of risk. By democratic relations of definition, we offer the following principles:

1) The official experts, technicians, etc. should be free from any real or apparent conflict of interest, especially in relation to polluters (i.e., mining companies), etc. For obvious reasons, they should never be paid directly or indirectly by one polluter.
2) No actor(s) should dominate the relations of definition and the management of risk through the use of undemocratic measures such as repression, exclusion or intimidation.
3) Special attention should also be paid to the undemocratic effects of the unequal access to basic resources such as knowledge, money, and time. Counter-experts, citizens, and activists, should have enough resources to participate in the relations of definition in an efficient and fair manner.
4) Risk-bearers, or at least their elected representatives, should be involved in the risk assessment from the beginning to the end (when the official experts are chosen, when the samples are collected, when different methods are selected, etc.).
5) Risk assessments should not be founded on the traditional gap between the experts who know and the lay-people who are invited to listen to and believe them. As McBurney and Parsons (2000: 21) noted, "systems have generally used argumentation as if from the perspective of an omniscient observer, rationally comparing cases for and against a claim, and combining these to produce a single coherent argument. But science policy debates are usually contentious, with different participants having different interests, values and preferences, and even different modes of reasoning." Following Habermas' principles for dialectical argumentation, McBurney and Parsons proposed the formation of "risk agora" "as a normative framework for debates about the risks of new substances and technologies" (2000: 21). They even proposed a list of rules for the risk agora (2000: 22), defining who may participate, how and when to withdraw, how to present a definition, a claim, etc. We do not have the space to discuss these rules and do not want to reduce democratic risk assessments to only rules, but this type of thinking is related to what we have in mind when we talk about the necessity of democraticizing relations of definition of risks.
6) By following Latour's works on science (2004, 1987), democraticizing definitional struggles and the management of risks also implies inviting all the non-human actors that should be part of the study. Any metal, chemical product, etc., that is reasonably suspected of affecting human health or the ecological system should be integrated into the study. As well, this principle should be applied to human actors such as retired workers, childbearing age women, and so on.

Complying by the first principle should be the main target of social movements and citizens in what Beck (1992) called "subpolitics." In their study of the rhetoric of risk in the California low-level radioactive waste debate, Bedsworth, Lowenthal and Kastenberg (2004) found out that three "policy frames" shaped this debate. "Policy frames" were developed by different actors being involved in debates around the perception and management of risk. They are ways "of selecting, interpreting, and organizing information to construct a policy argument" (2004: 414). The choice of one frame "leads to different characterizations of the risks (...), of the need for action, and of the appropriate course of action" (2004: 408). The legal and scientific story (or frame) "places all value in the legitimacy of science and reason" (2004: 420); pragmatic story (or frame) "respects the scientific evidence but recognizes that it may not be sufficient to overcome political and social obstacles" (2004: 420); and cautious and skeptical story (or frame) "views the legal and scientific basis of regulation with a wary eye" (2004: 420). Our point here is that any real or apparent conflict of interest has the potential of reinforcing one cautious and skeptical frame to the detriment of other possible stories or perceptions. In the case studied by Bedsworth, Lowenthal and Kastenberg, many people though that "scientists who work on these issues are biased by funding from the DOE and from the nuclear industry, and the regulation is the product of a captured regulator" (2004: 420). We have shown that there are many signs indicating that this is also what many people think in Sudbury. Clearly, this type of frame destroys the legitimacy of the experts and of the whole risk assessment.[4]

In addition to avoiding conflicts of interest, there is other good reason to decrease the role of private businesses in risk assessments. Previous negative stories where governments and industries failed to protect the public from toxic residuals have raised the public consciousness and mobilized environmental movements to pay attention to the importance of democratic relations of definition (see, for instance, Fletcher, 2003). The public evaluation of risk assessments does not only depend only on the green image projected by industries, supposed open decisional processes and consultations, and scientific facts. The public evaluation of risk assessments is also influenced by the memory of the actors, a priori perceptions, various interests, and different modes of reasoning (Bedsworth, Lowenthal and Kastenberg, 2004; McBurney and Parsons, 2000; Rouse, 2004). The Chernobyl nuclear power plant accident in 1986 is now part of the collective memory and influences our evaluation of the safety of nuclear plants. Any terrible accident where people were victims of polluting industries and the lack of diligence of governments, such as the Love Canal affair or the Bhopal accident, leaves negative memory traces to other similar people involved in risk assessment.

In Sudbury, the evaluation of the SSS is also dependent on the collective memory regarding Vale Inco and Xstrata Nickel. Let us use the example of Vale Inco. In spite of all the real efforts made by this firm in the last decades to improve its image and its past investments to restore the environment in the area of Sudbury, its perception is tainted by memories of environmental destruction and previous conflicts with residents. In 2001, for instance, residents from Port Colborne, Ontario launched a class-action lawsuit against Vale Inco, seeking $750 million in damages to health, property value, and quality-of-life. Along with Shell and Dow Chemicals, Vale Inco has been a large, visible firm that has been "at the center of environmental controversy" (Macdonald, 2007: 18). Evaluation of local actions by

---

[4] Kasperson (1986: 277-278) explained that the believability of risk information (social trust) is multidimensional: i) it comprises "a judgment of competence about the agency which is conducting the risk assessment," ii) "the agency must be seen as unbiased, able to conduct its activities uncompromised by hidden agenda or undue influence by particular interests," and iii) "people must have confidence that the agency cares about those whom it serves, that it will provide adequate opportunities for individuals to make their concerns known."

Vale Inco is also influenced by the global reputation of the company. In 2006, Inco was removed from the FTSE4Good (index company) for failing to meet their human rights criteria (outside of Canada).[5] By funding and playing a major role in the SSS, Vale Inco took the risk of damaging the level of trust towards the results of the study. Greening the image of private firms and private investments in environmental restoration is not enough to allow companies to participate in risk assessments. Trustworthy risk assessments have to rely on independent risk assessors like democratic political institutions have to rely on independent politicians.

In this sense, the policy of deregulation based on voluntarily participation must be revised. This policy has been advocated by Canadian governments and industries from the 1980s until today (Macdonald, 2007). Deregulation was a reaction against the establishment of the regulatory system (1956-1993) in Canada. Generally speaking, deregulation was supported by the private sector and opposed by environmentalists. The structure of the SSS and the major role played by the two mining industries reflect the policy of deregulation. Once again, funding is provided by Vale Inco and Xstrata Nickel who participated on a "voluntary" basis. Generally speaking, voluntarism refers "to government initiatives intended to bring about improvements in the environmental performance of the firm, without putting in place either financial incentives or legally binding requirements" (Macdonald, 2007: 138-139). As we have previously mentioned, voluntarism is usually connected to efforts made by industries for "greening" their image. By voluntarily accepting the assessment of health effects and ecological impacts of their industrial activities, private firms also hope to: i) avoid being regulated by governments, and ii) improve their legitimacy as good members of the community. The reports published by the SARA Group are full of statements and examples showing this strategy. Without denying that some past practices of companies were destructive in many ways, the reports insist that the recent tendency has been to move from degradation to restoration:

> Sudbury's legacy of mining and smelting activities is also the story of environmental degradation, and more recently, environmental restoration and rehabilitation. As the hills of the Sudbury region turn green once more, it is important to consider the whole story, from roast yards to emission controls, and from environmental degradation to ecological restoration and rehabilitation so Sudbury and its people can move decisively into the future. (SARA Group, 2008b: EC-ii)

The two mining companies have played a major role as voting members on the technical committee of the SSS. Once again, since decisions were made on a consensual basis, the two companies had some sort of a veto right. Other "partners" from the community and the provincial government were also voting members on the technical committee. The two mining companies did not run the technical committee by themselves. The structure of stakeholder linkages was presented "to ensure that the study is conducted in a transparent manner" (SARA Group, 2008b: EC-i).

The Public Advisory Committee, composed of 10 to 12 community members, was in charge of "reviewing and contributing to all study communications and consultation materials and initiatives" (SARA Group, 2008b: EC-v). Citizens were invited to comment at several public meetings and through phone and email. One independent process observer was in

---

[5] See http://www.ftse.com/Indices/FTSE4Good_Index_Series/Downloads/FTSE4Good_March_2006_Review.pdf

charge of reporting to the community. Union representatives were invited as observers. International experts conducted peer reviews of the report and two additional scientific reviewers evaluated the work of the SARA Group. Obviously, many efforts were made in terms of consultation and review; however, as we saw earlier, the study created significant skepticism. This skepticism arose mainly because many people were worried about the health effects of the contaminants and saw the two mining industries as being in a conflict of interest. Concerns were raised about the choice of the scientific reviewers and the independency of the process reviewer. Many thought that few people would have the autonomy, capacity and/or the courage to significantly challenge big mining companies, which are the economic lungs of this part of the world. It was also mentioned that secret deals can be done and implicit or explicit pressure can be applied and placed outside the formal processes of consultation and supervision.

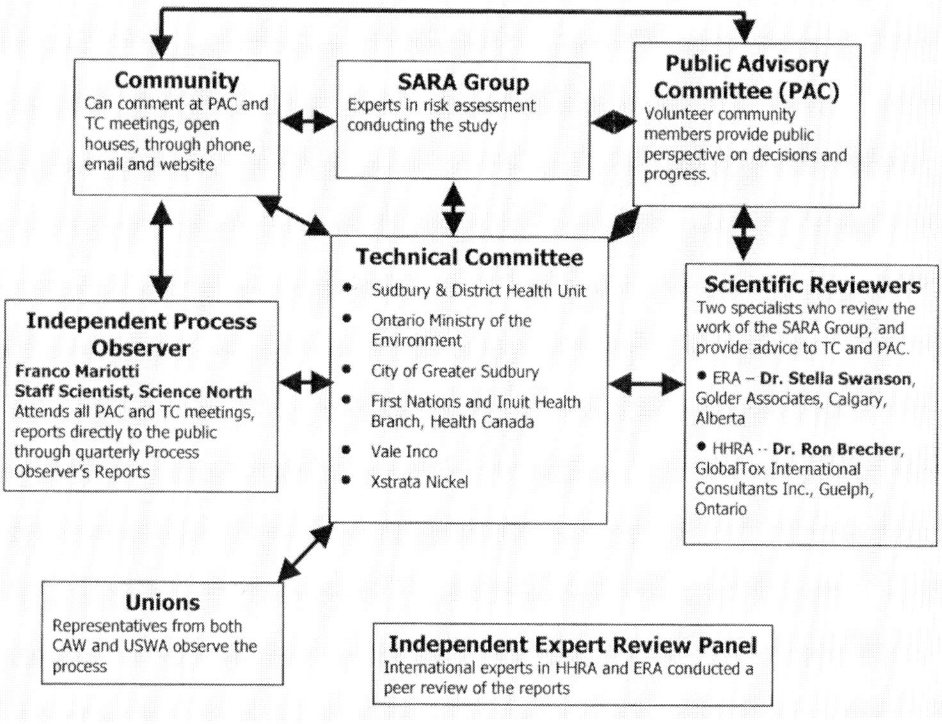

Figure 3. Stakeholder Linkages within the Sudbury Soils Study.

Many people also mentioned that they did not feel they had the capacity to ask relevant questions during the public consultations. Part of the problem was that they did not feel they were invited to ask questions. But this issue is also related to the fact that discussions can hardly lead to mutual understanding and public judgment when they are based on unequal access to resources such as time, money, and knowledge. Usually, activists and citizens do not have enough resources to evaluate and test the research conducted by the official experts. They may also have a hard time understanding the expert's technical explanations. This makes it difficult for activists and citizens to ask relevant questions and/or understand the answers. They are asked to trust the experts without being able to understand and test their works. They also have to believe that the scientific reviewers and laboratories are actually independent, that the governments are actually protecting the public, etc. With this logic, enlightened acceptance of risks founded on some real understanding and discussion of risk assessments is usually replaced by passive tolerance or skepticism:

> Since most risks are imposed on a less than fully informed risk-bearer, the response is more properly thought of as tolerance or acquiescence rather than acceptance. With limited choice and imperfect knowledge, the individual does not resist the imposition of the risk. (Kasperson, 1983: 16)

Confronted with their own lack of knowledge and various tools, many lay-people avoid any implication by being passively skeptical, retreating, and/or simply and blindly believing the officials' explanations and results. Others challenge the risk assessments. In this logic, active and informed public participation should be seen as a solution rather than a problem by the rulers of the risk assessment. Unfortunately, the public is often seen as a potential threat that should be controlled, educated, etc.:

> Thus, agency-designed participation efforts characteristically recognize goals such as "correcting misperception," "educating the public," "reducing conflict," "easing implementation," or "increasing legitimacy." (Kasperson, 1986: 276)

This type of elitist strategy is usually based on maintaining the gap between people who really know the "facts" and "irrational" people; however, there is another possible social dynamic based on what Scuthfield, Ireson and Hall called "public judgment":

> Experts have created a culture of technical control that results in elites talking with other elites when problems arise. An alternative to this contemporary form of elitism would be citizens coming to public judgment, which means, "the state of highly developed public opinion that exists once people have engaged an issue, considered it from all sides, understand the choices this leads to, and accept the consequences of the choices they make." (2004: 198)

Again, we do not have the space to propose and compare technical solutions that could help the real democratization of risk assessments (for instance, see Kamrin, 1998, Kasperson, 1986). But if risk assessments are not democratic, they can raise a lot of skepticism. This skepticism can be raised even when there is no proof that experts were co-opted, corrupted, manipulated, controlled, or intimidated. The goal is not to improve the level of trust of the public; the goal is to produce more democratic and efficient risk assessments. As one researcher on the U.S. chemical weapons disposal program wrote:

> Comparing the practical consequences of both decision-making approaches suggests that increasing collaboration among citizens and experts can produce decisions that are both politically viable and technically safer than more insular, adversarial forms. (Futrell, 2003: 451)

Good risk assessments are not managed only by experts. They are based on the genuine invitation and participation of many other actors (Latour, 2004). Improving risk assessments requires constant ongoing and retrospective evaluation of procedures, structures, norms, etc. used by governing bodies. It is reflexive modernization against scientism. Bettering risk assessment procedures involves the democratization of economic and techno-scientific activities that are beyond the traditional spheres of politics (parliaments, political parties, etc.), and engage risk-bearers. This can be further accomplished through relations between sub political actors such as activists, counter-experts, and citizens. At the end of the day, it is all about building networks of concerned people who refuse to be believers in a modern society. It is trying to be finally modern in modernity (Latour, 1993) by enforcing reflexive modernity (Beck, 1999; 1992). In this respect, social scientists have a major role to play in this emerging social movement by: i) explaining that risk is relational, and ii) insisting on the need and the possibility of democratic sciences.

REFERENCES

Beck, U. (1999). *World Risk Society*. Cambridge, UK: Polity.

Beck, U. (1992). *Risk Society. Towards a New Modernity*. London: Sage Publications.

Bedsworth, L.W., Lowenthal, M.D. and Kastenberg, W.E. (2004). Uncertainty and regulation: The Rhetoric of Risk in the California Low-Level Radioactive Waste Debate, *Science, Technology, & Human Values*, vol. 29, no. 3: 406-27.

Fletcher, T.H. (2003). *From Love Canal to Environmental Justice. The Politics of Hazardous Waste on the Canada-U.S. Border*. Peterborough, Ontario: Broadview Press.

Futrell, R. (2003). Technical Adversarialism and Participatory Collaboration in the U.S. Chemical Weapons Disposal Program, *Science, Technology, & Human Values*, vol. 28, no. 4: 451-82.

Higgins, V. and Natalier, K. (2004) in White, R. (Ed by). Controversies in Environmental Sociology. Cambridge, Uk.: Cambridge University Press: 77-90.

Kamrin, M. (1998). Environmental Health, Risk Assessment, and Democracy, *Environmental Health Perspectives*, vol. 106, no. 5: A216.

Kasperson, R.E. (1986). Six Propositions on Public Participation and Their relevance for Risk Communication, *Risk Analysis*, vol. 6, no. 3: 275-281.

Kasperson, R.E. (1983). Acceptability of Human Risk, *Environmental Health Perspectives*, vol. 52: 15-20.

Latour, B. (2005). *Reassembling the Social. An Introduction to Actor-Network-Theory*. New York: Oxford University Press.

Latour, B. (2004). *Politics of Nature. How to Bring the Sciences into Democracy*. Cambridge, Mass.: Harvard University Press.

Latour, B. (1993). *We Have Never Been Modern*. Cambridge, Mass.: Harvard University Press.

Latour, B. (1987). *Science in Action. How to Follow Scientists and Engineers Through Society*. Cambridge, Mass.: Harvard University Press.

Macdonald, D. (2007). *Business and Environmental Politics in Canada*. Peterborough, Ontario.: Broadview Press.

McBurney, P. and Parsons, S. (2000). Risk Agoras: Using Dialectical Argumentation to Debate Risk, *Risk management*, vol. 2, no. 2: 17-27.

Rouse, M.T. (2004). Knowledge Translation and Risk Management, *Risk Management*, vol. 6, no. 2: 9-15.

Ruckelshaus, W.D. (1983). Science, Risk, and Public Policy, *Science*, Vol. 221, no. 4615: 1026-28.

SARA Group (2008a). *Sudbury Soils Study. Summary of Volume II: Human Health Risk Assessment* (www.sudburysoilsstudy.com)

SARA GROUP (2008b). *Volume 1: Background, Study Organization and 2001 Soils Survey* (www.sudburysoilsstudy.com)

Scutchfield, F.D., Ireson, C. and Hall, L. (2004). The Voice of the Public in Public health Policy and Planning: The Role of Public Judgment, *Journal of Public Health Policy*, vol. 25, no. 2: 197-205.

Slovic, P. (2000). Perceived Risk, Trust and Democracy, in Slovic, P. *The Perception of Risk*. London: Earthscan.

Slovic, P., Flynn, J. and Layman, M. (2000). Perceived Risk, trust and the Politics of Nuclear Waste, in Slovic, P. *The Perception of Risk*. London: Earthscan.

St. Pierre, D. (2009). Activists Doubt Soils Study is Valid, *The Sudbury Star*, no. 44, p. A1.

Starr, C. (1985). Risk Management, Assessment, and Acceptability, *Risk Analysis*, 5: 97-102.

Strydom, P. (2002). *Risk, Environment and Society*. Philadelphia: Open University Press.

Welch, M., Rivera, R., Conway, B., Yonkoski, J., Lupton, P. and Giancola, R. (2005). Determinants and Consequences of Social Trust, *Sociological Inquiry*, vol. 75, no. 4: 453-473.

Zinn, J.O., Ed. (2008). *Social Theories of Risk and Uncertainty. An Introduction*. Oxford, UK: Blackwell Publishing.

In: Health and Environment: Social Science Perspectives ISBN 978-1-60876-216-3
Editors: Helen Kopnina and Hans Keune

Chapter 4

THE ENVIRONMENT AND THE PERCEPTIONS OF THE GLOBAL HEALTH CRISIS

Steven Ney

ABSTRACT

Few people would argue that health and the environment are intricately related. It stands to reason, one may think, that as long as we depend on the environment or, better, environments, for our sustenance, both physical and social, then the state of these environments is likely to have an impact on our health. And yet, the precise ways in which the environment affects health—and vice versa—remains shrouded in uncertainty. One thing, however, does seem clear: whatever the relationship, it is unlikely to be straightforward.

Given this degree of uncertainty and complexity, how do stakeholders define and frame the relationship between health and environment? Using narrative analysis, this chapter systematically compares the stories contending policy actors tell about health and the environment. The chapter compares contending policy narratives in terms of how divergent underlying assumptions of the relationship between health and the environment structures leads to very different and, ultimately, policy recommendations.

By analysing policy positions of key stakeholders at national and international levels, the chapter identifies three contending policy stories archetypes. Each narrative is based on a specific set of assumptions and identifies a particular causal relationship between health and the environment, as well as recommends specific policy measures. The first policy story, one told by market-oriented stakeholders, argues that human health—infinitely adaptable and flexible as it is—thrives when we efficiently and effectively make use of the environment. The second narrative, associated with environmental groups, tells us that human health is an integral and irretrievable part of the environment: in this view, it is the degradation of the natural world that makes us sick. The third policy discourse tells a story of how human health and the environment are in a delicate balance that can only be maintained by rational and expert-driven management of the sort only well-organised institutions can provide.

In the final section, the chapter explores the implications for policymaking of this argumentative and conflictual policy process. In particular, the chapter explores ways of harnessing the creative and innovative energies of this policy conflict while avoiding the tendency for policy impasse as debate deteriorates into a unresponsive shouting match.

INTRODUCTION

Few people would argue with the contention that health and the environment are intricately related. It stands to reason, one may think, that as long as we depend on the environment or, better, environments, for our sustenance, both physical and social, then the state of these environments is likely to have an impact on our health. And yet, the precise ways in which the environment affects health—and vice versa—remains shrouded in uncertainty. One thing, however, does seems clear: whatever the relationship, it is unlikely to be clear-cut.

Since the relationship between health and the environment is complex and uncertain, policy actors will make sense of the impact of the environment on health in terms of general perceptions of global threats to human health. This chapter, then, outlines and compares how different policy actors conceive of and construct emergent global threats to human health. While most policy actors at the international level agree that a range of global pressures presently threaten human stakeholders, they cannot seem to agree on what this "global health crisis" actually entails.

This chapter uses narrative analysis to reconstruct the way stakeholders grapple with the global health crisis and, in this way, frame the relationship between environment and health. After briefly outlining the method, the article compares three narratives or stories told by key actors in the health policy domain. Each of these narratives tells a plausible yet contentious story about global health challenges. Each of the narratives sets out from a particular set of premises that form the setting of the stories. Each of the three narratives identifies causes for global health problems and apportions blame—these are the villains of the story. And each of the narrative points to preferred solutions for global health challenges—the heroes of the tales. Each of the narratives is based on facts and objective data. And yet, each of the narratives is based on values and ideas to weave these facts into coherent and plausible arguments.

The upshot of the argument is that agreement on how best to tackle global health challenges, and with it the link to the environment, is likely to remain elusive. The way policy actors make sense of global health crisis means that conflict between contending political communities in the process is endemic: conflicts are unlikely to be resolved either by new facts or by bargaining. Thus, the broad implications of this analysis for policymaking are that policymakers need to embrace this frame-based conflict. At the same, policy actors must ensure that this conflict does not degenerate into what Martin Rein and Donald Schoen (1993, 1994) have called "intractable policy controversies."

Messy Problems and Narrative Analysis

Narrative analysis provides insight into policy processes that emerge around a certain class of policy problems that commentators have called "messy" (Rhodes, 2997) or "wicked" (Rittel and Weber, 1973). As early as 1973, Horst Rittel and Melvin Webber (1973) recognized that policymakers could probably not approach these "wicked problems" with the same toolbox that had served them so well for more orderly issues. On the one hand, messy challenges are complex. Contemporary policy problems no longer have clear causes but rather a whole host of loosely connected and interrelated factors. Developments in one, seemingly unrelated policy field, say education, may impinge in unpredictable and intricate ways on policy realities of another policy field, say, health policy. In terms of health policy, for example, the World Health Organization (WHO) (2000) and World Bank (1993) argue that health outcomes in developing countries will depend, in part, on the education of girls. On the other hand, messy policy problems are shrouded in uncertainty. While we have an awful lot of data about issues such as climate change, health care or transport policy, much about the basic variables and factors, let alone how they relate to one another, is simply not known with any degree of certainty.

The problem with messy policy challenges is that they generate persistent and divisive conflict about how best to solve them. The two sociologists Martin Rein and Donald Schön (1993, 1994) have nicely captured the spirit of these conflicts in the term "intractable policy controversy." These types of conflicts, they continue,

> ...are immune to resolution by appeal to the facts. Disputes of this kind arise around such issues as crime, welfare, abortion, drugs, poverty, mass unemployment, the Third World, the conservation of energy, economic uncertainty, environmental destruction and resource depletion, and the threat of nuclear war. Disputes such as these tend to be intractable, enduring, and seldom finally resolved (Rein and Schön, 1994, p.4).

The danger with intractable policy controversies is that they all too easily degenerate into what the two policy scientists Paul Sabatier and Hank Jenkins-Smith (1993) call a "dialogue of the deaf"—essentially a shouting match between contending policy actors. It is these dialogues or, more precisely, the failure of dialogue that leads to the policy impasse described as Reformstau.

How, then, do messy policy problems generate intractable policy controversies?

In part, intractable policy controversies are a product of the institutional spaces in which policymaking takes place. Policy scientists such as R.A.W. Rhodes (1990, 1997) argue that the expansion of state responsibilities in the past 40 years or so has also fragmented government. As parts of government specialize on particular policy areas, they need to acquire—either through a process of "buying-in" or "outsourcing"—the requisite technical knowledge to deal with an issue are such as the environment. This creates new types of dependencies between governments and non-state actors (Ansell, 2000). Moreover, policymaking today involves a wide range of actors and organizations. Instead of being one monolithic institution, the state is scattered across a range of discrete institutional networks that focus on particular policy problems. The collection of these "policy networks" (Rhodes, 1990, 1997), policy communities (Richardson and Jordan, 1987), issue networks (Heclo, 1978), or policy subsystems (Sabatier and Jenkins-Smith, 1993, 1999) is what Chris Ansell

(2000) has called the "networked polity." And it is in these pluralist policy networks that policy-actors, including social entrepreneurs, today have to deal with messy issues.

In part, however, messy issues generate intractable policy controversies because they are complex and uncertain. Despite the availability of reams and tera-bytes worth of facts and evidence on any particular issue, this knowledge is not always helpful in determining what to do about messy policy problems. On the one hand, the volume of data on any given messy policy challenge is too large for any policymaker or group of policymakers to consider. On the other hand, scientific facts and evidence rarely address the issues that are salient for policymaking directly. That is why an analyst needs to select (what data is relevant) and interpret (how is this data relevant to the practical policy issue in question). What is more, since messy problems are complex and uncertain, the analyst must resort to rhetoric to be heard. All this requires judgment and is a far cry from the commonplace that "facts speak for themselves."

Judgment is guided by shared ideas, values and beliefs that Rein and Schön (1993, 1994) call "frames." As the term suggests, these frames help policy actors put evidence and data into a specific context. They provide the criteria for focusing on salient data. They also provide blueprints or templates about how best to interpret the data (Rayner, 1991).

In the networked polity, policy scientists have found that individuals tend to work together if they share a particular frame. These groups—sometimes called "advocacy coalitions" or "discourse coalitions"—consist of people with a wide range of professional and institutional backgrounds. Their common frame provides them with a shared policy enterprise. Advocacy coalitions try to influence policy in their particular subsystems. In this view, they pursue their common policy enterprise by crafting arguments for or against a particular policy. These arguments, produced by refracting data and evidence through their frame, aim to mobilize other policy actors and galvanize existing support.

Advocacy coalitions, through their arguments, tell stories about messy issues. What they are, why they are a problem, who is to blame and what we should do about them. But since advocacy coalitions centre on different frames, these stories of messy policy issues are likely to diverge considerably. What is more, these differences cannot be resolved by facts since frames determine what is to count as a fact and what facts are relevant. Neither is this type of conflict amenable to resolution by bargaining: since contending world views are at issue, there is no basis for negotiation. Policymaking about messy challenges, then, is an inherently argumentative process in which contending advocacy coalitions pit arguments—plausible and convincing accounts of what is and what should be going on—against each other. This is why conflict about messy issues is inevitably about values and beliefs. And that is also why frame-based conflict about messy issues is inherently intractable.

A useful way of comparing contending narratives about messy issues is to think of them as stories. These stories are based on specific assumptions (the setting), define the problem (the villains) and propose solutions (the heroes).

The approach compares these policy stories using the typology of frames inspired by the late Dame Mary Douglas (1970, 1982, 1996) and developed by Mike Thompson (Thompson et al., 1990; Thompson et al., 1999). This approach, sometimes called Cultural Theory, is based on the simple insight that the way we live shapes the way we think about the world. The social relations we enter into provide us with the normative and cognitive tools to make sense of messy policy issues. These tools also furnish us with norms to guide the interaction with others. In this view, social structures generate specific frames and interpretive templates. These frames, in turn, help legitimate the preferred social form.

Cultural Theorists identify five basic sets of social relations: hierarchy, the market, nested-bounded groups, isolation, and the hermit. Each of these social relations gives rise to a particular frame: hierarchism, individualism, egalitarianism, fatalism and autonomy. Each of these frames enables its members to select and focus on the salient aspects of messy issues. Here, salience is what helps reproduce and legitimate the preferred form of social relations.

We can use this typology to compare contending advocacy coalitions and their stories. Some coalitions are likely to impose clear and differentiated roles on their members. Members of these coalitions will value order, harmony and process. In other coalitions, members will freely negotiate their relations with one another. These coalitions will emphasize individual liberties, competition and the primacy of the bottom-line. Other coalitions will be well-defined groups that shun internal distinctions. Members of these coalitions will stress equality, holism and the ever-present need to protest. Members of the last two forms of social relations do not take part in policy debates. Fatalists, isolated as they are, see no reason to participate in politics since much of what they seem to do never amounts to much. Hermits, in turn, go out of their way to avoid any social interaction.

The typology of frames, then, is the scalpel for dissecting contending policy stories. By identifying and reconstructing contending policy stories, the typology gauges the distance between contending advocacy coalitions. This is the "scope of policy conflict." But since each advocacy coalition defines itself in contradistinction to the other, the contending coalitions are constantly probing, examining and exploring each other's stories. This leads to complicated patterns of agreement and disagreement across advocacy coalition boundaries. This is what we will call the "structure of policy conflict." Since each frame is a partial and selective account of any particular messy issue, each policy story has strengths and weaknesses. These, then, are the "potential impacts" of policy stories.

The Scope of Policy Conflict about the Global Health Crisis

The structured content analysis reveals (at least) three contending policy narratives about the global health crisis. Each story is based on a specific set of assumptions. Each narrative identifies different causes and culprits of the global health crisis. Most importantly, each story comes to very different conclusions about the solutions to the global health crisis as well as the relationship between the environment and health.

Choice and Health

The Choices Policy Story challenges public health systems across the globe. Its proponents include international organizations such as the World Bank, libertarian think-tanks such as the Adam Smith Institute or the Economist, as well as market actors in the pharmaceutical industry. The bureaucratic behemoths of public health care provision look tired in the today's fast-moving global economy. What is needed, they contend, are health systems that can keep up with dynamic and competitive economies. Here, the environment is, well, just that: the social or natural context in which entrepreneurs recognise and seize opportunities to innovate.

The Setting

Health statistics conclusively show that wealth begets health (Musgrove and Zeramdini, 2001). Economic growth generates the resources needed for developing the costly medical therapies that "...have brought huge benefits to the health and quality of life of millions of people over the last 100 years" (GSK, 2006). At the same time, health industries drive economic growth by creating employment as well as buying goods and services (GSK, 2006; IFPMA, 2007). So, the hard-nosed reality for any health system—private or public—is that economic growth

> ...has a large impact on health outcomes—both by directly improving outcomes and by generating increased resources that can be mobilized by governments for increased public spending (Gottret and Schieber, 2006, p.4)

However distasteful it may be to some, the Choices Story tells us that only firms pursuing profits in competitive markets can create the wherewithal to finance health care (IFPMA, 2007; GSK, 2006). In particular, this means being able to put the natural environment to the most profitable use possible.

The Villains

In this view, health care provision in both rich and poor countries suffers from rampant inefficiency and waste brought about by political mismanagement (World Bank, 1993; Anonymous, 2004a, 2004b, 2004c; Hansen, 2007). In developing countries, the World Bank (1993) argues that the misallocation and inefficiency creates gross health inequalities. Instead of funding low-cost but high-impact health interventions, public health systems waste resources on expensive but ineffective health services (World Bank, 1993). What people in the developing world need, so the argument goes, is affordable disease prevention such as vaccinations for common illnesses, safe food and water, as well as health education (World Bank, 1993). Moreover, they also need the capacity to use their environments in the most efficient and healthiest ways (Listorti and Dumani, 2001). What they are getting instead are high-tech health services, more often than not funded by public subsidies, whose location and price put them beyond reach for the neediest (World Bank, 1993). Worse still, poor regulation is distorting markets and is thereby promoting unhealthy and unsustainable use of the environment in poor countries (Listorti and Dounami, 2001). The effects are unnecessary suffering and excess mortality of the poor.

In the developed world, in turn, inefficiencies in-built into public health systems have sent health costs into orbit. The recent history of health sector reform in developed countries has been a protracted and largely unsuccessful battle against rising costs (OECD, 2004). This, proponents tells us, should not come as a surprise: reforms are little more than piecemeal adjustments unable to address the fundamental flaws of public health systems (Anonymous, 2004a, 2004b). As a result, health care expenditure has steadily grown over the past decades; since 1995 alone, health care expenditure has increased an average of 4% per year in OECD countries (OECD, 2007, p87). And, to add insult to injury, poorly conceived environmental policies are driving businesses and entrepreneurs into the ground (Cato Institute, 2001).

The reason, proponents contend, public health care systems are letting down patients around the globe is that they corrode individual responsibility (Anonymous, 2004d; Hansen, 2007). Socialized health care leave those lucky enough to enjoy this privilege secure in the knowledge that, whether they smoke, participate in risky sports or allow themselves to become obese, someone else will to foot the health care bill. Instead of taking responsibility for our health, Hansen (2007) argues, we rely on expensive therapies—paid for by others—to save us from our poor health choices. Additionally, public health systems allow providers of health services to determine the "need" and level of these services, a set-up that does not encourage restraint (Anonymous, 2004c). Providers, then, profit handsomely from irresponsible health behaviour. Thus, proponents argue, health systems reflect "...priorities of providers—doctors and hospitals—rather than consumers, and that there is plenty of scope to push for a better bargain from the huge amount of money now spent on medical care" (Anonymous, 2004, p4).

In the developing world, pervasive corruption exacerbates these design flaws of government-dominated health care. Political elites use the public health system to divert much-needed resources from the poor to pander to rich but politically influential middle-classes (World Bank, 1993). As Daniel Vasella (2004), the President and CEO of Novartis, points out,

> ...poverty is aggravated by political mistakes, such as spending on military instead of investing in education, hygiene and preventive medicine. Some governments seem to care more about lining their own pockets than caring for citizens they are supposed to lead and represent (p. 43).

The Heroes

The solution, argue proponents of the Choices Story, is to administer a curative dose of individual responsibility by expanding patient choice. Creating what the World Bank calls "health-enabling environments"—both social and natural—calls for fundamental changes to existing health systems. It will require overcoming entrenched and powerful political interests. It will require overcoming the ingrained misperceptions that health care is best provided by the public sector. Most importantly, it will require convincing consumers and producers of health care that preventing diseases is considerably cheaper than curing them.

Health-enabling environments offer patients cost-effective health services. In the developing world, such environments target basic health services to those who need them most. This implies excluding those who can afford to pay for their own health care from public provisions (World Bank, 1993). In order to free-up much-needed resources, the World Bank (1993) argues, health policymakers need to relocate health spending away from expensive HIV/AIDS treatment, curative treatment for terminal cancers, or intensive care for very premature babies. These resources, argue proponents of this discourse, would be more wisely spent on a limited but cost-effective package of "essential clinical services" for the poor. In the developed world, the standardization of common health services, so the argument goes, would go some way to create transparency about the costs of therapy. Ideas such as DRGs enable patients clearly to associate a set of health services with a particular price.

Standardizing health services in this way, proponents contend, allows patients to choose the most suitable and cost-effective therapies.

Over and above service provision, health-enabling environments must encourage cost-effective choices on part of the consumers. One way of doing this, proponents agree, is to provide information on health services and disease prevention. In the developing world, the World Bank (1993) points out, much can be achieved by strengthening basic education of the poor. Among other things, this will lead to a more efficient and healthy use of the immediate environment (Listort and Doumani, 2001). In rich countries, information technologies and the internet could enable patients to take responsibility for their health (Anonymous, 2004d; Hansen, 2007). What is more, rather than ineffective environmental regulation, governments of rich countries should concentrate on providing the right incentives for protection of one's environmental as well as one's health (Cato Institute, 200##).

In addition to information, proponents of the Choices Story argue, health systems need to create compelling incentives for cost-effective health choices. This implies a fundamental reform of existing financing mechanisms. On the revenue side, health care financing needs to reflect different risk profiles (Anonymous, 2004c; Hansen, 2007). Staggered insurance contributions or premiums financially reward people who choose to pursue a healthy lifestyle. In terms of expenditure, reforms will have to find ways to "pay doctors for doing the right things at the right time" (Anonymous, 2004c, p18). Introducing more outcome-orientation in the payment of health care providers, proponents contend, will go some way to restoring individual responsibility. However, the most effective way of rewarding actual performance and quality, maintain many adherents of this policy story, is to abandon third-party purchasing of health care (Tanner, 2006; Hansen, 2007). In this way, consumers will have complete control and responsibility over their health care expenditures.

Taking responsibility for health choices presupposes that there are meaningful choices to take. In both the developing and developed world, advocates of the Choices Story see scope for far more competition and diversity in health care provision. Governments, so the argument goes, need to concentrate less on providing actual health care services and more on building strong, sustainable health markets. In the developing world, the World Bank (1993) suggests that privatization and decentralization in health service provision frees-up public resources for core tasks of poverty-alleviation and regulation. Decentralization of service delivery also expands choice: for this reason, policymakers need to shift the emphasis of health care spending away from the expensive "apex" (sophisticated tertiary care facilities located in cities) towards the base (basic primary care facilities located in the villages). Privatization and decentralization also generate choices and encourage efficiency in the developed world. "Instead of trying to supplant the market," *The Economist* argues,

> ...governments should be striving to promote competition while upholding social values about equity in health care. Thus, in Britain, a Labour government now sees no contradiction between its historic commitment to free services by doctors and hospitals and the pursuit of an internal market in the NHS (Anonymous, 2004c, p.18)

Peoples' health depends on their wealth. That is why, the World Bank reminds us, "...[e]conomic policies conducive to sustained growth are thus among the most important measures governments can take to improve their citizen's health" (p.7). Health policymakers around the world would do well in understanding the implications of this simple truth. "What is needed," *The Economist* tells us,

> is a change in the structure of health-care systems so that competitive pressures push them in a more useful direction, enhancing power of purchasers and increasing competition in the supply of medical care. The traditional argument has been that health care is too important to leave to the market. The opposite holds true: it is too important not to be exposed to the market (Anonymous, 2004c, p17)

Health Is a Human Right: An Egalitarian Story

For the proponents of the Health Rights Story—NGOs, humanitarian organizations, new social movements but also critical physicians—global health inequities brought about by unjust and exploitative world order fuel the global health crisis. Abject poverty in the developing world bars billions from adequate health care, causing unnecessary suffering. In the developed world, profligate and unsustainable lifestyles drive people into expensive health care, causing unnecessary suffering. The health care mess we are in, they argue, is beyond technocratic fixes or market utopias. What we need—and we need it quickly—is fundamental change in the way we live. Significantly, this includes a fundamentally different relationship between humans and the natural world.

The Setting

The keystone of the Health Rights Story is the Alma-Ata Declaration of 1978. In outlining a comprehensive primary health care reform agenda, the Declaration defines health as "... a state of complete physical mental and social well-being, and not merely the absence of disease or infirmity..."(p.2).

This broad definition lays out the basic beliefs of the Health Rights Story. First, health is about a lot more than just disease. Understanding health, so the argument goes, requires looking at the full range of factors that shape human well-being. Apart from the prosaically biological, then, "*...the main determinants of health are social, economic and political*" (Werner, 2003, p.17, original emphasis).

Second, since all humans are equal, so, too, are their needs for health and well-being. This is why the Peoples' Health Charter (2000) pronounces that the

> ...attainment of the highest possible level of health and well-being is a fundamental human right regardless of a person's colour, ethnic background, religion, gender, age, abilities, sexual orientation or class (p.3)

It follows, proponents contend, that health is neither negotiable nor can it be traded. Health and well-being cannot be parceled off and sold, just as access to them cannot be rationed and prioritized. So fundamental is this right to health and well-being that proponents of this story believe it ought to be the primary focus for policymaking (PHC, 2001).

Third, this type of health implies holistic and participatory health policymaking. Health policy that does not address social, economic, political and ecological conditions can only ever alleviate with the symptoms (Sanderson, 2003). Effective health policy will need to integrate processes in a range of different policy domains, such as the transport, industrial policy or, significantly, the environment and actors, such as patients, traditional healers or critical social groups (UN and WHO, 1978; Sanders, 2003; Werner, 2003).

The Villains

Of the billions of U.S. dollars spent on medical R&D each year, Oxfam, VSO and Save the Children (2002) inform us that a mere 10% is committed to diseases that account for 90% of the global disease burden (p.19). The unjust economic and political world order responsible for these inequities, argue proponents of the Health Rights Story, is the root cause of the global health crisis.

Economic globalization has commodified and demeaned health as it has ravaged the natural environment (Banerji, 2003; Werner, 2003; Sanders, 2003). To make room for commercial health care provision, proponents of the Health Rights Story argue, neo-liberal economic policies, such as Structural Adaptation Programmes in the South and massive cuts to social budgets in the North, have all but destroyed public health systems (Banerji, 2003; Werner, 2003; Sanders, 2003). The perpetrators of this crime—political elites beholden to powerful business interests—have tried to conceal their machinations behind unfulfilled promises of efficiency, choice and quality (Schiff et al., 1994; PHM, 2000). As the ability to pay for health care crowds out the social right to health, billions of people are denied access to adequate health care.

Worsening inequities, argue proponents of the Health Rights Story, invariably punish the weak and vulnerable. In countries such as India or Bangladesh, de facto privatization of health services has closed down access to the most basic of health services for the vast majority of people (Islam and Tahir, 1999). Additionally, proponents contend, local health inequities grow as health becomes a factor in macro-economic and corporate production functions. The more health care focuses on the economically active, the more health needs of children, the old, the disabled or pregnant women fall by the wayside (PHC, 2000; Benson, 2001; John, 2003). In this systems, people and nature are little more than resources to be brutally exploited by the rich and powerful.

The profit-motive also corrupts the quality of health care for people in rich countries. Even if you are not among the unlucky 39 million U.S. citizens without health insurance coverage, all the exorbitant health insurance premiums buy you is dependency on high-tech drugs, flashy diagnostic gadgetry and macho surgical procedures (Robinson, 2008, p.1). The dominant curative approach to health care does little for your real health needs, but works wonders for "shareholder profit, inflated CEO salaries, and top-heavy administration" (Bündnis 90/ Die Grünen, 2002; Robinson, 2008, p.1). Quality, the American organization Physicians for a National Health Program (PNHP) argues,

> ...is distorted when the ability and willingness to pay become the criteria for determining which services are provided. Marginally effective or even harmful treatments for the well-insured affluent take priority over more needed and appropriate services (Schiff et al., 1994, p.3)

If poverty caused by economic inequity is the "...mother of all health problems" (Sivaraman, 2003, p21), then the father is political exclusion. Globally, the Health Rights Story argues, a "syndicate of rich countries and the ruling elite of poor countries" (Banerji,

2003, p.12) have consistently scuppered attempts at implementing Alma-Ata's transformative vision of health care. In its place, this cartel has installed the "...dominant inequitable paradigm of development" (Werner, 2003, p.18). A paradigm, no less, that degrades both humans and the environment.

Locally, the dominant interests of providers marginalize important stakeholders, such as patients, family-members or non-medical health professionals. Political exclusion invariably leads to discrimination of marginalized groups—e.g., the very old—in commercial and technocratic health systems (Peoples' Health Charter, 2000, Bündnis 90/Die Grünen, 2002). As a result, the International Alliance of Patients' Organisations (IAPO) contends,

> ...the patient's voice is not valued enough in policymaking and practice. Patient involvement is often merely tokenism; its influence on policymaking can be restricted by practical and financial structures, differing knowledge bases, cultural barriers and personal attitudes (IAPO, 2005, p.1).

The global economic and political system, argue proponents of the Health Rights Story, is as "undemocratic as it is unsustainable, [since] it promotes economic growth of the rich regardless of the human and environmental cost" (Werner, 2003, p17). Instead of liberating and empowering people, commodified health care imposes inappropriate "cures" on both the poor and the rich.

The Heroes

In this view, fighting the global health crisis means tearing down the social, economic and political barriers to equitable health care and a sustainable future. In the both the developing and developed world, proponents of Health Rights demand free and universal access to primary health care (PHM, 2000; Bündnis 90/ Die Grünen, 2002). Since the private sector cannot be trusted, proponents contend, this is a responsibility of the public and tertiary sectors (PHM, 2004). And since centralised and bureaucratic systems are oppressive, advocates of Health Rights call for the radical decentralization of health care provision (Sanders, 2006; PHM, 2000; Bündnis 90/Die Grünen, 2002).

Moreover, inclusive and empowering health care, proponents argue, must appreciate that people's health and the health of their environments (both social and natural) are intrinsically interconnected. This means "demystifying health and medical technologies" (People's Health Charter, 2000) by integrating traditional medicine into mainstream allopathic therapies. Most importantly, empowering health care requires a "strongly participatory strategy" (Werner, 2003, p.15). For this reason, civil society and communities, proponents contend, must play a central part in health policy processes (Sanderson, 2003; Werner, 2003; Bündnis 90/ Die Grünen, 2004).

However, argue the proponents of this discourse, if reforms are to treat more than just the symptoms, they must address global economic and political inequities. First, the reckless casino capitalism and its institutions—the WTO, the World Bank, the IMF and the Security Council—must go (PHC, 2000). In particular, the WHO needs to radically change to be able to insulate the poor from the commercialization of health care (PHC, 2000). Second, in order to end the ruthless exploitation of the weak and poor, global economic processes must be brought under democratic control. This not only implies the regulation of MNCs but also the taxation and control of global financial flows. Third, the growth-oriented paradigm, as dominant as it is wrong-headed, must give way to a renewed commitment to "equitable social investment" (PHC, 2000). Last, we must put an end to the destruction of nature caused by the profligate consumption of the rich (People's Health Charter, 2000; Bündnis 90/ Die Grünen, 2004). That is why the Peoples' Health Charter (2000) calls for activists to pressure "wealthy and industrialized countries to reduce their consumption and pollution by 90%" (p.6).

A just world order is the only feasible response to the global health crisis. Anything else, the proponents of the Health Rights Story argue, inevitably leads to exploitation, suffering and disease. For this reason, the People's Health Charter (2000) proclaims:

> Equity, ecologically-sustainable development and peace are at the heart of our vision of a better world—a world in which a health life for all is a reality; a world that respects, appreciates and celebrates all life and diversity; a world that enables the flowering of people's talents and abilities to enrich each other; a world in which people's voices guide the decisions that shape our lives.

Stewardship and Health: A Hierarchical Story

Whatever else it may be, providing health care is always an immense administrative challenge. In the past, proponents of the Stewardship Story contend, only public organizations have been equal to this challenge. And this will remain the case in the foreseeable future. Neither the visions of glitzy high-tech medicine nor the touchy-feely PHC utopias can dispel the fundamental truth: health care as well as environmental protection need competent administration, oversight and, yes, control.

The Setting

Health, advocates of stewardship argue, is an unusual thing.

In certain respects, health is a commodity. Not only, the European Commission (2007) tells us, is health "...important for the well-being of individuals and society, but a healthy population is also a prerequisite for economic productivity and prosperity" (p.5). And yet, unlike a pair socks, health affects us in an uncomfortably immediate and inescapable way. After all, the consumer of health care "...is also the physical object to which all such care is directed" (WHO, 2000, p.4). Unlike assets such as education, health is subject to large and unpredictable risks that are independent from one another. Since many health choices are made in the context of fear, pain or bereavement, people cannot be expected to make decisions about health like they would about, say, living-room curtains (Blunden and Smith, 2005; BMA, 2004). This is why, proponents argue,

> ...markets work less well for health than for other things, why there is need for a more active and also more complicated role for the state, and in general why good performance cannot be taken for granted (WHO, 2000, p.4).

In other respects, then, health also is a right. Apart from fulfilling the health care wants of those who can afford it, health systems "have an additional responsibility to ensure that people are treated with respect, in accordance with human rights" (p.4). And yet, like any other right, health is not an absolute right but must be weighed against other human rights

(Hunt, 2004). For example, promoting health rights through encouraging healthy lifestyles must not infringe on the right of self-determination.

In this view, then, health care is about striking a balance between health as a commodity and health as a right. The art of striking and sustaining this balance is called Stewardship: a "...function of government responsible for the welfare of the population, and concerned about the trust and legitimacy with which its activities are viewed by the citizenry" (Saltman and Ferousier, 200#, p735). Stewardship, the WHO (2000) tells us, involves "...the tasks of defining the vision and direction of health policy, exerting influence through regulation and advocacy, and collecting and using information" (p.xiv). Moreover, proponents argue, the best way of thinking about the environment in this context is also in terms of Stewardship: like human health, the environment requires careful and judicious balance that only experts can achieve and maintain (Douglas et al., 1998).

The Villains

Health policy challenges in the developing world are diametrically opposed to problems in industrialized countries. In the developing world, health systems are out of balance. As a result, they can neither keep people healthy nor provide fair finance. In countries of the developed world, in turn, ageing and globalization ask awkward questions of health policymakers that, if not answered with the utmost care, threaten the balance these high-performance health systems.

The root cause of health system imbalance in the developing world, proponents contend, is poor stewardship. First, poor stewardship impairs health policy vision so that policymakers lose sight of the population's needs and demands (WHO, 2000). Further, policymakers are unable to "see" and engage with other important actors, such as the private sector. Significantly, poor stewardship causes an inability to recognise poor practices: as a result, "...stewardship is subverted; trusteeship is abandoned and institutional corruption sets in" (WHO, 2000, p121).

Second, poor stewardship leads to policy paralysis. Many developing countries have no explicit policies for the health sector (WHO, 2000). The health policies that do exist are often inappropriate for the given level of resources. Policy paralysis also closes down pathways of health system development. For example, privatization only creates more capacity for health care provision in conjunction with regulatory capabilities. Yet, privatization under weak stewardship exacerbates imbalances (WHO, 2000).

Third, blinded and immobile, the patient becomes vulnerable to political abuse and exploitation. Weak stewardship leaves health systems exposed to the irrationality typical of politics in poor countries.

In the developed world, the Stewardship Story tells us, demographic ageing threatens the equipoise of high-performance health systems. As we have seen, ageing populations demand expensive health services (OECD, 1998; European Commission, 2007). Inopportunely, this is happening at a time when globalization is squeezing public coffers. Maintaining performance will require health policymakers to recalculate and redistribute the societal burdens of health care provision.

Regrettably—and this is what really afflicts developed health systems—governments in affluent countries have tried to control health care costs by introducing competition (Freeman, 1998). This narrow focus on the market and choice, however, undermines the basis for effective health care planning and management (BAK, 2003; BMA, 2004). Since many health choices are made in the context of fear, pain or bereavement, people cannot be expected to make decisions about health like they would about, say, living-room curtains (Blunden and Smith, 2005; BMA, 2004).

What is more, competition in health systems actually reduces choices by destroying capacity (BMA, 2004; BAK, 2007). On the one hand, health care providers who cannot "attract" customers face closure. In this way, competition "...will destabilize a vital public service and lead to inequalities in service provision from area to area that will fragment care and be very costly" (BMA, 2004, p.6). In Germany, reforms have led to hospital closures, the rationing of health services and the geographical concentration of facilities (BAK, 2006). On the other hand, "ruinous competition" (BAK, 2003) undermines the autonomy of health professionals by empowering health care insurers. External cost controls mean that accountants, lawyers and, worse still, politicians determine therapies rather than competent professionals. Invariably, the quality of health care suffers since targets and standards "...that are unrelated, too numerous and politically driven distort clinical care." (BMA, 2004, p12). This, argue proponents of the Stewardship Story, is driving competent health professionals from health systems that force them to pick up the tab for poor management and persistent under funding (BAK, 2007). How the resulting shortage of qualified health professionals in the developed world contributes to patient choice, proponents of Stewardship ask, is anyone's guess.

The Heroes

Health policymakers anywhere have their work cut out for them. In the developed world, health systems must start providing basic health services to all who need them while finding ways of financing health care that protect the vulnerable (WHO, 2000; ILO, 2007). In health systems of the developed world, the issue is how best to maintain the high quality of health care in the face of changing demographic and socio-economic circumstances.

The precondition for solving either set of challenges, however, is getting Stewardship right. Stewardship enables policymakers to balance the many things in health care provision that pull in opposite directions: things such as public and private interests, compassion and hard-nosed technical competence, choice and guidance, or patient rights and professional autonomy. And because stewardship must square these contradictory pressures, "...the ultimate responsibility for the overall performance of a country's health system must always lie with government" (WHO, 2000, p119). Governments are the only societal actors, so the argument goes, at the right vantage point to ensure "coherence and consistency" across different departments and sectors (WHO, 2000).

This, proponents of the Stewardship Story argue, does not necessarily mean a return to bureaucratic planning of health care. On the contrary, effective stewardship, the WHO (2000) argues, implies that states' "...key role is one of oversight and trusteeship—to follow the advice of 'row less and steer more'" (p.119).

Rowing less and steering more implies three tasks. First, Stewardship must clear a legal, political and socio-economic space in which other policy actors can deliver health care services (ILO, 2007). In the developing world, creating these spaces means fighting persistent corruption by insulating health policy from politics (WHO, 2000, p.xvi). Institutionalising "social dialogue"—the deliberation between accredited stakeholders in a "prepolitical space" (ILO, 2007)—goes a long way towards cutting out malfeasance. Good stewardship also needs strong administrative institutions that take an active role in shaping health care provision as well as health needs.

Second, good stewards must take responsibility for defining the vision and general direction of health systems (WHO, 2000). For the German Medical Association (BAK), this means that governments "...determine the type, the volume and the level of social protection in case of disease and in terms of prevention" (p.2). Instead of formulating elaborate and detailed health sector plans that cannot be implemented, health policymakers in developing countries should concentrate on designing flexible health policy frameworks (WHO, 2000).

Third, good stewardship implies exerting the right kind of influence on providers and patients through coordination, regulation and education. It involves striking "... a balance between avoiding regulatory capture by private interests and maintaining productive dialogue with them to ensure that regulatory frameworks are realistic" (WHO, 2000, p.129). For the European Commission (2007), good stewardship means identifying and managing the synergies between health systems and other policy sectors (p.6). Similarly, the future role of health insurance, the ILO (2007) argues, involves the management of networks and intersectoral partnerships: here, "[c]oordination and collaboration are key in a multi-actor environment and crucial for avoiding duplication and inefficiencies" (p.6)

Effectively tackling the global health crisis, fundamentally a crisis of health system Stewardship, requires a new mindset. Leaving health care entirely to the market or to participative primary care, the Stewardship Story tells us, is a recipe for failure. What is needed today in both the developing and the developed world is an approach that exploits the synergies of both extremes while avoiding their pitfalls. This, then, is what the WHO (2000) refers to as "New Universalism":

> Rather than all possible care for everyone, or only the simplest and most basic care for the poor, this means delivery to all of high-quality essential care, defined mostly by criteria of effectiveness, cost and social acceptability. It implies explicit choice of priorities among interventions, respecting the ethical principle that it may be necessary and efficient to ration services, but that it is inadmissible to exclude whole groups of the population. (p.xiii)

Figure 1 provides an overview of the scope of policy conflict in the debate about the global health crisis.

The Structure of Policy Conflict: Irreducible Plurality and Conflict

Table 1 points to an irreducible plurality of policy narratives about the global health crisis. All three policy stories provide contradictory yet plausible accounts of the global health crisis and the role of the environment. All three stories identify a different culprit and suggest divergent but credible policy solutions. None of the stories gets it completely wrong. And, more importantly, none gets it absolutely right, either. In short, the stories do not reduce onto one another. Contestation about the global health crisis, then, takes place in a discursive space delimited by the three narratives. We can visualise this space as a triangle as in Figure 2.

	Choices	Rights	Stewardship
Setting	Wealth means health Health is what you think it is Health care provision is about satisfying health wants The environment is the context in which entrepreneurs seize opportunities	Health is a Right Broad definition of health Health care provision is about meeting basic health needs The environment is an integral part of human well-being and health	Health is not quite a commodity and it is not quite a right Health care provision is about striking a balance between contradictory tendencies Like human health, the environment needs to be in balance
Villains	Public health care systems do not encourage efficient consumption of health goods; not only ineffective but also unfair Poor countries: inefficiency, misallocation of resources, corruption; Rich countries: skyrocketing costs All this is exacerbated by wrong-headed environmental policy	Global inequality cause disease of humans and the environment; Inequitable economic system has eroded access to health for billions and has degraded their natural, healthy environments Inequitable political system; exclusion at all levels	Different worlds, different problems Poor: health systems are imbalanced because of poor stewardship; Rich: balanced systems are being undermined by ideological commitment to choice
Heroes	Create real choices in health care provision; Health enabling environments; Empower consumers viz. providers (information, education); Right incentives for individual responsibility in health care and environmental protection; encourage diversity and plurality.	Eliminate inequities; Break down financial, geographical, ideological and political barriers in health systems; Equitable and unsustainable global governance; New global development paradigm;	Strengthen Stewardship; Steer more, row less; Create appropriate public sector institutions: insulate from politics; Processes for goal setting; Skillful day-to-day management in health care and environmental policy: education, management of networks, guidance; good regulation

Figure 1. The Scope of Policy Conflict about the Global Health Crisis.

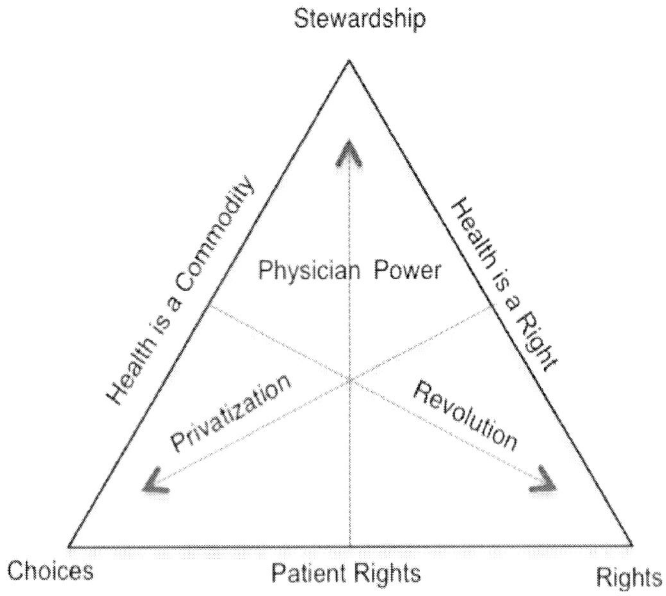

Figure 2. The Triangular Policy Space.

In this view, conflict about the global health crisis is likely to persist for two reasons. First, judgment, guided by incompatible frames, transports knowledge about the global health crisis into the policy debate. For example, the statistics of global health disparities are essentially undisputed. Yet, the Choices Story takes them as evidence for public sector inefficiency (Vasela, 2004; World Bank, 1993), the Rights Story uses the same data to highlight private sector greed (PHM, 2000) and the Stewardship Story takes from the data a pressing need for more regulation (WHO, 2000; ILO, 2007). As we have seen, objective data on the global health crisis is not in short supply. Rather what is missing is a common key for deciphering the policy relevance of this data for both health care and environmental policy.

Second, the triangular form of the debate about global health suggests there is no common basis for political bargaining. This is not to say that there is no agreement among contending coalitions (see Table 2). Yet agreement across coalitions is likely to be fragile. For one, agreement between any two stories is mounted on a more fundamental rift over basic principles. For example, both the Choices and the Health Rights Story champion patient rights. However, the former sees these rights empowering consumers while the latter understands rights as a means of emancipating citizens.

Moreover, the triangular constellation of narratives ensures that there is always someone to argumentatively probe and poke tenuous agreements across advocacy coalition lines. In this triangular policy space, then, agreement is too tenuous

The inherently moral nature of deliberation about the global health crisis may proceed into a "dialogue of the deaf." In the debate about the global health crisis, accusations of incompetence and bad faith abound. Advocates of Health Rights and Stewardship accuse of proponents of the Choices Story of willfully destroying health systems for profit. Policy-actors associated with the Choices and Stewardship narratives dismiss the Health Rights vision as irresponsibly utopian (WHO, 2000; World Bank, 1993). Last, the advocates of Health Rights and Choices Stories blame proponents of Stewardship for denying patients their rights. The danger that the current policy debate turns into an shouting match is captured by a proponent of the Choices Story:

> Whenever pharmaceutical corporations are denounced as greedy, irresponsible, and socially insensitive crowd of "couldn't-care-less" capitalists, it becomes difficult to expect that the same corporations will queue up to commit their funds and technology. (Leisinger, 2006, p. 7)

	Choices - Stewardship	**Rights-Choices**	**Stewardship-Rights**
Agreement	Health implies wealth; costs need to be controlled	Patient rights; measures to strengthen patient rights	Health is a social right
Mutual Rejection	Sever link between wealth and health	Centralised and hierarchical systems of health provision	Commercialisation and privatisation of health
Disagreement	Market vs. management	Social vs. consumer rights	Empowering Participation vs. Competent administration

Figure 3.

THE IMPACT OF POLICY CONFLICT

One way of avoiding this shouting match would be to exclude the annoying veto-players (Tsebelis, 2001) that inconveniently and persistently contest what others perceive to be rational policy. However, complexity and uncertainty of global health issues means that solutions proposed by contending narratives are, at best, partial. Each policy story, then, features inherent weaknesses and blind spots.

The unquestioning trust proponents place in the efficiency of the market leads them to downplay the adverse impacts on equity and service provision that the privatization of health care may bring about. The Health Rights Story's confidence in communitarian self-determination leaves it vulnerable to free-riding. Convinced of the superiority of their moral argument, proponents of health rights are prone to intolerance. Finally, the Stewardship Story believes that only management skills of health professionals and sturdy governance institutions can solve contemporary health challenges. By concentrating almost exclusively on the management of processes, the Stewardship Story would leave health systems open to bureaucratic inertia, ineffectiveness and corruption.

Without frame-based conflict or, more precisely, the institutions to encourage frame-based contestation, the inherent weaknesses of proposed solutions in each individual narrative are likely to go uncontested, if not undetected. The resulting "unforeseen" consequences—unforeseen only to proponents of the particular policy story—are likely to undermine proposed policy solutions. For example, Benson (2001) points out that the privatisation of health provision in Tanzania has resulted in a concentration of services on urban males in employment: as a consequence, access of women, children and the old in both the rural and urban areas has declined. Similarly, the WHO (2000) contends that the excessive focus on providing equitable health care meant that the primary health care movement, for all their good intentions, provided poor and inadequate services to everyone. The World Bank (1993), in turn, points to inequalities and inefficiencies in health care provision brought about by the inherent corruption and ineffectiveness of public health care provision and environmental regulation.

CONCLUSION

The chapter has shown that policy actors tell at least three diverging policy stories about the global health crisis that also construct the relationship between health and the environment. All three policy stories tell a plausible tale about the global health crisis. All three stories relate health and the environment in different ways. All three stories outline credible, if divergent, policy solutions. Yet, left to their own devices, each would bring about highly undesirable impacts. Since contending advocacy coalitions frame the global health crisis in fundamentally different ways, conflict is inevitable. While this undoubtedly makes policy processes in the health domain rather unsightly, it enables the critical appraisal of policy arguments. Each of the policy frames is a useful lens through which policy actors critically refract the policy proposals of contending policy stories. The triangular model of the contestation about the global health crisis, then, suggests that conflict, particularly the frame-based and value-driven sort, is a crucial yet volatile resource for global health policymaking. On the one hand, narratives are the main vehicle for transporting knowledge about global health issues into the policy debate. It is the conflict articulated in terms of these stories that enables policy-oriented learning. On the other hand, frame-based conflict about the global health crisis also risks degenerating into a "dialogue of the deaf." Here, no learning or change is possible.

Rather than obstructing the policy process, this is the way policy stories inform the public debate on global health and the environment. In a very real sense, policy stories are the vehicle for transporting scientific knowledge into the policy debate. Policy frames allow actors to make sense of the uncertainties and complexities of health care provision. It is policy frames that enable actors to tell stories by using scientific knowledge to build plausible scenarios about health and health care provisions. Stories provide some answers, albeit partial and partisan, to the central questions that science helps us formulate but which it cannot resolve. How does globalization affect health systems? What will demographic ageing mean for health care provision? What are policymakers to do about old and new infectious diseases? How can we address local and global health disparities? What are the advantages and disadvantages of privatizing health care? What are we to do about increasing health care costs? The divergent answers to these questions—embedded in contending stories—and the policy conflict this gives rise to create the concepts and strategies that form the basis for health care and environmental policy.

REFERENCES

Angell, C. (2004). Excess in the Pharmaceutical Industry, *Canadian Medical Association Journal,* vol.171, no. 12, pp1451-1453.

Anonymous (2004a). Survey: The Health of Nations, *The Economist*, vol. 372, no. 8384, p3.

Anonymous (2004b). Survey: No Reverse Gear, *The Economist*, vol. 372, no. 8384, p5.

Anonymous (2004c). Survey: Keep Taking the Medicine, *The Economist*, vol. 372, no. 8384, p17.

Anonymous (2004d). Survey: Treating the Symptoms, *The Economist*, vol. 372, no. 8384, p14.

Banerji, D. (2003). Reflections on Twenty-Fifth Anniversary of the Alma Ata Declaration, in R. Narayan and Unnikrishnn PV (Eds.) *Health for All Now—Revive Alma-Ata*, Books for Change, Bangalore.

Benson, J. (2001). The Impact of Privatisation on Access in Tanzania, *Social Science & Medicine*, vol. 52, no. 12, pp1903-1915.

Blank R.H. and Burau, V. (2004). *Comparative Health Policy*, Palgrave, Basingstoke.

Blunden, F. and T. Smith (2005). What Does Choice Mean? How Can it be Made More Meaningful, *Health Policy Review*, vol. 1, no. 1, pp16-35.

Bonoli, G. and Palier, P. (2001). How Do Welfare States Change? Institutions and their Impact on the Politics of Welfare State Reform in Western Europe, *European Review*, vol. 8, no. 3, pp333 –352.

BMA (2004). British Medical Association written submission to the Public Administration Select Committee Inquiry—Choice and voice in public services, www.bma.org.uk/ap.nsf/Content/choiceandvoice~defining, accessed 1 October 2008.

BAK (2003). Gesundheitspolitische Grundsätze der deutschen Ärzteschaf', http://www.bundesaerztekammer.de/page.asp?his=0.5, accessed 23 February 2008.

BAK (2004). Kriterien der duetschen Ärzteschaft zur Finanzierung einer patientengerechten Gesundheitsversorgung, http://www.bundesaerztekammer.de/page.asp?his=0.5, accessed 23 February 2008.

BAK (2008). Gesundheitspolitische Grundsätze der Ärzteschaft—Ulmer Papier, www.bundesaerztekammer.de/page.asp?his=0.5, accessed 1 October 2008.

Bündnis 90/ Die Grünen (2002). The Future is Green: Party Program and Principles, www.gruene.de/cms/files/dokbin/145/145643.party_program_and_principles.pdf, accessed 1 October 2008.

Epstein H. (2007). *The Invisible Cure: Africa, the West and the Fight Against AIDS*, Viking/ Penguin Books, New York, NY.

European Commission (2007). *Communication from the Commission to the Council, the European Parliament, the Economic and Social Committee and the Committee of the Regions on the Health Strategy of the European Community*, European Commission, Brussels.

EFPIA (2007). *Medicines in Developing Countries*, EFPIA, Brussels.

Freeman, R. (1998). Competition in Context: the Politics of Health Care Reform in Europe, *International Journal for Quality in Health Care*, vol. 10, no. 5, pp395-401.

Gilson, L. (1995). Management and Health Care Reform in Sub-Saharan Africa, *Social Science & Medicine*, vol. 40, no. 5, pp695-710.

GSK (2006). A Human Race: Corporate Responsibility Report 2006, Corporate Responsibility Report, http://www.gsk.com/responsibility/index.htm, accessed 2 October 2008.

Gottret P. and G. Schieber (2006). *A Practitioner's Guide: Health Care Financing Revisited*, The World Bank, Washington, D.C.

Hansen, F. (2007). Self-Care—Essentials of 21st Century Health Care Reform, http://www.adamsmith.org/think-piece/health/selfcare-%11-essentials-of-21st-century-health-care-reform-2007111640/, accessed 2 October 2008.

Hunt, P. (2004). The Right to Health: What Do I Expect from a Pharmaceutical Corporation? in K. Leisinger (ed) *The Right to Health: A Duty for Whom? International Symposium Report 2004*, Novartis Foundation for Sustainable Development, Basel, pp68-73.

Ikegami, N. (2005). Japan, in R. Gauld (ed) *Comparative Health Policy in the Asia-Pacific*, Open University Press, Maidenhead.

IAPO (2005) Policy Statement—Patient Involvement, http://www.patientsorganizations.org/showarticle.pl?id=810;n=109, accessed 2 October 2008

IFPMA (2007). *The Pharmaceutical Innovation Platform Meeting Essential Global Health Needs The Pharmaceutical Innovation Platform Meeting Essential Global Health Needs The Pharmaceutical Innovation Platform: Meeting Essential Global Health Needs*, IFPMA, Geneva.

ILO (2007). *Social Health Protection: an ILO Strategy towards Universal Access to Health Care*, International Labour Office, Geneva.

ISSA (2007). *Investing in People's Health: Towards Strengthening Health Promotion and Prevention in Social Health*, International Social Security Associations—Technical Commission of Medical Care and Sickness Insurance, Geneva.

Islam A. and M. Zaffar Tahir (1999). Health Sector Reform in South Asia: New Challenges and Constraints, *Health Policy*, vol. 60, pp151-169.

John, P.C. (2003). Whatever Happened to Alma-Ata, in R. Narayan and Unnikrishnn PV (Eds.) *Health for All Now—Revive Alma-Ata*, Books for Change, Bangalore.

Marshall, T.C. (1950). *Citizenship and Social Class and other Essays*, Cambridge University Press, Cambridge.

Meek, J. (2003). Trillion Dollar Disease, *London Review of Books*, vol. 25, no. 15.

Mullan, F. (2006). The Metrics of the Physician Brain Drain, *The New England Journal of Medicine*, vol. 353, pp1810-1818.

OECD (2004). Towards High-Performing Health Systems, OECD, Paris.

OECD (2006). Future Budget Pressures Arising from Spending on Health and Long-Term Care, *OECD Economic Outlook*, vol. 2006, no. 1, pp145-156.

OECD (2007). *Health at a Glance: OECD Indicators*, OECD, Paris.

OECD(1998). *Maintaining Prosperity in an Ageing Society*, OECD, Paris.

Oliver, J. E. (2006). *Fat Politics: The Real Story behind America's Obesity Epidemic,* Oxford University Press, Oxford.

Oxfam, VSO and Save the Children (2002). *Beyond Philanthropy: the Pharmaceutical Industry, Corporate Responsibility and the Developing World*, Oxfam, VSO, Save the Children, London.

PHM (2000). People's Charter for Health, www.phmovement.org/charter/pch-index.html, accessed 2 October 2008.

PHM (2004). Mumbai Declaration, http://www.phmovement.org/md/md-english.html, accessed 2 October 2008.

Robinson (2008). Mythbusting Canadian Healthcare Part II: Debunking the Free Marketeers, http://www.ourfuture.org/blog-entry/mythbusting-canadian-healthcare-part-ii-debunking-free-marketeers, accessed 11 February 2008.

Saltman, R. B. and O. Ferrousier-Davis (2000). The Concept of Stewardship in Health Policy, *Bulletin of the World Health Organisation*, vol. 78, no. 6, pp732-739.

Sanders, D. (2003). Twenty-Five Years of Primary Health Care: Lessons Learned and Proposals for Revitalisation in R. Narayan and Unnikrishnn PV (Eds.) *Health for All Now—Revive Alma-Ata*, Books for Change, Bangalore.

Schiff, Bindman and Brennan (1994). A Better-Quality Alternative: Single-Payer National Health System Reform, *JAMA -The Journal of the American Medical Association*, vol. 272, September.

Sivaraman, S. (2003). "Fuzzy Words, Sharp Bullets"—Media, Globalisation and Health in R. Narayan and Unnikrishn PV (Eds.) *Health for All Now—Revive Alma-Ata*, Books for Change, Bangalore.

Tanner, M. (2006). No Miracle in Massachusetts: Why Governor Romney's Health Reform Won't Work, *Cato Briefing Papers*, June, no. 97, p1-13.

Vasella, D. (2004). The Right to Health: The Novartis Position in K. Leisinger (Ed.) *The Right to Health: A Duty for Whom? International Symposium Report 2004*, Novartis Foundation for Sustainable Development, Basel, CH.

Walker, A. (2002). A Strategy for Active Ageing, *International Social Security Review*, vol. 55, no. 1, pp121-139.

Werner, D. (2003). The Alma-Ata Declaration and the Goal of "Health for All" 25 Years Later—Keeping the Dream Alive in R. Narayan and Unnikrishnn PV (Eds.) *Health for All Now—Revive Alma-Ata*, Books for Change, Bangalore.

WHO (2000). *The World Health Report 2000: Health Systems—Improving Performance*, WHO, Geneva.

WHO (2002). *Active Ageing: A Policy Framework*, WHO, Geneva.

WHO (2006). *World Health Report 2006: Working together for Health*, WHO, Geneva.

WHO (2007). *World Health Statistics 2007*, WHO, Geneva.

WHO and UNICEF (1978). *Primary health care: report of the International Conference on Primary Health Care, Alma-Ata*, USSR, 6-12 September, 1978, World Health Organization, Geneva.

WTO Secretariat and WHO (2002). *WTO Agreements & Public Health*, WHO, Geneva.

World Bank (1993) *World Development Report 1993: Investing in Health, World Bank*, Washington, D.C.

PART II. SOCIAL SCIENCES PERSPECTIVES

In: Health and Environment: Social Science Perspectives ISBN 978-1-60876-216-3
Editors: Helen Kopnina and Hans Keune

Chapter 5

From Environmental to Ecological Sociology: Implications for Health

Gary Bowden

Abstract

This chapter briefly traces the rise and institutionalization of environmental sociology in order to answer the following questions: A) Has the emergence of environmental sociology forced a fundamental restructuring of sociological theory? B) If not, what have been the blockages to such a development and do they still exist? C) What would an ecological (as opposed to environmental) sociology look like? And D) what are the implications of ecological sociology for our understanding of environment and health issues?

Introduction

The rise and institutionalization of environmental sociology represents one of the most significant changes to the discipline of the past quarter century. Underlying this success, however, is a deeper question: Do these developments represent the first step down a longer path toward more fundamental theoretical change, or have the changes to date delivered us to the Promised Land? Riley Dunlap (2002), one of the sub discipline's founders and most tireless advocates, recently argued the latter position. The first three sections of this chapter combine to argue the former, that fundamental theoretical change is still required. The first section distinguishes between environmental and ecological sociology and advocates the latter. The second section examines current sociological theorizing about the environment and concludes that the label "environmental sociology" accurately describes the current state of affairs. The third section identifies three conceptual blockages that have, in the past, prevented the development of ecological sociology. Recent scholarship has removed each of these blockages, thus paving the way for the development of ecological sociology, the characteristics of which are briefly described in the fourth section. The fifth and final section examines the implications of ecological sociology for our understanding of health.

Distinguishing Between Environmental and Ecological Sociology

A quarter century ago, William Catton and Riley Dunlap wrote a series of articles defining environmental sociology (Catton and Dunlap 1978a; 1978b; 1980; Dunlap and Catton 1979; 1983; 1994). Frustrated with perceived deficiencies in contemporary sociological practice, they argued that contemporary sociologists should be, but weren't, focused on the relationship between society and the biophysical environment. They accounted for the oversight by examining the taken-for-granted assumptions of mainstream sociology and explaining how those assumptions led the discipline to ignore the biophysical environment. To rectify the situation, they advocated a new paradigm based on an alternative set of background assumptions.

A close reading of Catton and Dunlap's publications shows two distinct definitions of environmental sociology: one methodological and the other paradigmatic. In contrast to Dunlap and Catton, I do not see the two definitions as alternative ways of rendering the same thing. Instead, I wish 1) to draw a distinction between environmental sociology (defined methodologically) and ecological sociology (defined paradigmatically) and 2) to argue that present thinking conflates the two.

Methodologically, Catton and Dunlap (1978a: 44) defined environmental sociology as the "study of interaction between the environment and society." Studies using social variables as both cause and effect, for example the examination of social class distinctions in public opinion toward the environment, merely transferred the gaze of traditional sociological theory to a new substantive area—attitudes about the natural environment. In contrast, Dunlap and Catton argued for an environmental sociology in which theories would include both environmental and social variables as cause and/or effect. Such studies, for example the examination of the impact of social class on environmental degradation or of energy shortages on society, they viewed as "real" environmental sociology.

Recognizing that such studies were exceptionally rare in the late 1970s, they argued that the disciplinary assumptions of sociology emphasized social causes and effects. According to Catton and Dunlap (1978a; 1978b; 1980), western society shares a variety of background assumptions that they termed the Dominant Western Worldview (see column 1 of Figure 1). Traditional sociology emerged out of this tradition and, hence, shared a set of related background assumptions, the Human Exceptionalism Paradigm (see Column 2 of Figure 1). In other words, they described a sociological paradigm based upon a shared anthropocentrism that—irrespective of particular theoretical orientation (Marxist, functionalist, symbolic interactionist, or whatever)—led sociologists to treat modern societies as "exempt" from ecological constraints. In contrast to these existing paradigms, they advocated an environmental sociology that embodied a new set of assumptions (Column 3 of Figure 1). Thus, the second definition defines a set of background assumptions comprising the paradigmatic structure of environmental sociology.

	Dominant Western Worldview (DWW)	Human Exemptionalism Paradigm (HEP)	New Ecological Paradigm (NEP)
Assumptions about the nature of human beings	People are fundamentally different from all other creatures on Earth, over which they have dominion.	Humans have a cultural heritage in addition to (and distinct from) their genetic inheritance, and thus are quite unlike all other animal species.	Despite exceptional characteristics, humans remain one among many species that are interdependently involved in the global ecosystem.
Assumptions about social causation	People are masters of their destiny; they can choose their goals and learn to do whatever is necessary to achieve them.	Social and cultural factors (including technology) are the major determinants of human affairs.	Human affairs are influenced not only by social and cultural factors, but also by intricate linkages of cause, effect, and feedback in the web of nature
Assumptions about the context of human society	The world is vast, and thus provides unlimited opportunities for humans.	Social and cultural environments are the crucial context for human affairs, and the biophysical environment is largely irrelevant.	Humans live in and dependent on a finite biophysical environment that imposes potent physical and biological restraints on human affairs.
Assumptions about constraints on human society	The history of humanity is one of progress; for every problem there is a solution, and thus progress need never cease.	Culture is cumulative; thus technological and social progress can continue indefinitely, making all problems ultimately soluble.	Although human inventiveness may appear to temporarily extend carrying capacity limits, ecological laws cannot be repealed.

Figure 1. Major Assumptions of the Dominant Western Worldview, Sociology's Human Exemptionalism Paradigm, and the New Ecological Paradigm. (adapted from Dunlap, 2002:333).

Figure 1 highlights three significant differences between the two conceptualizations of "environmental sociology." First, the paradigmatic definition is more elaborate and precise. In addition to the methodological element of the first definition, the assumption about social causation, the paradigmatic definition involves additional assumptions related to human nature and the context of and constraints on society. Second, the nature of those additional assumptions—particularly those about society—significantly shift the focus from environment to ecology (i.e., from the biophysical realm to a particular systemic conceptualization of the biophysical realm). This led Catton and Dunlap, who had originally labeled the NEP the "New Environmental Paradigm" (Catton and Dunlap, 1978a), to re-label it the "New Ecological Paradigm" (Dunlap and Catton, 1979:250), thus emphasizing its focus on the ecological foundation of society. Third, by casting their discussion in the language of paradigms and paradigm change, Catton and Dunlap invoked Kuhn's famous work *The Structure of Scientific Revolutions* (1962).

To summarize, Catton and Dunlap's work defined three distinct visions of sociological practice: traditional sociology, environmental sociology (the methodological definition) and ecological sociology (the paradigmatic definition). The conceptual differences among them can be clarified by a simple visual representation. Think of two colored circles, one yellow and one blue, separated by an empty space. The yellow circle represents the social sphere; the blue circle represents the natural sphere. Traditional sociology (the HEP) operates like a set of colored glasses that obscures the blue circle and, hence, sociological theorizing ignores the biophysical environment. Environmental sociology, the methodological definition imploring sociology to take the biophysical environment (and its interactions with society) into account, removes those glasses. We now see both circles and, in addition, the arrows of causation (representing interactions between the two spheres) flowing between them. Note, however, that expanding the perspective to include the natural sphere does not necessitate a fundamental overhaul of the way the social is theorized (it continues to exist as an autonomous, identifiable yellow circle).

In contrast, the paradigmatic definition (ecological sociology) argues not that the social sphere is separate from the natural but, rather, that each sphere is, in part, embedded within the other. This can be visualized as two overlapping circles, one yellow and one blue. Embedding the spheres within each other has two significant implications. First, the shapes of the yellow and blue circles are altered (they each have a bite taken out of them by the overlap). Second, a new region (the green area created by the blending of the yellow and blue in the overlapped area) is created. Stated in theoretical terms, paradigm change involves a significant modification of a discipline's conceptual apparatus (the change in shape of the yellow area). In addition, seeing the two spheres as embedded (rather than interacting) focuses attention on the possibility of a new type of emergent phenomenon (e.g., phenomenon such as global warming that emerge from the blending of the two spheres) represented by the green area.

THE CURRENT STATE OF ENVIRONMENTAL SOCIOLOGY

Which of the three representations best describes current practice? Looking over the developments of the past 25 years, several things are evident. First, sociological studies of the nature-society connection have been institutionalized as part of a recognized sub discipline, complete with journals and scholarly associations. Second, as a series of recent theoretical reviews have noted (Dunlap et al., 2002b; Redclift and Woodgate, 1997; Buttel, 1996; Dunlap, 1997), theoretical diversity among self-identified environmental sociologists has increased over time. Indeed, the variety of theories in environmental sociology closely approximates the various theoretical traditions present in the discipline as a whole: Marxist, Durkheimian, Weberian, world-systems, social constructionist (for examples of each of these, and others, see Dunlap et al., 2002a). As environmental matters moved from the periphery to the center of social consciousness, they were similarly integrated into the fabric of existing sociological theory. Third, there exists a general consensus among scholars about how that occurred; through a process of "greening" previously existing theories (Dunlap et al., 2002b;

Redclift and Woodgate, 1997; Buttel, 1996; Dunlap, 1997). In other words, the core of each of the previously existing theories has remained fundamentally intact. The modifications have occurred through a process of extension—i.e., integrating previously ignored concerns into the fabric of the existing theories. This is the equivalent of the second alternative (environmental sociology) described above. The basic outline of sociology's theories—be it social constructivist approaches, Marxist perspectives, cultural studies, or whatever—as represented by the yellow circle remains unchanged.

Without denying the significance of these developments, however, it is erroneous to claim, as Dunlap (2002) does, that the greening of sociological theory constitutes a Kuhnian paradigm shift (see Bowden, 2004, for an elaboration of this point). On the other hand, self-identified environmental sociologists tend to embrace the assumptions of the NEP and, as a result, see themselves as working within a paradigm that differs from that of traditional sociology.

How can we explain this apparent contradiction: the objective fact that the emergence of environmental sociology has not caused a paradigm revolution within sociological theory juxtaposed against the belief of environmental sociologists that they employ a different paradigm? A close examination of Column 3 of Figure 1 provides an answer. The defining characteristic of the NEP is a concept borrowed from biology—the notion of carrying capacity—or the existence of biophysical limits on human society. By removing sociology's colored glasses, thus enabling a view of both colored circles, environmental sociologists have become aware of, and adopted, ideas from outside the traditional discipline, specifically the biologist's concept of carrying capacity.

As Catton and Dunlap correctly note, the carrying capacity perspective is antithetical to the assumptions embedded in the "Dominant Western Worldview." Indeed, it has generated a long running disciplinary dispute between economists and biologists (see Homer-Dixon, 1999, for an insightful summary of this debate). Sociology has been, at most, a peripheral participant. The concept at the core of the debate, scarcity, while central to both economic and ecological theory, is not a central concept in sociological theory. Thus, environmental sociologists have embraced biology's orientation toward scarcity (in contrast to the economic conception traditionally held by the discipline), while "greening" sociological theory rather than radically revising it.

Just because the emergence of environmental sociology did not bring about a paradigm change does not mean that such a change is not needed. Several recent empirical studies have reached the same conclusion: that systems combining people and nature are not the same things as the interaction of autonomous social and natural systems (Berkes and Folke, 2000; Scheffer et al., 2000). More specifically, they argue that social and natural systems are involved in a process of co-evolution (Holling, 1994). This concept implies the need for an integrative framework that encompasses both the natural and the social, i.e., for ecological sociology. But such a theoretical development has yet to occur. Why?

THREE BARRIERS TO PARADIGM CHANGE AND THE DEVELOPMENTS THAT REMOVED THEM

Three major factors have acted as barriers to the paradigm change necessary for the emergence of ecological sociology: the division between economic optimists and neo-Malthusians, the division between realists and social constructivists, and the inadequate and discredited conception of system present in sociology. In each case, recent developments have removed existing impediments to true paradigm change.

As noted above, the distinction between the HEP and the NEP reflects a division between economic and biological conceptions of scarcity. The economic optimists argue that social arrangements—typically free markets, science and liberal democracy—provide the institutional arrangements that allow humans to solve any problem they confront. Economic markets raise capital and provide incentives for entrepreneurs to solve problems; science provides knowledge about the functioning of the natural world, knowledge that can be applied to the problem; and democracy allows participation by wider segments of the public, thus enhancing the probability that a solution will be found. In contrast, the neo-Malthusian tainted NEP holds that the demand for resources, stemming from population growth and the increasing standard of living, grows exponentially while the ability to provide them grows arithmetically. As a result, there exist strict biophysical limits on resource supply—the carrying capacity of an ecosystem or the biosphere.

This argument was recently resolved by Homer-Dixon (1999). Where the economic optimists and neo-Malthusians argued past one another, holding that the key claims of the other side were false, Homer-Dixon recognized that both groups offered partial truths and developed a theory integrating the insights from both. Like the economic optimists, Homer-Dixon recognized the deficiencies of a theory based on the strict limits of carrying capacity. Through the application of ingenuity, humans can substitute for resources that are becoming physically exhausted (e.g., shift from oil to another energy source) or develop technologies that expand productivity and, hence, the "limits" of growth. On the other hand, Homer-Dixon argues that ingenuity cannot necessarily solve all problems—there are limits to our ability to solve problems. Specifically, he argues that a) a variety of processes, for example the rapidly increasing scale of human modification of the environment, are contributing to generate problems of greater and greater complexity and b) the ability of humans to comprehend and respond to problems of such complexity are dependent on a variety of factors affecting human capacity to understand and address them. Thus, the real problem is not the biological limits of carrying capacity but, rather, the "ingenuity gap" that emerges when the requirement for ingenuity brought about by problems of greater and greater complexity cannot be met by the level of ingenuity a society is capable of supplying.

A second roadblock flows from sociology's debate between realist and constructionist ontology. Chronologically paralleling the rise of environmental sociology was another development—the sociology of scientific knowledge—which treated scientific conclusions as just another socially constructed knowledge claim (Latour and Woolgar, 1979). Many environmental sociologists, however, took a strong realist view; science legitimized the existence of environmental problems and the need for remedial action. These debates culminated in the science wars (Parsons and Long, 2003). Since that time, researchers on both sides have moved toward a common middle ground. See, for example, the "critical realism" of Collier (1994) or the "realist constructivism" of Murphy (2002). Even Bruno Latour now claims that "things strike back" and, in recognition of this, has added "a realist social philosophy of science" to the "old tired theme of social construction (Latour, 2000: 115, 114).

The third blockage involves sociology's attitude toward system thinking. In the late 1970s, when Dunlap and Catton were advancing their ideas, sociological theorists were engaged in a Oedipal rejection of functionalism, the dominant tradition of an earlier generation of theorists. In the process of throwing out the functionalist bath water, the discipline also threw out the baby—any discussion of system. Functionalism treated social systems as a self-regulating equilibrium in which, when confronted with pressure to change, a mechanism acts to restore the system to a state of balance. This approach, exemplified in the highly influential work of Parsons (1951), was beset with a number of problems; an excessively rigid relationship between parts and the whole and an inability to deal with either a) sudden change or b) diversity. Unable to overcome the objections, systems theorizing in sociology came to a halt.

That was not the case in other disciplines. Versions of systems theory that overcome the objections leveled against traditional systems thought in sociology, most notably the idea of complex adaptive systems (Byrne, 1998; Holland, 2000; Kauffman, 1995), are now being applied to social systems. The most promising such theory, Holling's (2001) notion of a panarchy, outlines a common framework for the explanation of social, economic and ecological systems while recognizing the distinctive nature of social systems (Westley et al., 2002).

The implications of the above developments for sociological understanding of environmental and health issues have affected the discipline at an uneven pace. Most appreciated is the resolution of the realist/constructivist debate and its implications for our understanding of the nature/society divide (Murphy, 2002). However, while correctly defining the object of study (an emergent, conjointly constituted social/nature hybrid), appreciation of the object does little to articulate a theoretical appreciation of *how the object operates*. Similarly, while a number of sociologists have argued for the application of complex adaptive system theory, their main object of study has been social systems (e.g., Walby, 2007) rather than social/ecological hybrids. In contrast, a large number of ecologists, led by Holling, have developed a complex adaptive systems approach to ecosystems, which they have used to theorize a more general conceptualization of systems operation (Gunderson and Holling, 2002). The key point is that understanding a hybrid socio-ecological system requires a fundamentally interdisciplinary approach and, hence, a deep understanding of both social and ecological processes. To date, the most impressive synthesis of these emerging trends can be found in the work of Thomas Homer-Dixon (1999, 2001, 2006).

Conceptualizing Ecological Sociology

As a result of this new theorizing—particularly in relation to way systems are conceptualized and the replacement of "scarcity" with "ingenuity" as the limiting factor on human adaptability—the vision of ecological sociology that emerges at the end of this process is not identical to Dunlap and Catton's paradigmatic characterization. With the disputes resolved, the guiding principals of an ecological sociology become apparent:

1. An understanding of the social/nature relationship as an *emergent nature/society hybrid.*
2. The recognition that such an object is best theorized as a *conjoined socio-ecological system.* As suggested by Dunlap and Catton, this implies the need to integrate sociological thinking with ecological thinking. However, in contrast to their approach (with its emphasis on the limitations of carrying capacity), current research suggests attention to the processes of ecological systems. Broadly speaking, this draws our attention to a) processes of biogeochemical flows—such as the water and carbon cycle and b) the necessity for increasing amounts of net energy input into the system in order to support increasing levels of complexity.
3. The recognition that such socio-ecological systems exhibit the characteristics of *complex adaptive systems* (e.g., positive and negative feedback loops, coupling, thresholds effects, multiple equilibria, etc.).
4. There are two general approaches to understanding the operation of complex adaptive systems. Most researchers explain the observed systems properties through reference to a large number of interacting causal processes. In contrast, Holling's (2001) notion of *panarchy* argues for a model based on a set of hierarchically ordered and interacting adaptive cycles where each level's adaptive cycle is controlled by a small number of causal factors. For present purposes, this approach highlights the need to attend to causal factors across a range of spatial-temporal levels (from short duration, localized "micro" processes to long duration, globalized "macro" processes).
5. A *model of human nature*—humans as problem-solvers—predicated upon what is ecologically unique about our species: we are adaptive generalists rather than adapted to the specifics of a particular ecological niche (Homer-Dixon, 2001:191-220) and an appreciation of the role of human perception in relation to our dealings with nature. Of particular relevance are the inherent limitations on human understanding of the dynamics of the system operations captured in the distinction among known knowns, known unknowns and unknown unknowns (Homer-Dixon, 2001: 171-187).
6. A broad understanding of these processes allows humans to make plausible projections or envision likely scenarios for the future (as is done by climate change modelers) though the limit to make accurate predictions is limited by a) *limitations on our knowledge* (the inability to incorporate unknown unknowns), b) the possibility of what Perrow (1984) labeled "*system accidents*" and c) the human capacity for agency.
7. These scenarios, however, are limited not only by limitations on our understanding, but also by the fact that the system itself is a moving target. To borrow Holling's (1994a) phrase, we have entered an era of "*nature evolving*" characterized by the mutual evolution of nature and society.
8. The conjoined nature-society system is evolving at an *increasing speed.* Larger human populations, higher resource consumption, and more powerful technologies sharply raise the density, intensity, and pace of our interactions with each other and the environment. These factors have all exploded since the industrial revolution, leading to a profound increase in the speed at which the conjoined socio-ecological system is evolving (Homer-Dixon, 2001).

ECOLOGICAL SOCIOLOGY AND HEALTH

What implications does the shift from environmental to ecological sociology have for the way we conceptualize health and its relation to the environment? Rather than attempt an encyclopedic discussion of the consequences, the remainder of this chapter develops a small number of examples designed to illustrate key characteristics of the perspective.

Our first example uses one of Canada's worst public health disasters, the E. coli outbreak in the small farming community of Walkerton, Ontario, to illustrate the deeper understanding of the events provided by the ecological sociology perspective. In late May, 2000, more than 2,300 people became ill and seven people died as a result of consuming water from the town's municipal supply—water that had been contaminated by cattle manure containing the deadly E. coli O157:H7 bacterial strain. Most investigations of the event, both journalistic (CBC, 2004) and official (Walkerton Inquiry Commission, 2002) treated it as an anomalous occurrence resulting from the combination of a) unexpected flooding of pastureland resulting from heavy rains that let contamination into the water system via b) a shallow well covered with porous rock that was vulnerable because of c) operator error (failure to maintain proper chlorine levels) made possible by d) poor policies and practices implemented by a variety of different government and non-government agencies that were, in part, a result of budgetary pressures and neo-liberal regulatory reforms.

The difference in understanding afforded by ecological sociology can be seen by contrasting Ali's (2004) account of the incident. As summarized in Figure 2, Ali examines a variety of social and ecological factors that operated across a number of spatial-temporal scales (micro, meso, macro). Notably, this leads Ali to identify the relevance of a variety of factors that the other accounts ignore. Take, for example, the role of human-induced environmental changes in the evolution of the E.coli bacterium (the macro level cell in the ecological dimension of Figure 2). The bacteria affecting Walkerton residents took a circuitous ecological pathway: "from the intestines of cattle, through the surface water pathways, through the soils into the ground water pathways and through the constructed drinking water system to be ultimately consumed by the unfortunate human victim." How was the bacterium able to survive as it passed through these various environments with such diverse ecological characteristics? According to Whittam et al. (1998), E. coli O157:H7 is particularly well adapted to survival in extreme environments because of its resistance to acid, salt and chlorine. They hypothesize this resistance is an evolutionary adaptation to human-induced environmental changes (specifically, the acidification of water brought about by acid rain). Similarly, Ali (2004: 2605) notes that factory farming contributes to the proliferation and spread of virulent pathogens in two ways: 1) by providing the pathogens with suitable habitat-niches and 2) by placing the cattle under stress; leading to the need for "maintenance doses" of antibiotics to keep them healthy and the resultant selection pressure for virulent strains that are better able to develop antibiotic resistance. Of course, without the factory farm, ecological niches and characteristics specific to the factory farm would not exist. Thus, one cannot account for the opening of these new niches and pressures without reference to those changes in the international political economy that account for the shift from traditional livestock farming methods to the factory farm (the macro cell of the social dimension).

This example illustrates several key features of the ecological sociology perspective. First, rather than treating the outbreak as an anomalous event, it renders it as the emergent, conjointly constituted product of an extended set of social and ecological processes. Second, rather than focus on a single spatial-temporal scale, the processes involved operate across the entire spatial-temporal spectrum, from small and fast (micro) to large and slow (macro). As a result, the framework focuses attention on factors that other accounts tend to ignore. Despite the exceptionally thorough public examination of the Walkerton case, involving numerous investigative journalists and a federal inquiry costing 9 million dollars, Ali's (2004) social-ecological analysis identified a variety factors that all previous studies had missed. Third, by viewing the socio-ecological system as a product of a diverse set of processes across a variety of scales, the perspective highlights a) the complexity of the event, b) the potential limits on human understanding of the phenomenon (exemplified in the distinction among known knowns, known unknowns, and unknown unknowns), and c) the potential for system accidents (Perrow, 1984) resulting from unanticipated interactions among various factors. Fourth and finally, the example draws attention to the increasing level of systemic risk. Instead of treating it as a unique event, the perspective renders the outbreak as a systemic property of the current socio-ecological system and, hence, potentially repeatable in a variety of other locations. Indeed, by locating key causative features in long-term social-ecological processes (e.g., climate change, the transformation of farming) the perspective draws attention to the increasing levels of risk and, hence, the increased likelihood of similar outbreaks in the future.

As the Walkerton example illustrates, the perspective of ecological sociology provides a powerful new lens for conceptualizing the emergence of disease and other forms of toxic exposure. It also suggests an alternative focus on the meaning of health itself. Most commonly, in our day-to-day lives, human health is discussed at an individual level: how healthy are you? Epidemiology, in contrast, studies the factors that affect the health and illness of populations. A related area, population health, recognizes that there is no such thing as a single population (e.g., that income inequality and other social factors affect health and lead to differential health outcomes). This perspective emphasizes the social determinants of health and, hence, argues that successful health promotion policy needs to reflect a collectivist rather than individualist ethos (Low and Teriault, 2008). In short, standard perspectives on human health typically examine humans as objects, existing either as individuals or groups, affected by their interactions with other aspects of the social and natural systems.

These approaches contrast with the more systemic and relational focus associated with ecological sociology. Tracing its conceptual roots more to ecology and its study of ecosystem health than to traditional studies of human health, this work takes a systemic view in order to integrate concerns for human, wildlife and ecosystem health within a holistic perspective. The approach addresses how human health and environmental quality are determined by complex relationships among the diverse components of an ecosystem with the implication that human health can be protected and improved through more sustainable ecosystem management. Key examples of this approach can be found in Millennium Ecosystem Assessment (2005), Brown, et al. (2005), Lobel (2003), Gunderson and Holling (2001) and the journal *EcoHealth.*

Two final examples, focusing on the role of flows, illustrate other key implications of the shift from environmental to ecological sociology. We begin with a brief review of one strand of research in environmental sociology, research focusing on human exposure to toxics. A large segment of this research, building on classic works like Bullard's (1990) *Dumping in Dixie,* focuses on the existence of social inequities in the exposure to toxic substances. These works argue that race and class play a prominent role in the location of hazardous waste dumps, factory farms, chemical refineries, and other kinds of infrastructure responsible for large amounts of potentially dangerous contaminants. Thus, the poor and people of color are more likely to be exposed to toxins that may negatively affect their health. Much of the power of their argument comes from maps that visually and dramatically depict positive correlations between the percentage of poor people or people of color living in an area and the number and location of various kinds of hazardous waste producing infrastructure. Simply put, this research argues not only that "race matters" but also that "place matters." (For a recent review of this literature, see Bullard, et. al., 2007).

	Micro Level	Meso Level	Macro Level
Ecological Dimension	Individual Water Well; Extreme Weather Event	Regionalized Factory Farm Setting	Evolution of new and (Re)Emerging Diseases; Processes of Global Climate Change
Social Dimension	Operator Failure; Individual Farmer	Public Utilities Company; Government Agencies (MOE, public health); Private Lab Testing	International Political Economy (Neoliberalism); Agribusiness

Figure 2. A Socio-ecological examination of the factors responsible for the E. coli outbreak in Walkerton, Ontario, Canada (adapted from Ali, 2004:2603).

Without contesting the basic insight of this research, it is important to note that "environment" is reduced to geographic location, and distance from the hazardous waste site is treated as a proxy for exposure. We know this is overly simplistic. Thus, for example, the level of exposure to hazardous chemicals buried in Love Canal showed little correlation with how close a person's residence was to the dump site. The chemicals traveled along underground rivers, radiating out like spokes along those paths, rather than seeping equally in all directions (Levine, 1982). Thus, a more fine-grained appreciation of the dynamics of exposure requires attention not only to "place" but also to ecosystem processes affecting the flows of toxic chemicals.

But, attention to flows shouldn't be treated simply as a nuance related to determining the likelihood of exposure for those living within a threatened area. Sometimes the individuals most affected live great distances from the origins of the pollutants. Thus, for example, the indigenous peoples of the Arctic show some of the highest body burdens ever measured for a number of persistent organic pollutants (e.g., DDT, PCBs, lindane) despite the fact that such chemicals aren't used in the Arctic. The explanation for this rests in the "grasshopper effect" (Cone, 2006). Chemicals evaporate out of the soil in warmer countries where they are still used, and travel in the atmosphere toward cooler areas, condensing out again when the temperature drops. The process, repeated in hops, can carry the chemicals thousands of kilometres in a matter of days. Snow, in comparison to rain or low temperature, is particularly good at scrubbing the chemicals from the atmosphere and, hence, areas with high snowfalls are particularly vulnerable. Once deposited, the chemicals travel up the food chain, accumulating in the fatty tissue of predatory animals including traditional food sources of the indigenous people. In short, attention to the ecological flow—through both the air and the food chain—is necessary to understand how the pollutants came to reside within the bodies of the Inuit.

The above example was introduced as a way of illustrating a difference between environmental and ecological sociology. It also, however, illustrates the way ecological sociology draws on an understanding of ecology different from that used in the paradigmatic vision of Catton and Dunlap (1978a). Specifically, where Catton and Dunlap take carrying capacity as the dominant ecological concept of relevance, contemporary ecological sociology builds instead on ecology as process and, hence, places emphasis on factors such as energy inputs and biogeochemical flows. The final example links the discussion of flows with the idea of a conjoined socio-ecological system.

Over the past two decades, a large body of research has emerged documenting the potential health consequences of climate change. Among them are the following: a) changes in temperature-related morbidity and mortality, b) the health effects of extreme weather events, c) changes in the distribution of vector-born zoonotic diseases, d) health effects resulting from food and water shortages, and e) many others (McMichael et. al., 2003). Less appreciated, however, is the fact that climate change, at its most fundamental level, results from human alteration of the carbon cycle. Carbon that has been locked up in trees and underground in oil and gas deposits is being released into the atmosphere via deforestation and the burning of fossil fuels. Put simply, the flow of the "natural" carbon cycle is being disrupted by human influence. Or, stated another way, the carbon cycle is no longer a strictly natural process; it is now the product of a conjoined socio-ecological system. Significantly, this view varies significantly from conventional wisdom, which continues to think of natural resources as a stock to be exploited, rather than one location in a biogeochemical flow.

Furthermore, the carbon cycle is only one such flow and, indeed, it is in some ways less perturbed than the flows of other cycles. Klee and Graedel (2004) have produced the first systematic attempt to assess the relative contributions of natural and anthropogenic mobilizations to elemental flows within the biosphere. Analyzing the flows of 77 of the first 92 elements on the periodic table, they found that human activities dominate (i.e., anthropogenic activities are responsible for more of the flow than are natural activities) or strongly perturb (i.e., anthropogenic activities contribute between 15 and 50% of the total flow) the cycles of 66 of those elements. To put this in context, human activities contribute only 9% of the total carbon flow, though the overall magnitude of the total flows of carbon significantly dwarf those of the other elements.

To render these findings in a graphic form, picture a variety of pizzas. The one extremely large pizza (carbon), has a relatively small slice (9%) that is the result of human activities while most of the other pizzas (which range in size from large to very tiny) have human-produced slices that make up substantially larger proportions of the total (from 30 to nearly 100%). In short, human activities (particularly mining and industrial production) are dramatically transforming the flows of most of the elements through the biosphere. As such, we have become co-creators of the biogeochemical flows that underpin the operation of the biosphere. While a) it is impossible to know the scale and scope of what this implies at this time and b) the carbon cycle is one of the most basic and, hence, alterations to it would be expected to have substantial consequences (as documented in McMichael et. al., 2003), there is little doubt—given the significant impact that alteration of the carbon cycle has had on health—that these developments also have substantial health implications.

Conclusion

Environmental sociology has established itself as an institutionalized sub discipline and, as a result, may be entering a phase more interested in protecting established sub disciplinary boundaries than in radical theoretical change. However, shifting the focus from the analysis of autonomous but interacting social and ecological systems to a framework that truly embeds humans in ecosystems and envisions both systems as the product of a common co-evolutionary framework will require fundamental theoretical change. Fortunately, recent developments have removed the roadblocks to the development of ecological sociology, and it is now possible to sketch the outlines of such an approach. This approach advances our understanding of human health and the health-related challenges facing us in a variety of ways.

References

Ali, H. (2004). A socio-ecological autopsy of the *E. coli* O157:H7 outbreak in Walkerton, Ontario, Canada. *Social science & medicine*, *58*(12), 2601-2612.

Berkes, F. & Folke, C. (2002). Back to the future: Ecosystem dynamics and local knowledge. In L. Gunderson & C. Holling (Eds.), *Panarchy: Understanding transformations in human and natural systems* (pp. 121-146). Washington: Island Press.

Brown, V., Grootians, J., Ritchie, J., Townsend, M. & Verrinder, G. (2005). *Sustainability and health: Supporting global ecological integrity in public health.* London: Earthscan.

Bowden, G. (2004). The need for a real paradigm shift: Moving from environmental to ecological sociology. Paper presented at the Annual Meeting of the Canadian Sociology and Anthropology Association, Winnipeg, Manitoba, June 6.

Bullard, R. (1990). *Dumping in Dixie: Race, class and environmental quality.* Bolder, Co.: Westview Press.

Bullard, R., Mohai, P., Saha, R. & Wright, B. (2007). (PDF), *Toxic wastes and race at twenty: 1987-2007*, United Church of Christ, http://www.ejrc.cau.edu/TWART-light.pdf, retrieved on 2009-03-06.

Buttel, F. (1996). Environmental and resource sociology: Theoretical issues and opportunities for synthesis. *Rural sociology, 61*, 56-76.

Byrne, D. (1998). *Complexity theory and the social sciences: An introduction.* London: Routledge.

Canadian Broadcast Corporation, 2004

Catton, W. & Dunlap, R. (1978a). Environmental sociology: A new paradigm. *The American sociologist, 13*, 41-49.

Catton, W. & Dunlap, R. (1978b). Paradigms, theories, and the primacy of the HEP/NEP distinction. *The American sociologist, 13,* 256-259.

Catton, W. & Dunlap, R. (1980). A new ecological paradigm for post-exhuberant sociology. *American behavioral scientist, 24,* 15-47.

Collier, A. (1994). *Critical realism.* London: Verso.

Cone, M. (2006). *Silent snow: The slow poisoning of the Arctic.* New York: Grove Press.

Dunlap, R. (1997). The evolution of environmental sociology: A brief history and assessment of the American experience. In M. Redclift & G. Woodgate (Eds.), *The international handbook of environmental sociology* (pp. 21-39). London: Edward Elgar.

Dunlap, R. (2002). Paradigms, theories and environmental sociology. In R. Dunlap, F. Buttel, P. Dickens, & A. Gijswijt (Eds.), *Sociological theory and the environment: Classical foundations, contemporary insights* (pp. 329-350). Lanham, Md.: Rowman & Littlefield.

Dunlap, R., Buttel, F., Dickens, P & Gijswijt, A. (Eds.) (2002a). *Sociological theory and the environment: Classical foundations, contemporary insights.* Lanham, Md.: Rowman & Littlefield.

Dunlap, R., Buttel, F., Dickens, P & Gijswijt, A.(2002b). Sociological theory and the environment: An overview and introduction. In R. Dunlap, F. Buttel, P. Dickens & A. Gijswijt (Eds.) *Sociological Theory and the Environment: Classical Foundations, Contemporary Insights* (pp. 3-32). Lanham, Md.: Rowman & Littlefield.

Dunlap, R. & Catton, W. (1979). Environmental sociology. *Annual review of sociology, 5,* 243-273.

Dunlap, R. & Catton, W. (1983). What environmental sociologists have in common (Whether concerned with "built" or "natural" environments). *Sociological inquiry, 53,* 113-135.

Dunlap, R. & Catton, W. (1994). Struggling with human exemptionalism: The rise, decline and revitalization of environmental sociology. *The American sociologist, 25,* 5-30.

Freudenberg, W., Frickel, S. & Gramling, R. (1995). Beyond the nature/society divide: Learning to think like a mountain. *Sociological forum 10* (3), 361-392.

Gunderson, L.H. & Holling, C.S. (2002). *Panarchy: Understanding transformations in human and natural systems.* Washington: Island Press.

Holland, J.H. (2000). *Emergence: From chaos to order.* Oxford: Oxford University Press.

Holling, C.S. (1994). An ecologist's view of the Malthusian conflict. In K. Lindahl-Kiessling & H. Landsberg (Eds.), *Population, economic development, and the environment* (pp. 79-103). Oxford: Oxford University Press.

Holling, C. S. (1994a). New science and new investments for a sustainable biosphere. In A. Jansson, M. Hammer, C. Folke & R. Costanza (Eds.), *Investing in natural capital: The ecological economics approach to sustainability.* Washington DC: Island Press.

Holling, C.S. (2001). Understanding the complexity of economic, ecological, and social systems. *Ecosystems, 4,* 390–405.

Homer-Dixon, T. (1999). *Environment, scarcity and violence.* Princeton: Princeton University Press.

Homer-Dixon, T. (2001). *The ingenuity gap: Can we solve the problems of the future?* Toronto: Knopf.

Homer-Dixon, T. (2006). *The upside of down: Catastrophe, creativity and the renewal of civilization.* Toronto: Knopf.

Kauffman, S.A. (1995). *At home in the universe: The search for laws of self-organization and complexity.* London: Viking.

Klee, R. & Graedel, T. (2004). Elemental cycles: A status report on human or natural dominance. *Annual Review of Environment and Resources. 29,* 69-107.

Kuhn, T. (1962). *The structure of scientific revolutions.* Chicago: University of Chicago Press.

Latour, B. (2000). When things strike back. *British journal of sociology, 51,*107-123.

Latour, B. & Woolgar, S. (1979). *Laboratory life: The construction of scientific facts.* Beverley Hills: Sage.

Levine, A. (1982). *Love Canal: Science, politics and people.* New York: D.C. Heath.

Lobel, J. (2003). *Health: An ecosystem approach.* Ottawa: IDRC Books.

Low, J. & Theriault, L. (2008). Health promotion policy in Canada: lessons forgotten, lessons still to learn. *Health Promotion International 23 (2),* 200-206.

McMichael, A., Campbell-Lendrum, D., Corvalán, C., Ebi, K., Githeko, A., Scheraga, J. & Woodward, A. (2003). *Climate change and human health: Risks and responses.* Geneva: World Health Organization.

Meyrowitz, J. (1985). *No sense of place: The impact of electronic media on social behavior.* New York: Oxford University Press.

Millennium Ecosystem Assessment. (2005). *Ecosystems and human well-being: Health synthesis.* Geneva: World Health Organization.

Murphy, R. (2002). The internalization of autonomous nature into society. *Sociological Review, 50,* 313-333.

Parsons, K. & Long, R. (2003). *The science wars: Debating scientific knowledge and technology.* Amherst, NY: Prometheus.

Parsons, T. (1951). *The social system.* New York: Free Press.

Redclift, M. & Woodgate, G. (Eds.) (1997). *The international handbook of environmental sociology.* London: Edward Elgar.

Scheffer, M., Westley, F., Brock, W. & Holmgren, W. (2002). Dynamic interactions of societies and ecosystems: Linked theories from ecology, economy and sociology. In L. Gunderson & C. Holling (Eds.), *Panarchy: Understanding transformations in human and natural systems* (pp. 195-240). Washington: Island Press.

Walby, S. (2007). Complexity theory, systems theory, and multiple intersecting social inequalities. *Philosophy of the social sciences, 37* (4), 449-470.

Walkerton Inquiry (2002). Brief of part one report of the Walkerton inquiry. http:// walkertoninquiry.com/ The Walkerton Inquiry, Dennis R. O'Connor, Commissioner.

Westley, F. (2002). The devil in the dynamics: Adaptive management on the front lines. In L. Gunderson & C. Holling (Eds.), *Panarchy: Understanding transformations in human and natural systems* (pp. 333-360). Washington: Island Press.

Whittam, T., McGraw, E. A., & Reid, S. D. (1998). Pathogenic E. coli O157:H7: a model for emerging infectious diseases. In R. M. Kause (Ed.), *Emerging Infections* (pp. 163–183). San Diego: Academic Press.

In: Health and Environment: Social Science Perspectives ISBN 978-1-60876-216-3
Editors: Helen Kopnina and Hans Keune

Chapter 6

BIOMEDICAL AND ENVIRONMENTAL HEALTH PERSPECTIVES: THE EXAMPLE OF CONFRONTING BREAST CANCER

Lori Baralt

ABSTRACT

This chapter explores the dominant biomedical paradigm and the challenges that social scientific environmental health perspectives pose to this model. The two predominant charity and advocacy approaches to breast cancer are used as examples to demonstrate the conflict between the two perspectives outside of the realm of social science and the potential role of social scientists in furthering environmental health perspectives. While both approaches to breast cancer are anthropocentric in that they maintain the ultimate goal of the eradication of breast cancer as an illness, the dominant charity form of breast cancer advocacy relies on a biomedical model of health, which ignores the environment altogether, while the alternative environmental breast cancer model is a hybrid advocacy model that highlights the inextricable connections between human health and human-induced environmental harms. Embedded within a sociological framework, I begin by presenting an overview of the main tenets of the biomedical paradigm of health and illness. I then describe challenges posed to the biomedical paradigm by the sociological field of environmental health studies, which has developed at the intersection of medical and environmental sociology. The divergent charity and advocacy approaches to breast cancer are then explored within this context. Finally, the role of social scientists in contributing to positive social change for the well-being of the environment and human health are then discussed.

> As we tinker with and change the environment, we know, or suspect, that it is tinkering with and changing us, too often from moderately healthy people to sick ones. (Brown et al., 2000, 17)

> We live in a toxic culture that degrades human and environmental health, particularly in communities of the poor and people of color. In workplace, home, communities, and recreational areas, exposure to an expanding array of toxic conditions in the air, water, and soil poses an increasing long-term threat to public health. However, what makes our society a toxic culture are the social arrangements that encourage and excuse the deterioration of the environment and human health. (Hofrichter, 2000, 1)

INTRODUCTION

Emphasis on genetics, individual risk factors, and medical diagnosis and treatment are the elements of the dominant health and illness discourse. Often times, in fact, it may seem that other ways of understanding health and illness outside of the biomedical paradigm are nonexistent, but this is not the case. The field of environmental health in the social sciences,[1] developing out of the increasing dialogue between medical and environmental sociologists, provides an alternative, albeit one fraught with controversy, to the biomedical paradigm (Brown, 2000). As an environmental health sociologist, I draw connections between human-induced environmental harms and the subsequent negative effects on human health in my research. I also remain critical of strictly biomedical approaches to human health that removes humans from the political, economic, social and environmental context in which they exist. In this chapter, I will provide an overview of the biomedical paradigm, followed by a presentation of the alternative environmental health perspective. The two predominant charity and advocacy approaches to breast cancer are then discussed. While both types of breast cancer advocacy are humanistic-focused, in that both maintain the ultimate goal of the eradication of breast cancer as an illness, the dominant charity form of breast cancer advocacy relies on a biomedical model of health, which ignores the environment altogether, while the alternative breast cancer advocacy model highlights the inextricable connection between breast cancer and human-induced environmental harms. Social scientists have the potential to play an important role in drawing connections between human health and the environment. Within sociology, environmental sociologists and sociologists of health and illness have come together in order to draw connections between human health and the environment. Sociological environmental health research and theory has challenged the dominant biomedical paradigm and encouraged medical sociologists to broaden their understanding of human health and illness.

THE BIOMEDICAL PARADIGM

Currently, issues of health and illness are primarily understood through the biomedical paradigm, sometimes simply referred to as the medical model of health (Bury, 2005). This approach is entirely anthropocentric because it prioritizes human health and does not place

---

[1] The field of environmental health spans the social and natural sciences; however, for the purposes of this chapter, I am limiting my focus to the social sciences, and particularly to the development of environmental health studies at the intersection of medical and environmental sociology.

humans in a broader environmental context, but rather focuses on individual bodies at the level of genetics. Medical sociologists, or sociologists of health and illness[2] as they are more commonly referred to now, have paid a great deal of attention to understanding the biomedical paradigm. While scholars have focused on various aspects of this paradigm, in what follows, I will focus on the key elements.

Firstly, the biomedical model of health and illness may be described as a "negative model" of health (Bury, 2005, 2). Within this model, health is typically conceptualized as the absence of illness (Bury, 2005; Annandale, 2002). Practices of promoting positive health, therefore, are not emphasized in this paradigm. Furthermore, "in adopting what is termed a reductivist approach, biomedicine assumes that health and disease are natural phenomena which exist in the individual body, rather than in the interaction of the individual and the social world" (Annandale, 2002, 6) (emphasis in original). The body, therefore, is often viewed as separate from the environment in which it exists, as well as separate from the mental state and experience of the individual. In other words, based on this paradigm, a doctor or medical scientist (for example, doing genetic research) is able to assess a patient's health (or if disease is present, lack of health), without reference to anything beyond the individual's body (or genes).

A related aspect of the biomedical paradigm is the "*doctrine of specific etiology*" (Annandale, 2002, 6) (emphasis in original). This characteristic, as defined by Dubos (1960, 87 as quoted by Annandale, 2002, 6), is the view that "disease could be produced at will by the mere artifice of introducing a single specific factor—a virulent micro-organism—into a healthy animal." Again, disease is conceptualized very narrowly and as a biological process that occurs within an individual body without reference to the environment in which it exists.

Finally, the biomedical model posits that medical research and practice are scientifically neutral and objective. According to Annandale (2002, 7), the claim is that "medicine can be rational, objective, and value-free, treating each individual according to their need and irrespective of any sense of moral worth." Many medical sociologists have challenged the medical model's claim to scientific objectivity and neutrality (e.g., feminist critiques of the biomedical model, for example, see Annandale, 2002).

As the biomedicalization of society (Clarke et al., 2005) increases, the biomedical paradigm continues to gain strength, despite its inability to address many people's concerns about health and illness. Conrad (1992, 2007) argues that medicalization refers to processes through which aspects of life previously defined outside of medicine come to be constructed as medical problems (e.g., menopause, ADHD, erectile dysfunction). Clarke and her colleagues (2005, 442) argue that "recently, medicalization is intensifying in new ways. Biomedicine is being reorganized through the remaking of the technical, informational, organizational, and hence institutional infrastructures of the life sciences and medicine, largely via the incorporation of computer and information technologies." Biomedicine, which Clarke et al. (2005, 442-42) define as "the increasingly technoscientific, complex, multi-sited, multi-directional processes of medicalization" is transforming the practices of medicine and understandings of health and illness in the twenty-first century.

---

[2] There are persistent debates within the field regarding the use of the term "medical sociology" or embracing the "sociology of health and illness" identification. Some sociologists argue that the term medical sociology too closely associates the field with medicine, and the field needs to develop an independent position, not based on the medical model, but based on the social aspects of health and illness. For more on this debate, see (Annandale, 2002).

Building on the "medical industrial complex" concept developed in the 1970s, Clarke and her colleagues (2005, 446) argue that the "Biomedical TechnoService Complex, Inc." typifies the political economy of the current era of biomedicalization. The corporatized and privatized research, products, and services made possible by technoscientific innovations that further biomedicalization are emphasized by this concept. "Through its sheer economic power, the Biomedical TechnoService Complex, Inc. shapes how we think about social life and problems" (Clarke et al., 2005, 446). Within the political economy of biomedicalization, which consists of multinational corporations that are spreading the biomedical paradigm and processes of biomedicalization around the world, solutions to health problems are increasingly corporatized and commodified.

Processes of biomedicalization also transform the concepts of health, risk and surveillance. Whereas processes of medicalization redefined social problems as medical illnesses, disabilities and/or diseases, the biomedicalization of health has redefined health so that the lack of symptoms does not necessarily indicate the lack of illness or potential illness. Therefore, health must be maintained through biomedical surveillance and risk assessment (Clarke et al., 2005, 446). Individuals who may otherwise appear and feel healthy are defined as "at risk" based on family history, personal behaviors, and/or results from genetic testing. Risk is then managed through "processes of surveillance, screening, and routine measurements of health indicators" (Clarke et al., 2005, 448).

As the biomedical model continues to spread around the world through transnational processes of biomedicalization, alternative approaches to understanding human health are increasingly marginalized. Social scientists have played a role in critiquing the myopic view embraced by the biomedical model of health and illness and have argued that health must be understood in a broader social, political, and economic context (e.g., Brown, 1995; Brown & Kelley, 2000; Brown, et al., 2003; Budrys, 2003; Davis & Webster, 2002; Kawachi, Daniels, & Robinson, 2005; Lahelma, 2001; Link & Phelan, 1995; Reich, 2000). The increasing dialogue between environmental sociologists and sociologists of health and illness, in particular, has led to a more developed critique of the narrowness of the biomedical model and the ever-increasing need to understand human health and environmental harms as interrelated.

Bringing the Environment into our Understanding of Human Health: Challenging the Biomedical Paradigm

While the biomedical paradigm developed within the medical field, environmental health perspectives often develop within grassroots social movements that recognize the need to connect issues of health, illness, and the environment (Brown, 2002). Environmental health perspectives remain anthropocentric, in that the focus remains on human health; however, they bring the environment into the picture and seek to alleviate both human illnesses and environmental harms, as an integrated process. The toxic waste movement, various environmental justice movements, and the environmental breast cancer movement are just a few examples of social movements that have made connections between the health of the environment and the health of humans and have worked to enact public policy, raise awareness, confront polluters, and challenge medical and epidemiological sciences (Brody &

Rudel, 2003; Brown, 1995, 2002; Brown, et al., 2006; Couch & Kroll-Smith, 2000; Kroll-Smith & Floyd, 1997; Shostak, 2004). These movements have demonstrated the need for sociologists of health and illness, environmental sociologists, social scientists in other disciplines, and natural and medical scientists to join together, embracing an interdisciplinary approach, to address issues of health, illness and the environment as integrated phenomena.

The social scientific field of environmental health counters the assumptions and priorities of the biomedical paradigm. Environmental health perspectives stem from a vastly different understanding of health and illness. The biomedical paradigm defines health and illness in an individualized medical framework in which an individual is either healthy or ill, and health is largely defined as the absence of disease, or, increasingly, relatively low levels of individual risk factors for disease. Environmental health perspectives provide a broader, contextualized definition of health and illness. According to Brown (2002, 7), the broadest definition of environmental health

> would include the totality of unhealthy living and working conditions: bacteria and viruses in human waste; animal vectors for infectious diseases; surface water and groundwater pollution; air pollution from fires, vehicle exhaust, and incineration; chemical and petroleum product spills and explosions; and disasters, such as floods, hurricanes, and fires (which may be either natural, human caused, or human exacerbated).

Because this definition is particularly broad and all-encompassing, Brown (2002, 7) suggests that environmental health researchers, activists, and policymakers have rightly focused on "the health effects caused by toxic substances in people's immediate or proximate surroundings (soil, air, water, food, and household goods)." Even with a narrower focus on health effects caused by chemical-related, air pollution-related, and radiation-related symptoms and diseases, the field of environmental health challenges the myopic biomedical definition of health.

Social scientific environmental health perspectives locate individuals within broader social and environmental contexts in order to understand their experiences of health and illness. This challenges the biomedical paradigm, which tends to strip the individual of his or her environmental context, at times not even focusing on the whole person but merely on a particular diseased body part or genetic trait without reference to the context in which the individual, body part, and/or gene exists. Steingraber (1997, 137) illustrates this point about the biomedical paradigm in the following recounting of her experiences with medical professionals when she had bladder cancer at a young age,

> In all the years I have been under medical scrutiny, no one has ever asked me about the environmental conditions where I grew up, even though bladder cancer in young women is highly unusual. I was once asked if I had ever worked with dyes or had ever been employed in the rubber industry (no and no). Other than those two questions, no doctor, nurse, or technician has even shown interest in probing the possible causes of my disease—even when I have introduced the topic. From my conversations with other cancer patients, I gather that such lack of curiosity in the medical community is usual.

The biomedical paradigm does not include making connections between diseases and environmental factors. Steingraber's illness was explained within the biomedical model as a "genetic fluke" that was detected by medical technologies, treated through surgery, chemical interventions, and pharmaceutical drugs, and her body would be continually monitored for reoccurrences through medical surveillance (e.g., blood work, x-rays, etc.).

In addition to challenging the narrow biomedical definition of health and illness, the field of environmental health within the social sciences counters the "doctrine of specific etiology" of the biomedical paradigm that focuses on a specific factor (e.g., bacteria, virus, genetic defect, etc.) as the exact cause of each disease (Annandale, 2002). The biomedical paradigm explains diseases as developing within otherwise healthy individuals due to a single particular cause, in which either something that was already in the body (e.g., a genetic defect), or something introduced to the body (e.g., bacteria) led to the manifestation of a particular disease, and it is the role of medical professionals to identify which disease it is and medically treat it accordingly. Environmental health scholars and activists argue that many diseases cannot be completely explained or understood within the biomedical paradigm. This is due to the fact that elements beyond the scope of the biomedical paradigm—such as environmental pollutants and toxins that people are continuously exposed to at varying levels depending on their geographic location, occupation, race, class, and gender—affect their health in myriad ways (Brown, 1995; Brown & Kelley, 2000; Thomas, 2001). Without taking these environmental factors into account, the biomedical paradigm provides insufficient understandings of numerous illnesses (e.g., asthma, breast cancer) and ignores others all together (e.g., Gulf War Syndrome,[3] multiple chemical sensitivity, see Brown, 2002).

Social scientific environmental health perspectives and environmental health activists conceptualize prevention, risk, and treatment/solutions in terms of environmental pollutants and toxins that contextualize the disease manifesting itself within an individual or within multiple individuals in a community. Because the field of environmental health focuses on environmental causes of illnesses, prevention is conceptualized as removing or limiting exposure to environmental disease-causing chemicals, rather than as early detection and "chemo-prevention" (Brown, 2000a; Eisenstein, 2001; Steingraber, 1997).

Additionally, risk is not understood primarily on a genetic or individual level; rather, it is conceptualized in a broader context of exposure, sensitivity, and susceptibility to environmental pollutants and toxins. Social scientific environmental health perspectives focus on public health and the health of communities, often minority and lower-class communities that are exposed to higher levels of illness-inducing chemicals (Brown, 1995; Brown et al. 2000; Capek, 2000; Eyles, 2000). Risk, therefore, is understood as environmental risk, external to the individual's body or genetic make-up, although often interacting with genes in ways that produce disease (Shostak, 2003). Finally, in locating the causes of illness largely outside of the biomedical framework, social scientific environmental health perspectives challenge the biomedical model of treatment as well. Environmental health perspectives emphasize the importance of identifying, regulating, and reducing (or eliminating) harmful pollutants and chemicals in the environment in order to prevent further increases in environmentally induced diseases.

The shift from a traditional model of biological or natural science research to a participatory research model represents another challenge to the biomedical paradigm posed by environmental health perspectives. The biomedical paradigm's claim to scientific

[3] After years of pressure from Gulf War veterans suffering from health problems, in 2008, a U.S. congressionally mandated panel, the Research Advisory Committee on Gulf War Veterans' Illnesses, confirmed that Gulf War illness, as reported by veterans since the war in 1991, is a real disease that should be treated as such by doctors and veterans' health and disability benefits (Maugh & Engel, 2008).

neutrality and objectivity is challenged by environmental health perspectives. Traditional scientific approaches, those used in medical, biological and natural science research in most cases, rely upon "a narrowly circumscribed set of assumptions about causality, the political and public role of scientists, and corporate and governmental social responsibility" (Brown, 2000a, 371). This narrow view depicts medical and scientific research as existing outside of politics and as being devoid of its own internal assumptions. Fosket (2000, 19) asserts that

> in practice, cultural, embodied and experiential knowledges are always drawn upon in the construction of biomedical knowledge just as in other knowledges, even when at the same time such "illegitimate" sources of knowledges are dismissed as unreliable or illegitimate ways of constructing biomedical knowledge.

Brown (2000a, 371) argues that it is useful to draw on both the political-economic perspective—which presents scientific inquiry as tied to corporate, political, and foundation interests that "direct research and interpretation toward support for the status quo" —and the social constructivist perspective—which "shows us the internal workings of the scientific community." The field of environmental health is interested in science and scientific methodology not only as it relates to affecting understandings of illnesses for medical purposes, but also, and more importantly, for its role in encouraging government agencies to act on regulating environmental health hazards.

The field of environmental health has drawn on the experience of activists in critiquing the supposed value-neutrality of science. "Activists disagree that epidemiology is a value-neutral scientific enterprise conducted in a sociopolitical vacuum" (Brown, 2000a, 372). Environmental health advocates argue that traditional scientific practices lack value-neutrality because their endeavors are shaped by "social forces such as media influence, economic interest, political pressure, and social movement activism" (Brown 2000a, 372). Additionally, science does not typically take direction from the lay public.

Environmental health advocates argue that the overemphasis on methodological and scientific certainty does not always lead to "good" science (Brown, 2000a). Lay people in communities where there is clear evidence of contamination and visible health effects often question the methodology and validity of science that is either unwilling to address the issue altogether or, in conducting research, finds there to be no statistically-significant connection between the two (Brown, 2000a).

Social scientists in the field of environmental health, therefore, play a role in supporting the active involvement of lay people in the development of scientific questions and research methods. According to Brown (2000a, 377), lay involvement, which he refers to as "popular epidemiology" produces "good science," by changing the nature of scientific inquiry. Lay involvement produces "good science" by identifying cases of "bad science," "e.g., poor studies, secret investigations, failure to inform local health officials" (Brown, 2000a, 377). Additionally, lay involvement draws attention to the limitations of science as it is traditionally practiced with regard to "opposing lay participation in health surveys, demanding standards of proof that may be unobtainable or inappropriate, and being slow to accept new concepts of toxic causality" (Brown, 2000a, 377).

Popular epidemiology, because it develops out of lay peoples' experiences of the relationship between environmental hazards and health effects, often presents medical scientists with challenges that they are unwilling or unable to adequately address. However, by developing lay methods of data collection and inquiry into environment-health connections, lay people have been able to confront medical scientists, challenge their priorities and methodology, and offer new ways of working with lay people to produce "good science."[4]

Kroll-Smith et al. (2000) frame the current state of environmentally induced health problems within the context of a post-natural world. According to the authors, in a post-natural world, "nature is increasingly mediated by human actions, and disasters are less likely to be natural and more likely to be the results of human error, malfeasance, or simply bad planning" (Kroll-Smith et al., 2000, 3). In the U.S., every year approximately 1,000 new industrial chemicals are introduced into the environment (Kroll-Smith et al., 2000). These chemicals are introduced into the environment at such a rate that most of them have not even had a chance to be tested for health-related effects. Of the 70,000 chemicals that were used in manufacturing in 1995, "only 2% of them were tested for human health effects" (Kroll-Smith et al., 2000, 2). The field of environmental health, by drawing connections between industrial practices and health consequences, opens up a contentious political debate (that the biomedical paradigm avoids), by challenging "our chemical dependency, reckless use of radiation, gas-guzzling and particulate-emitting vehicles, and genetically modified organisms" (Brown, 2002, 8). In linking ill-health to corporate industrial practices, environmental health perspectives challenge "agribusiness, the chemical industry, auto production, energy companies, garbage disposal firms, and many other large sectors of the economy" (Brown, 2002, 8). The political ramifications of the environmental health approach, as opposed to the biomedical approach, are severe. According to Brown (2002), while other scholars and scientists choose not to put politics at the forefront of their science and scholarship, environmental health scholars and scientists do. Brown (2002, 8-9) states,

> ...Action on environmental issues is always propelled by citizens who believe that there are dangerous activities being conducted by industry, along with insufficient government regulation and enforcement...Environmental science leads to worldwide debates, conferences, and agreements on ozone depletion, greenhouse gases, and endocrine disruptors, where the political stakes are often quite high and where the United States often looks bad in its failure to go along with the growing pro-environmental approach. And, of course, the politics of oil—so central to environmental preservation and health—has long embroiled us in Middle Eastern politics based on an environmentally destructive use of fossil fuels.

Environmental health scholars demonstrate the ways in which promoting environmental and human health is incompatible within the current global corporate capitalist political economy (Dauvergne, 2005; Davis & Webster, 2002; Seis, 2001; Thomas, 2001). According to Seis (2001, 138), "most environmental problems and ineffective environmental policy can be linked directly to the inherent contradictions of attempting to reconcile the goal of unrestrained capital accumulation within finite ecosystems." The biomedical paradigm, in addressing health and illness as purely individual medical issues, refrains from directly confronting the current capitalist system. The medical focus on diagnosis, treatment, and

[4] For an example of a project representing this form of "good science" that includes lay people, social scientists, epidemiologists and bench scientists, see http://bcerc.org. The National Cancer Institute and the National Institute of Environmental Health Sciences in the U.S. funded a number of Breast Cancer and Environment Research Centers to investigate environmental links to breast cancer.

genetics, while failing to account for all aspects of many illnesses, leaves the current political and economic system unchallenged. When prevention is addressed within the biomedical model, it is usually in terms of unhealthy lifestyles without attention to the effects of systemic environmental factors. According to Dauvergne (2005, 9), "[medical] diagnosis and treatment rely on technology and science for solutions while protecting a society from systemic ecological changes [that] by definition requires political action" (emphasis in original). In her discussion of the interests, alliances, and partiality of biomedical knowledges, Fosket (2000, 19) quotes a woman who has dealt with breast cancer as stating,

> It's much easier to say, "Well, you know, alcohol consumption appears to be related to breast cancer and so women really need to think about reducing their alcohol consumption" than saying, "Gee...it looks like dioxin may actually be an oestrogenic chemical and may increase the risk of breast cancer." I mean, we're talking about how we live our lives, our production processes related to everything. And that...that's politically very contentious.

This statement demonstrates the political contrast between the biomedical paradigm, which provides solutions for prevention that challenge individual lifestyles and behavior, and environmental health perspectives that challenge our current social, political and economic structures.

Critical to the lack of political challenge posed by the biomedical paradigm is corporate funding of medical research and the integral role of the pharmaceutical industry in biomedicine. According to Dauvergne (2005, 10), "the lack of research on cancer and system-wide ecological change reflects corporate indifference to funding research on the health consequences of production and consumption." Furthermore, drug treatments are big business. While the medical field often presents medical developments as purely scientific and "value-free," in reality, they are value-laden and often driven by corporate profit motives. For example, "the top-selling cancer drug in history, Taxol, sells for 10 times the manufacturing cost for Bristol-Myers Squibb. Genetic research is potentially a corporate goldmine with 1,300 human-gene patents now in force" (Dauvergne 2005, 10).

Environmental health perspectives have drawn attention to and critiqued the corporate involvement in medicine and the role of the pharmaceutical industry in stifling research and funding of environmental causes of disease. Environmental health perspectives of breast cancer have been particularly apt in challenging the biomedical approach to breast cancer with regard to the role of corporate interests. According to Eisenstein (2001, 100), "the breast cancer establishment—which houses the different institutions that do research, study detection, initiate treatment, and provide advocacy—exist alongside and inside the petro/chemical-pharmaceutical/cosmetic complex." Some of the same corporate interests that pollute the environment and negatively affect human health simultaneously profit from the development of drug treatments. Zones (2000, 120) argues that "profitability is a fundamental requirement of a capitalist economy, and cancer has many profit centers—detection, treatment, prevention, and even advocacy." Pharmaceutical manufacturing is currently the most profitable industry in the U.S. (Klawiter, 2002; Seis, 2001; Zones, 2000). Scholars and activists espousing environmental perspectives on breast cancer causation and the need for prevention have been particularly critical of AstraZeneca, the multinational pharmaceutical company that develops and sells breast cancer drug treatments (Eisenstein, 2001; King, 2006; Zones, 2000). AstraZeneca is presented as an example of the conflict of interest involved in the manufacturing and distribution of cancer treatments by profit-driven corporations. According to Zones (2000, 144), "While its best-selling drug, tamoxifen, brings in approximately a half-billion dollars a year, Zeneca (which merged with Sweden-based Astra Pharmaceuticals to become AstraZeneca in 1999) also has sold pesticides and herbicides, some of which are thought to be carcinogens." As the case of breast cancer demonstrates, the political economy of health and illness, while often ignored by the biomedical paradigm, shapes the research priorities (emphasis on treatment rather than prevention) and the incompatibility of corporate capitalism with concerns for environmental causes of disease. As Eisenstein (2001, 124) states,

> Breast cancer is big business. Many of the same corporations that contaminate our bodily environments sell the drugs that are supposed to prevent malignancy. Zeneca manufactures pesticides at the one end and markets tamoxifen at the other. The post-industrial-medical/beauty industry complex researches and markets breast cancer at the same time. This just might be a deadly combination: moneymaking and women's health.

The debate surrounding biomedical and environment health perspectives is not limited to the world of social scientists. Rather, charity and advocacy organizations engage in framing and challenging various environmental and health issues. The case of breast cancer charity and advocacy is particularly interesting with regard to biomedical and environmental health perspectives, as discussed below.

Breast Cancer Charity and Advocacy: Biomedical and Environmental Models

Breast cancer charity and advocacy in the U.S. and Western Europe can generally be divided into two categories: biomedically focused and environmentally focused. The biomedical model of breast cancer charity is promoted by such large organizations as Susan G. Komen for the Cure and the Avon Foundation and is currently expanding to countries in the Middle East, Africa, Asia, Eastern Europe, and Central and South America (King, 2006). Much less widespread, the environmental model of breast cancer advocacy tends to develop in communities where "cancer clusters" or higher-than-average rates of cancer are noticed by residents (Brody & Rudel, 2003; Brody, Tickner, & Rudel, 2005; Brown, et al., 2006).

Biomedical breast cancer charity and advocacy organizations focus their attention on finding biomedical solutions to breast cancer, such as improved treatment techniques, pharmaceutical developments, genetic links to the disease, and ultimately a cure for breast cancer. These organizations raise significant amounts of money each year through charity events. The funds are then funneled into hospitals, university medical centers, and pharmaceutical research and development projects in the hope of eradicating the disease.

Despite these efforts, which have been growing stronger in the U.S. and Western Europe since the 1980s, breast cancer provides an example where the biomedical paradigm is inadequate in explaining the increasing incidence and specific etiology of the disease. While physicians and medical scientists tout the strides made by biomedicine toward improved understanding of breast cancer, much about the disease remains unexplained (Eisenstein,

2001; Fishman, 2000; Fosket, 2000; Rosser, 2000; Steingraber, 2000). "As with most women's health issues, the biomedical model remains too restrictive for exploring causes, treatments, and prevention of breast cancer" (Rosser, 2000, 248). Because the biomedical model is focused on individual bodies and body parts at the cellular, hormonal and genetic levels, the context beyond this level of understanding is absent. Additionally, in this "human genome era, biomedical emphasis is primarily focused on genetic understandings of disease" (Rosser, 2000). For breast cancer research, medical attention shifted toward isolating breast cancer genes, and, despite the success of identifying the BRCA1 and BRCA2 gene mutations linked to breast cancer, "the reality is that only 5-10% of breast cancer is inherited, and that the BRCA1 and BRCA2 genes are responsible for only half of the inherited cases" (Eisenstein, 2001; Rosser, 2000, 250). Medical doctors supplement this account of breast cancer causation, which fails to explain up to 95% of breast cancer cases, with a focus on "lifestyle factors," (e.g., diet, exercise, age at first childbearing, breastfeeding practices) to explain the rest of the occurrences (Fosket, 2000; Zavestoski et al., 2004). The increasing rate of breast cancer, particularly in (but not limited to) more developed countries, is explained within the biomedical paradigm as an artifact of improved medical detection and increased screening (Aronowitz, 2007).

Processes of biomedicalization are evident in the case of breast cancer advocacy with regard to how prevention, risk, and treatment are conceptualized and promoted. Within biomedical breast cancer charity and advocacy, breast cancer "prevention" is conceptualized as both early detection through medical technology (e.g., mammogram) and as "chemoprevention," in terms of tamoxifen and raloxifene in women who are labeled "high risk" based on genetic testing and other individualized risk factors (Fosket, 2004; Klawiter, 2002). Processes of biomedicalization have increased the reach of the medical model, by not only dealing with diseases once they manifest themselves, but by promoting routine medical surveillance that classifies all (healthy) women as "low, average, or high risk" for breast cancer (Fosket, 2004). Biomedical charity and advocacy organizations encourage women to undergo routine medical screening in addition to performing monthly breast self-exams as a means of breast cancer "prevention."

The environmental model of breast cancer advocacy focuses on particular environmental links to breast cancer and de-emphasizes the relatively minor role of genetics (Brown et al., 2000; Davis & Webster, 2002; Eisenstein, 2001; Krimsky, 2000; Potts, 2004; Steingraber, 1997). Due to the lack of certain knowledge regarding the causes of breast cancer, environmental breast cancer advocates, along with environmental health scholars, conceptualize the disease as a "contested illness" (Brown et al., 2000). Environmental breast cancer advocates, such as those working with Breast Cancer Action in the U.S. and Canada, call for research into possible links between environmental toxins and breast cancer incidence and have pursued the link between environmental factors and breast cancer incidence. In particular, xeno-estrogens, or estrogens that come from sources external to the body, have been hypothesized to play a role in breast cancer incidence (Brody & Rudel, 2003). "These estrogen mimics are found in many chemicals, fertilizers, pesticides, and foods. Synthetic estrogens are potentially able to disrupt the body's endocrine system, interfering with its ability to deliver hormones to the blood stream" (Brown et al., 2000, 21). Environmental breast cancer advocates have pursued this link between xeno-estrogens and breast cancer by calling for increased funding for research into environmental causes of breast cancer (Brody et al., 2005).

Prompted by pressure from environmental breast cancer advocates, currently the National Institute of Environmental Health Studies (NIEHS) and the National Cancer Institute (NCI) in the U.S. are funding research on environmental links to breast cancer. The research design is participatory, in the sense that epidemiologists, biological scientists and breast cancer advocates are working together to define the research questions and translate and disseminate the findings. Social scientists play a critical role in evaluating this type of participatory approach to research and providing feedback to improve upon the process (McCormick & Baralt, 2009; McCormick, Brody, Brown, & Polk, 2004a, 2004b).

Based on their alternative scientific studies, environmental breast cancer advocates challenge government and industry to regulate toxins, despite uncertainty. Environmental breast cancer advocates draw clear connections between science, policy and ethics (Zavestoski et al., 2004). In doing so, they try to influence scientists and policymakers to look beyond issues of certainty, recognize their own biases, and regulate toxins because it is ethical, regardless of the lack of certain science in some cases.

Environmental breast cancer advocates are not only committed to furthering research on environmental links to breast cancer, they also promote the precautionary principle. The precautionary principle is based on the simple, yet politically controversial, premise that "public and private interests should act to prevent harm before it occurs" (Steingraber, 2000, 270). According to this principle, it is the indication of harm, rather than proof of actual harm, that should require political action, particularly if delay may cause "irreparable damages" (Steingraber, 2000, 270). In other words, rather than waiting for certainty regarding the relationship between certain chemicals and disease, chemicals should be regulated based on the indication that the chemicals are harmful, without certainty, in order to protect human life.

Related to the precautionary principle is the "principle of reverse onus" (Steingraber, 2000, 270). According to this principle, the burden of proof regarding certainty of environmental hazards and risks needs to be reversed. Rather than waiting for scientists to prove that particular chemicals are, in fact, toxic, the burden of proof should lie with the industries that want to introduce the chemical into the environment to prove that it is safe for the environment and human health (Steingraber, 2000).

> For most industrial chemicals, no firm requirement for advance demonstration of safety exists. But chemicals are not citizens. They should not be presumed innocent unless proven guilty, especially when a verdict of guilt requires some of us to sicken and die to demonstrate the necessary evidence (Steingraber, 2000, 270).

Actually, even with people becoming sick and dying, certainty regarding the causation of harm from particular chemicals is very difficult to assess. In other words, if the regulation of chemicals relied on certainty regarding environmental hazards and risk, very few hazards would be regulated.

Relatedly, environmental breast cancer advocates also promote the "principle of the least toxic alternative" (Steingraber, 2000, 271). This principle "presumes that toxic substances will not be used as long as there is another way of accomplishing the task" (Steingraber, 2000, 271). In other words, in the case of ridding crops of weeds and pests, if non-toxic alternatives exist, which they do, the government should mandate that they be used in place of toxic herbicides and pesticides.

Unlike biomedical breast cancer charities that align with the current political economic system by promoting and contributing to biomedical research, the principles promoted by environmental breast cancer advocates radically challenge the current capitalist political economy, based on profit-motive, which often means the devaluing and harming of the environment and human health. The European Union provides an example of a political body that has elected to engage with the precautionary principle. On April 13, 1999, the Council of the European Union adopted a resolution urging the EU Commission "to be in the future even more determined to be guided by the precautionary principle in preparing proposals for legislation and in its other consumer-related activities and develop as priority clear and effective guidelines for the application of this principle" (Commission of the EU, 2000, 8). The EU Commission adopted this resolution based on public concern regarding environmental health issues. The commission stresses the importance of scientific evaluation and risk assessment as the primary means of dealing with environmental health risks. Risk assessment consists of "hazard identification, hazard characterization, appraisal of exposure and risk characterization" (Commission of the EU, 2000, 14), however, limits to scientific knowledge often lead to uncertainties that make risk assessment more difficult. In cases where there is public concern and scientific uncertainty regarding environmental hazards and risk remain, the EU Commission has agreed that the precautionary principle should be relied on in order to move forward despite uncertainty.

In some cases, environmental breast cancer advocacy organizations directly target particular industries or companies and publicly pressure them to remove a particular toxic chemical from their products, bypassing the issue of uncertainty altogether. For example, environmental breast cancer advocacy organizations and the campaign for safe cosmetics recently drew public attention to the fact that OPI Products, Inc. nail polish contained toxic chemicals, such as formaldehyde, toluene, and dibutyl phthalate. These citizen groups drew particular attention to the fact that OPI sells a safer reformulated product in Europe in order to comply with the European Union Cosmetics Directive, but failed to offer the same safer product in the U.S. (Breast Cancer Action, 2006). Based on widespread public pressure, OPI, Orly International, and Sally Hansen all pledged to remove toxic chemicals from their nail polishes. Through this campaign, environmental health advocates were able to challenge and ultimately change corporate practices based on citizen knowledge and concern, avoiding the issue of scientific uncertainty.

While biomedically oriented breast cancer charity and advocacy organizations are completely anthrocentric in their approach to the disease and narrowly focused in that they do not tend to work with other advocacy organizations, environmentally oriented breast cancer advocacy organizations have formed alliances with environmental organizations for the betterment of the environment and human health. These types of hybrid environmental health organizations and alliances have much potential to alleviate environmental ills and human diseases.

Conclusion

The health of the environment and human health are inextricably linked.

Social scientific perspectives on the connection between human health and the environment have challenged the biomedical paradigm in significant ways. Environmental health advocates have played a crucial role in challenging medicine and science to address issues of environmental causes of disease. Social scientists play a significant role in contextualizing human health and illness within the broader political economy and demonstrating that the current profit-driven capitalist political economic system is not sustainable environmentally or in terms of human health. Valuing the environment as well as human health, social scientists in the field of environmental health, as well as environmental health advocates, emphasize the importance of creating an environment that promotes health, as opposed to the status quo, which values profit-maximization over environmental and human health.

Shostak (2003) argues that with biomedicine's increasing focus on genetics and environmental health's focus on environmental causes of illness, there is much potential for the future of understanding gene-environment interactions. According to Shostak (2003, 2338), the increasing "geneticization" within biomedicine could lead to the "entrenchment of genetic reductionism and individualized definitions of risk" as the core orientations of the public health field. On the other hand, if the field of public health takes advantage of the incorporation of genetic knowledge and molecular technologies, there is potential for "the discipline to attend to the production of 'local biologies,' health, and illness by attending to historically and geographically situated dialectics of biology and culture" (Shostak, 2003, 2238). If this occurs, according to Shostak (2003, 2238), "public health politics and practices would be increasingly oriented to the social, political, and economic relations that enable environmental exposures, gene-environment interactions, and their effects." Social scientists need to continue to draw connections between particular environmental problems (e.g., global warming, toxic dumping, and pollution) and human health problems (e.g., asthma, breast cancer, and multiple chemical sensitivity). While biological scientists work to elucidate these connections in order to improve the understanding and treatment of illnesses, the role of social scientists must be that of elaborating on the contribution of environmental factors to human illnesses with the goal of societal change. Social scientists play a crucial role in facilitating the connection between environmental justice and health activists and biological researchers. By encouraging a participatory scientific paradigm in which biological scientists, social scientists, and environmental health advocates all play a role, social scientists have the potential to alter "science as usual" in pursuit of politically relevant solutions to environmental problems and environmentally induced illnesses.

References

Action, B. C. (2006). Take Action, 2007.

Annandale, E. (2002). *The Sociology of Health & Medicine: A Critical Introduction*. Malden, MA: Polity Press.

Aronowitz, R. A. (2007). *Unnatural History: Breast Cancer and American Society*. Cambridge: Cambridge University Press.

Breast Cancer Action. (2006). Making Nail Polish a Little Safer. Newsletter 93, October/November. Accessed on 12 May 2009 at http://bcaction.org/index.php?page =newsletter-93j.

Brody, J., & Rudel, R. A. (2003). Environmental Pollutants and Breast Cancer. *Environmental Health Perspectives, 111*(8), 1007-1019.

Brody, J., Tickner, J., & Rudel, R. A. (2005). Community-Initiated Breast Cancer and Environment Studies and the Precautionary Principle. *Environmental Health Perspectives, 113*(8), 920-925.

Brown, P. (1995). Race, Class, and Environmental Health: A Review and Systematization of the Literature. *Environmental Research, 69*, 15-30.

Brown, P. (2000a). Popular Epidemiology and Toxic Waste Contamination: Lay and Professional Ways of Knowing. In S. Kroll-Smith, P. Brown & V. J. Gunter (Eds.), *Illness and the Environment: A Reader in Contested Medicine* (pp. 364-383). New York; London: New York University Press.

Brown, P. (2000b). Preface to Illness and Environment: A Reader in Contested Medicine. In S. Kroll-Smith, P. Brown & V. J. Gunter (Eds.), *Illness and the Environment: A Reader in Contested Medicine* (pp. xi-xii). New York: New York University Press.

Brown, P. (2002). Preface to Special Issue on Environmental Health. *The Annals of the American Academy, 584*(1), 7-12.

Brown, P., & Kelley, J. K. (2000). Physicians' Knowledge, Attitudes, and Practice Regarding Environmental Health Hazards. In P. Brown, S. Kroll-Smith & V. J. Gunter (Eds.), *Illness and the Environment: A Reader in Contested Medicine* (pp. 46-71). New York: New York University Press.

Brown, P., Kroll-Smith, S., & Gunter, V. J. (2000). Knowledge, Citizens, and Organizations: An Overview of Environments, Diseases, and Social Conflict. In S. Kroll-Smith, P. Brown & V. J. Gunter (Eds.), *Illness and the Environment: A Reader in Contested Medicine* (pp. 9-28). New York: New York University Press.

Brown, P., Mayer, B., Zavestoski, S., Luebke, T., Mandelbaum, J., & McCormick, S. (2003). The Health Politics of Asthma: Environmental Justice and Collective Illness Experience in the United States. *Social Science & Medicine, 57*, 453-464.

Brown, P., McCormick, S., Mayer, B., Zavestoski, S., Morello-Frosch, R., Altman, R. G., et al. (2006). "A Lab of Our Own": Environmental Causation of Breast Cancer and Challenges to the Dominant Epidemiological Paradigm. *Science, Technology & Human Values, 31*(5), 499-536.

Budrys, G. (2003). *Unequal Health*. New York: Roman & Littlefield.

Bury, M. (2005). *Health and Illness*. Malden: Polity Press.

Capek, S. M. (2000). Reframing Endometriosis: From "Career Woman's Disease" to Environment/Body Connections. In P. Brown, S. Kroll-Smith & V. J. Gunter (Eds.), *Illness and the Environment: A Reader in Contested Medicine* (pp. 345-363). New York: New York University Press.

Clarke, A. E., Shim, J. K., Mamo, L., Fosket, J., & Fishman, J. (2005). Biomedicalization: Technoscientific Transformations of Health, Illness and U.S. Biomedicine. In P. Conrad (Ed.), *The Sociology of Health & Illness: Critical Perspectives* (7th ed., pp. 3-22). New York: Worth Publishers.

Conrad, P. (1992). Medicalization and Social Control. *Annual Review of Sociology, 18*(209-232).

Conrad, P. (2007). *The Medicalization of Society: On the Transformation of Human Conditions into Treatable Disorders*. Baltimore: Johns Hopkins University.

Couch, S. R., & Kroll-Smith, S. (2000). Environmental Movements and Expert Knowledge: Evidence for a New Populism. In P. Brown, S. Kroll-Smith & V. J. Gunter (Eds.). *Illness and the Environment: A Reader in Contested Medicine* (pp. 384-408). New York: New York University Press.

Dauvergne, P. (2005). Cancer and Global Environmental Politics: Proposing a New Research Agenda. *Global Environmental Politics, 5*(3), 6-13.

Davis, D. L., & Webster, P. S. (2002). The Social Context of Science: Cancer and the Environment. *The Annals of the American Academy, 584*, 13-34.

Eisenstein, Z. (2001). *Manmade Breast Cancers*. Ithaca; London: Cornell University Press.

Eyles, J. (2000). Environmental Health Research: Setting an Agenda by Spinning Our Wheels or Climbing the Mountain? In P. Brown, S. Kroll-Smith & V. J. Gunter (Eds.), *Illness and the Environment: A Reader in Contested Medicine* (pp. 453-472). New York: New York University Press.

Fishman, J. (2000). Assessing Breast Cancer: Risk, Science, and Environmental Activism in an "at risk" Community. In L. Potts (Ed.), *Ideologies of Breast Cancer: Feminist Perspectives* (pp. 181-204). New York: St. Martin's Press.

Fosket, J. (2000). Problematizing Biomedicine: Women's Constructions of Breast Cancer Knowledge. In L. Potts (Ed.), *Ideologies of Breast Cancer: Feminist Perspectives* (pp. 15-36). New York: St. Martin's Press.

Fosket, J. (2004). Constructing "High Risk Women": The Development and Standardization of a Breast Cancer Risk Assessment Tool. *Science, Technology & Human Values, 29*(3), 291-313.

Hofrichter, R. (2000). Introduction: Critical Perspectives on Human Health and the Environment. In R. Hofrichter (Ed.), *Reclaiming the Environmental Debate: The Politics of Health in a Toxic Culture* (pp. 1-16). Cambridge: MIT Press.

Kawachi, I., Daniels, N., & Robinson, D. E. (2005). Health Disparities by Race and Class: Why Both Matter. *Health Affairs, 24*(2), 343-352.

King, S. (2006). *Pink Ribbons, Inc: Breast Cancer and the Politics of Philanthropy*. Minnesota: University of Minnesota Press.

Klawiter, M. (2002). Risk, Prevention and the Breast Cancer Continuum: The NCI, the FDA, Health Activism and the Pharmaceutical Industry. *History and Technology, 18*(4), 309-353.

Krimsky, S. (2000). Environmental Endocrine Hypothesis and Public Policy. In P. Brown, S. Kroll-Smith & V. J. Gunter (Eds.), *Illness and the Environment: A Reader in Contested Medicine* (pp. 95-107). New York: New York University Press.

Kroll-Smith, S., Brown, P., & Gunter, V. J. (2000). Introduction: Environments and Diseases in a Postnatural World. In S. Kroll-Smith, P. Brown & V. J. Gunter (Eds.). *Illness and the Environment: A Reader in Contested Medicine* (pp. 1-8). New York: New York University Press.

Kroll-Smith, S., & Floyd, H. H. (1997). *Bodies in Protest: Environmental Illness and the Struggle over Medical Knowledge*. New York: New York University Press.

Lahelma, E. (2001). Health and Social Stratification. In W. C. Cockerham (Ed.). *The Blackwell Companion to Medical Sociology* (pp. 64-93). Malden: Blackwell Publishers.

Link, B., & Phelan, J. (1995). Social Conditions as Fundamental Causes of Disease. *Journal of Health and Social Behavior* (Extra Issue), 80-94.

Maugh, T. H., & Engel, M. (2008, November 18). Report to Congress: Gulf War Syndrome is Real. *Los Angeles Times*,

McCormick, S., & Baralt, L. (2009). Lay-Expert Collaboration in Federal Environmental Breast Cancer Research: A Progress Report.

McCormick, S., Brody, J., Brown, P., & Polk, R. (2004a). Lay Involvement in Breast Cancer Research. *International Journal of Health Services, 34*, 625-646.

McCormick, S., Brody, J., Brown, P., & Polk, R. (2004b). Public Involvement in Breast Cancer Research: An Analysis and Model for Future Research. *International Journal of Health Services, 34*(4), 625-646.

Potts, L. (2004). An epidemiology of Women's Lives: The Environmental Risk of Breast Cancer. *Critical Public Health, 14*(2), 133-147.

Reich, M. R. (2000). Environmental Politics and Science: The Case of PBB Contamination in Michigan. In P. Brown, S. Kroll-Smith & V. J. Gunter (Eds.). *Illness and the Environment: A Reader in Contested Medicine* (pp. 430-452). New York: New York University Press.

Rosser, S. V. (2000). Controversies in Breast Cancer Research. In A. S. Kasper & S. J. Ferguson (Eds.). *Breast Cancer: Society Shapes an Epidemic* (pp. 245-271). New York: St. Martin's Press.

Seis, M. (2001). Confronting the Contradiction: Global Capitalism and Environmental Health. *International Journal of Comparative Sociology, 42*(1-2), 124-144.

Shostak, S. (2003). Locating Gene-Environment Interaction: At the Intersection of Genetics and Public Health. *Social Science & Medicine, 56*, 2327-2342.

Shostak, S. (2004). Environmental Justice and Genomics: Acting on the Futures of Environmental Health. *Science as Culture, 13*(4), 539-562.

Steingraber, S. (1997). *Living Downstream: A Scientist's Personal Investigation of Cancer and the Environment*. New York: Vintage Books.

Steingraber, S. (2000). The Environmental Link to Breast Cancer. In A. S. Kasper & S. J. Ferguson (Eds.). *Breast Cancer: Society Shapes an Epidemic* (pp. 271-302). New York: St. Martin's Press.

Thomas, C. (2001). Environmental and Health Research: Overlapping Agendas. Global Environmental Politics, 1(4), 10-17.

Zavestoski, S., Brown, P., & McCormick, S. (2004). Gender, Embodiment and Disease: Environmental Breast Cancer Activists' Challenges to Science, the Biomedical Model, and Policy. *Science as Culture, 13*(4), 563-586.

Zones, J. S. (2000). Profits from Pain: The Political Economy of Breast Cancer. In A. S. Kasper & S. J. Ferguson (Eds.). *Breast Cancer: Society Shapes an Epidemic* (pp. 119-152). New York: St. Martin's Press.

In: Health and Environment: Social Science Perspectives ISBN 978-1-60876-216-3
Editors: Helen Kopnina and Hans Keune

Chapter 7

THE ADHD REGIME AND NEURO-CHEMICAL SELVES IN WHOLE-SYSTEMS. A SCIENCE STUDIES PERSPECTIVE

Alexander Stingl

ABSTRACT

The present chapter explores the relation of Health and Environment from the general perspective of *Science Studies*. Science Studies are concerned with the history of science, society, medicine and technology as well as contemporary issues and applications of science. Science Studies utilize methodologies of sociology of knowledge, philosophy of science, medical anthropology, and applied ethics. This chapter focuses on "health and environment"—-aspects in a genealogical analysis of the discursive production of the concept *Attention* and its implications as a public, medical and psychological problem. Attention itself is, however, a vague concept, despite its frequent use as a medical and psychological category. It has a two-hundred year history, beginning with Kantian psychology/biology. In psychology, it occupies different roles, and so do its pathologies. Neuro-physiologically, "attention" is still a vague concept. Currently, it is assumed that there is a balance of different neurotransmitters that seems to play a role in the neurochemistry of attention. But while the concept itself is vague, doctors, parents, journalists, etc. use it in statements concerning normal and pathological forms of attention, as if it is self-evident and clear what attention and its pathologies are.

INTRODUCTION

This chapter is subsequently guided by three main questions:

- How does the discourse within science and between science and the public "construct" pathology and normalcy?

- What does this "construction" mean for the relation between health and environment in special regard to so-called lifestyle diseases, exemplified by case of attention-deficit and hyperactivity disorder (ADHD)?
- How need we effectively change current concepts of the human condition and our lifestyle?

The practices that construct the idea or concept of a pathology such as ADHD are part of "regimes of knowledge production" that conceive the human condition as that of the neuro-chemical self (Rose, 2007a,b). In the realm of discourse, the conception neuro-chemical self is found in different versions or interpretations among expert and layman discourses as

1. a *realist* scientific phenomenon that is a molecular system that requires optimization towards a positivist criterion of truth (King et al., 2006)
2. a popular science phenomenon that describes a somatic system (an organism/body) that must adequately fit in with the assumed natural laws of its environment, which define a baseline *normalcy* as a functional criterion within a system/environment wherein all deviations are inferred to be pathologies (see Rose, 2007b; Rabinbach, 1990)
3. a cultural phenomenon that understands pathologies such as ADHD as social constructions (Frazetto et al., 2007; Davis, 2008)

This chapter argues in favor of a "whole-systems"-oriented and presupposes that "parts and whole" cannot be understood separate from their mutual interactions. In other words, the brain—like other organs—cannot be comprehended in isolation from the entire human organism, nor can a human organism be comprehended without the interplay among its natural, cultural and social environments.[1]

Recent practices of knowledge production constitute a novel concept of the neuro-chemical self and its pathologies (such as ADHD) in separate natural, social, and cultural contexts.

In this context, we will have to review the role that different natural, social, and cultural environmental factors play in the use of conceptions in diagnosis and treatment: A major distinction should be made among causal, developmental and secondary causal (Developmental-turned-causal) factors.

The chapter focuses on a perspective that identifies the neuro-chemical self within the dialectic (or trialectic) of natural, social and cultural environment, and describes the triad "brain, organism, environments"[2] as a "whole system." Treating (Intervening in) a "whole system" requires a clear perspective on the triad and the interaction relations within it, taking into account the difference of causal (neuro-chemical system/heredity), developmental (environment) and secondary casual factors (developmental factors that dialectically shape the plastic elements of the neuro-chemical system).

---

[1] The term "social environment" refers to the people around us with whom we have communicative relationships, while the cultural environment refers to traditions, institutions and symbolic structures.

[2] A whole system in this perspective is the dynamic process of the mutual interactions among parts, whole, and environments of a system.

Picture 8. ADHD.

A whole system cannot be optimized nor restored to a "natural functionality." It can only be brought into a dynamic equilibrium, the criterion of which is "quality of life" (Varni, 2007 a,b; Matza et al., 2004; Klassen et al., 2004; Landgraf, 2002).

I. The Production of Knowledge about Normalcy and Pathology

a. ADHD and the Adaptation of the Human Organism to its Environment

When we speak about ADHD or similar disorders, we speak about a concept that describes a pathology that has been constructed in human interactions. In such interaction (scientific discourse is the primary example), the "interlocutors," those who are involved in the discourse, create an idea of what they think of as "normal." They try to categorize cases that deviate from that idea and classify them as pathologies. The French science historian Georges Canguilhem has introduced us to this argument with his seminal study *The Normal and the Pathological* in 1943 (Canguilhem, 2007).

For a "pathological condition" like ADHD, we can easily show a) that the classification is somewhat arbitrary, and b) it is largely dependent on the relation between the (human) organism and its environments.

The emergence of "attention disorders" can be reconstructed within the genealogy of Western psychology since at least 1800. Eventually, ADHD may have become institutionalized with its inclusion into the DSM III (*Diagnostic and Statistical Manual for Mental Disorders*), or the lectures on moral behavior by George Still in 1902, and the concept of "defects of inhibitory volition," or the teachings and writings of William James and Theodule Ribot in the 1880s and 1890s. A hundred years ago, the idea was established on the grounds of individuals who did not seem to fit into modern society. With the DSM III and the DSM IV,[3] a catalogue of symptoms was introduced that promotes the identification of ADHD in early childhood, allowing parents and doctors to intervene before adulthood (see Lakoff 2000; Rowland et al., 2002).[4] Finally, from the perspective of contemporary neuroscience and genetics, ADHD is the expression of genes that produce a neuro-chemical system (the brain) that does not function to optimum (Farah, 2002). We may ask, however, who defines this "optimum" as a "natural kind," how and why. Another perspective on ADHD is suggested by Eisenberg et al. (2008). Studying the lifestyle of the Ariaal, a Kenyan tribe, their research suggests that an "ADHD-like condition" is an evolutionary advantage towards survival:

"Increased impulsivity, ADHD-like traits, novelty-seeking-like traits, aggression, violence and/or activity levels may help nomads obtain food resources, or exhibit a degree of behavioral unpredictability that is protective against interpersonal violence or robberies."

Eisenberg and his team focused on "adaptive components" of the human organism. This perspective refers to the biological evolution of humanity. Therefore, it can be said that the Ariaal resemble a human condition that humanity occupied for the better part of its evolution. These adaptive components refer to an adaptation to a species' natural environment. This "optimum" is entirely different from the optimum that modern societies and their everyday tasks demand.

Harvard psychologist Ellen Langer (1998) has a different perception of pathology in mental health and human development. She assumes that many pathological categories are "made up" or at least "overdramatic." Instead, she argues, most people act and think "mindless," if they were trained to be "mindful," pathologies such as ADHD would literally disappear. Children, who are considered to have ADHD (by teachers or parents), are mostly bored in a "mindless" social environment and judged by people who make a "mindless" use of categories such as ADHD. In actuality, children diagnosed with ADHD prove to be better at multitasking or creative tasks. This would suggest that categories or "labels" such as ADHD often mistake talents or attributes, which differ from a "social norm," to be pathologies that need treatment or optimization, while they would actually need specific types of encouragement. Another Harvard psychologist, John Ratey (2008), has shown that lifestyle factors, above all physical activity, can assist or even replace treatment of several pathologies, including ADHD. Some authors argue that *lifestyle diseases* are the result of unhealthy diet (Gomez-Pinilla, 2008; Vaynman et al., 2004, 2006a,b,c). They point to the fact that the human organism has evolved with a different diet and a different level of physical activity

3 The DSM IV was published by the American Psychological Association in 1994. Information can be found on their website, including the progress on the next version DSM V, which is supposed to be finished by 2012.

4 While DSM IV and similar manuals suggest diagnosis and intervention starting with the age of six and above, there currently are clinical trials under way, testing the use of ADHD-pharmaceuticals on two and three year olds (see: Weill Cornell Medical College (2007, July 20). Pediatric Ritalin Use May Affect Developing Brain, New Study Suggests. *ScienceDaily*. Retrieved May 3, 2009, from http://www.sciencedaily.com /releases/2007/07/070719114451.htm).

than the one we encounter in modern lifestyles, reiterating that the "original" lifestyle was the result of the adaptation to the natural environment, whereas the current Western lifestyle is adaptation to the cultural environment of the last two centuries. This situation presents us with the question to what kind of "environments" the human organism must adapt to, and what kind of criteria conceptions of optimum, normality and pathology actually fulfill.

b. How Conceptions of the Normal and the Pathological Are Constructed

Before discussing the role of these categories and the adaptation of organisms to their environments any further, it is necessary to elucidate how—what we can call—*discursive practices* create the kind of knowledge represented in concepts of health and illness, normal and pathological, or efficiency and inefficiency. Michel Foucault has argued that *knowledge* should be understood more broadly. In order to understand *practices* or *disciplines,* we must first understand the interactions that take place among the interlocutors as well as their interactions with their environment. Introducing a heuristic model for analysis of the *regimes* of knowledge production and micro-climates of truth that have formed and continue to constitute those *practices* seems a reasonable course of action.

i. Knowledge Production

On a general level, we should distinguish between two kinds of elements: *productive/enabling* and *constraining/disabling* elements (Stingl, 2009). An interlocutor or actor is *enabled* by his *equipment* (Rabinow, 2003) to a certain practice or to make a certain statement or enunciation. An interlocutor is *constrained* in his actions and statements by members of his network, by bureaucracies, or conventionalizations (Bloor, 1997) in informal institutions such as a national or regional scientific community.

The constraining side (or the *negative selection regime*) can be tentatively called the *arrangement.* An *arrangement* consists of formal institutions, organizations, and bureaucracies on the one hand. These entities have their own histories, substrates (including paperwork, buildings, offices, etc.), and rules (or grammar) that are explicit and can be found in manuals, guidebooks, documented legislature, etc. These entities are *collective actors*, for as analytical units they appear as an acting agent themselves (see Oberschall, 1992). In the analysis of a health context, we would speak of the *Department of Health* as such a collective actor (distinguished from e.g., the actual caseworker Jones).

On the other hand, we also find *constellations* (Henrich, 1991). *Constellations*, as an analytical device, are formed by networks of people and informal institutions. Aside from the people that constitute them, they have no real world substrate. But they have, if you like, their own grammar or rules, even if these cannot be found in an explicit form. David Bloor's conception of *conventionalization* is an exemplification of this implicit grammar. *Constellations* are, therefore, also a *collectivity of actors* (see Parsons, 1951). As such, a collectivity depends on individual actions. In science, a "school" or an "invisible college" are examples of such *constellations*. *Collective Actors* can, therefore, also be considered to be part of the *external dimension* of an analysis, while *collectivities* are part of the internal dimension, because in the situation that is analyzed, there will be an actual representation of the *collectivity* by a person who is actually and internally involved in the collectivity. For the "grammar" of the internal dimension to be applied, a member who has internalized this "grammar" must be present, while this situation is different for an external entity. A doctor is not necessarily a member of the Department of Health, but he must adhere to revisions of the ethical guidelines, which he will be handed in written form or can look up in a book or online source. The difference can also be exemplified in the difference of sanction that the doctor will experience if s/he chooses to disregard the grammar (Foucault, 1966a,b).

Productive or *enabling factors*, on the other side, comprise *assemblages*.

We can utilize Paul Rabinow's Foucault-based terminology of *assemblages* and *equipment*. For the specific problems addressed here, it must, however, be extended. It can be argued that ADHD must not only be seen as such an *assemblage*. It additionally represents the very crisis of current neuro-science and the ideas of biological citizenship and biopolitics (Rose, 2006, 2008), for its two centuries of conceptual history and power over the life-course of children, it questions the presupposed "autonomy" of the idea of biological citizenship. It pays tribute to the fact that biological citizenship, like the idea of citizenship in general, is empty without some regulative idea of *civics*. The introduction of an environmental or whole-systems perspective in the discourses on biological citizenship and ADHD represents a first step towards a debate on *biological civics.*[5]

Heuristically, four components can be identified that constitute an *assemblage* along two ideal-typical dichotomies that represent them. The distinction between the dimension of the *intellectual climate* and the *equipment* on the one hand, and on the other, we find the temporal dimension distinguishing between synchronic and diachronic elements.

An *intellectual climate* (Stingl, 2009) circumscribes the field of possible conceptual *relations* an interlocutor can make. *Relations* include analogies, metaphors, equivocations, comparisons, etc.

The intellectual climate's diachronic aspect is circumscribed by a *thought-scape* or *Denkraum* (Dieter Henrich, 2004). A *thought-scape* represents a field or sphere of cognitively possible problems or *problematizations* (a term coined by Michel Foucault). *Problematization* describes a historical and social situation that constructs potential outcomes of truth-and-false selections in a web of possible solutions. This *problematization* is described as a "historical space of conditioned contingency" (Rabinow, 2003). In the historic progress of discourses, it so happens that a chain of discoveries leads to new problems that remain largely implicit and cannot initially be made explicit. They keep summing up and remain implicitly present but unresolved until they are concretized and rendered explicit (and largely public) by a string of works that thereby open a new *thought-scape*. Kantianism,[6] at the turn of the eighteenth to the

---

[5] Civics suggests that citizenship is attained by education and maturation and not bestowed by birth. With the rights of a citizen come responsibilities, which must be understood and accepted and which basically include conceptions of justice, government and ethics. Without civics, the concept of citizenship is passive to the point of emptiness. Biological civics suggests that education must include the understanding of biotechnologies of the self and of biopower, of the biological citizen in relation to the environment. The potentials of autonomy that biological citizenship supposedly offers, as its promoters claim, can only be realized if biological citizenship is complemented with *biological civics*. This is a relatively new aspect of the debate, which is gradually made explicit in part by the physical activity movement (John Ratey), the multiple intelligence movement (Howard Gardner) or mental health pluralism (S. Nassir Ghaemi). *Bio civics* will be at the centre of future discourse on biological citizenship, biopower and governmentality.

[6] Kantianism is a philosophical movement associated with the work of Immanuel Kant. Its basic premises are that experience is not an exact comprehension of the "thing-in-itself." The "thing-in-itself" is something we can never know. This uncertainty does not mean that we end in solipsism or that every theory about the world is arbitrary and random. What we know about the world attains a status of increasing certainty by the "tribunal of reason" and judgments we make in our continued communication with others. What we first need to look for

nineteenth century, represents the opening of a new *thought-scape*. With Kantianism, a new type of medicine was possible (Filho, 2009), wherein the relationship of mutual interaction, the idea of communication, and the premises of a science of life that is fundamentally different from the other natural sciences could be discussed and analyzed in ways that scholars could not even conceive of before (Stingl, 2009).

It *enabled* a new philosophy, biology, and psychology—including historic apriori conceptions that lead to our understanding of environment, the human condition and its pathologies.[7] We are largely within the boundaries of this Kantian thoughtscape today,[8] even if our conceptual languages have been drastically changed and even reduced. It accounts, however, for the temporal aspect that *thought-scapes* are diachronic and account for long-term and historic aspects of an *assemblage*.

A current and actual situation, on the other hand, enables conceptual relations also in a synchronic aspect. These aspects come as *dispositions* and are the actually present aspects or problems discussed in the intellectual climate. The discussion of reliability over validity in studies of the effect of pharmaceuticals in ADHD treatment would represent such a *disposition*.

The *equipment* accounts for the conceptual tools that are available. *Equipment* is distinguished in *theoretical/epistemological vernaculars* (synchronic) and *conceptual frames of reference* (diachronic).

The distinction on the temporal level is very important in this instance. The diachronic conceptual frames of reference do provide a stable fundament, while *vernaculars* are dynamic (yet also not arbitrary) structures. There can be, for example, a Kantian frame of reference. Whether an interlocutor is applying the physicalist-reductionist type of *vernacular* or the *biological vernacular:* there are common elements and concepts such as the idea of "apriorism," "historicity," or the "categorical approach."[9] These common elements are but one aspect; the other is the actual enunciation in the statements by an interlocutor. The interlocutor employs these forms in his enunciations in an *epistemological vernacular*. *Vernaculars* have dynamic histories and often "compete" for dominance within *constellations*—since the constraining and enabling levels are not independent of each other in concrete historic reality. With Kant and the birth of biology as a scientific *discipline*, for example, the biological *vernacular* shaped much of nineteenth-century science. In the late nineteenth century, the *vernacular* of physicalist reductionism emerged and gradually began to dominate even the life and social sciences of the twentieth century (Stingl, 2009).

These elements produce *assemblages* such as ADHD. The conceptual history and current practices that constitute and are constituted by ADHD have recently become the focus of the research of this author, because as an *assemblage*, ADHD represents more than just a disorder. It is a crucial focal point of the human condition and how we think of ourselves. The ADHD discourse illustrates this like few others do, because it brings together so much historical development and current conceptual crises.

---

in our effort of understanding (including science) are the conditions of how and why something is possible. (For a more detailed description see Stingl 2009).

[7] Kant is the name or address of the *author* (Foucault, 1969) whom we tentatively ascribe this achievement. The achievement-ascription to an individual person is a "genius"-type of closed accounts, which is the opposite of what I propose with a discursive model of knowledge production.

[8] Or as Foucault once said, that whether we realize it or nor, we are all Neo-Kantians now. (Foucault, 1963)

[9] Vandenberghe identifies Structuralism or Relationalism (as opposed to Substantialism) as the common ground of the Kantian frame of reference.

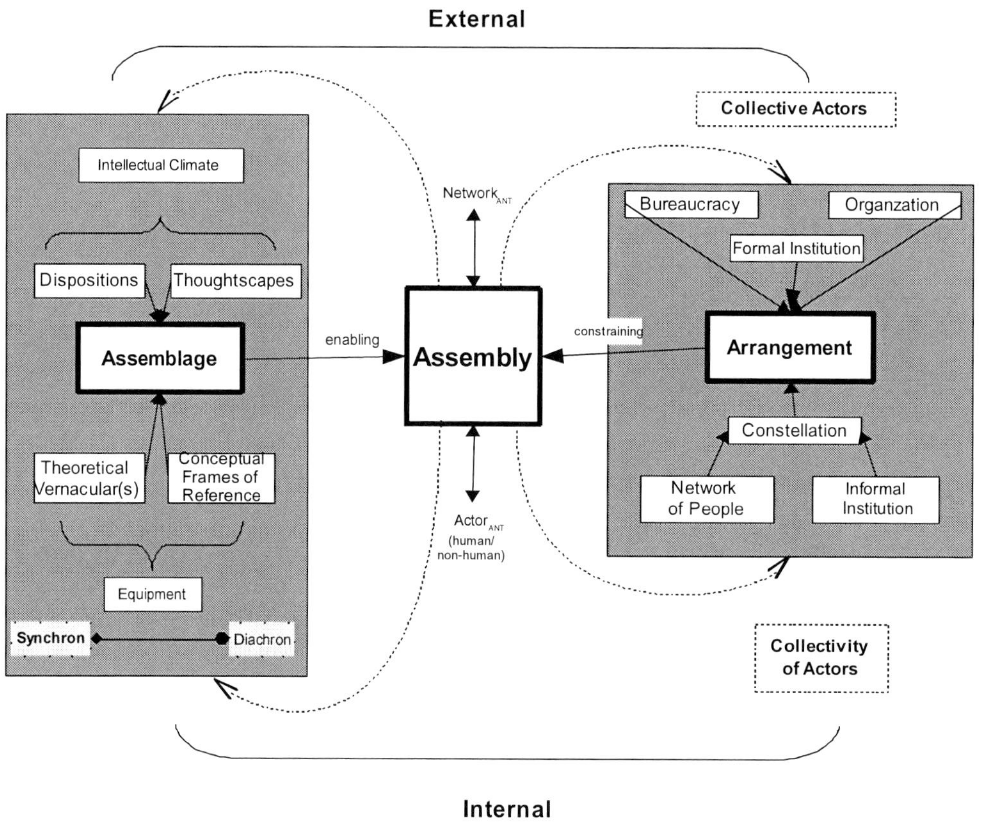

Diagram: Theoretical Model of a Truth Regime.

Figure 2.

Constraining factors are analogous to grammar or syntax and determine what is not permissible in a discourse, while *productive/enabling* factors are analogous to semantics and pre-structure what may be possible in discourse. The likelihood of an enunciation can be "equated" soliciting both factors. Progress usually occurs within a pre-existing thoughtscape and frame of reference. There are very few spaces in between that allow for degrees of freedom, where enunciations can be conceived that elude either. These occasions are historically rare, if they occur at all. In those cases, they are denoted to be either madness or real ingenuity by scholars such as Wilhelm Dilthey (1886; also Daston, 1998)). According to this model, most so-called geniuses have committed to progress, but not been *real* geniuses.

The analytical subject of this model is finally the *assembly*.

In each situation that we analyze, we find that *assemblages* and *arrangements* come together and produce and through selections constrain the *assembly* of a concrete entity. This entity or *assembly* requires two more analytical units or better yet points-of-view: the *network*$_{ANT}$ and the *actor*$_{ANT}$.

These are, as indicated by the indexicals, different from the actors and networks of the *arrangement*. Further, such a model includes Actor-Network-Theory (ANT) and the ideas of Bruno Latour (2007). Although the ends of the present research differ from ANT in so far that we still require ties to social aspects because that is the subject matter. But we should try to leave the *social* and *society* as explanatory categories behind to the most possible extent.

Latour's general criticism is that sociology has relied solely on the *social* as its reference point and explanatory device. He argues that this has resulted in severe limitations for explanatory power. The $actor_{ANT}$ and the $network_{ANT}$ are reference points for the analysis of *assemblies*. Such an *assembly* can be, very generally, a global assembly in politics (Sassen, 2008)—say, a global pharmacological initiative—or an actual child, Jones, who is diagnosed with ADHD, or the dopamine system in a physico-chemical system otherwise known as the brain of Mary and currently manipulated and analyzed under an fMRI *regime* by a Doctor Kadilak and Doctor Kerr at Boston MGH. The $actor_{ANT}$ in either case can be Mary or Peter, but it could likewise be the fMRI machine, the bench Mary is lying on while being analyzed or a specimen of a complex molecule that researchers call dopamine. The $network_{ANT}$ is very similar in presenting the other actors around that play a part in the actual moment of assembly. Environmental aspects have a substantial effect on and in the process of *assembly*. *Assembly,* therefore, happens all the time and all entities that are part of it have a history (most of which is hidden from us in *black boxes*), and it never leaves the entities unchanged. If read from a metaphysics point of view, this would mean that each and every thing, when seen as the point of view of the $actor_{ANT}$, is constantly re-assembled within the constraints of a contingent but not arbitrary sphere of possibilities.

ii. Translating between Different Approaches

The pragmatic "net gain" of this approach is that it enables a solution to the problem of "translatability" between scientific languages and cultures, such as Asian healing traditions (see Kaptchuk, 2000) and evidence-based medicine. Previously, this author (Stingl, 2009), among others, ascribed to serious doubts concerning the possibility of such translations, assuming the divergence between scientific *vernaculars* has progressed too far. However, there are at least three arguments that are reasons to assume that "bridges can be built":

(a) Thorough genealogical reconstruction can reveal common roots between different *vernaculars* that can indeed enable translatability.
(b) Translatability between different scientific cultures can be enabled by revealing analogous conceptions within the historical developments of these cultures.
(c) Since *vernaculars* themselves are a productive factor, in this theoretical perspective, these *vernaculars* are understood as intrinsically equipped with convertibility—similar to the very way that analytic philosophy assumes a *principle of charity.*[10] Therefore, the possibility of translation can be enabled by making explicit the historical constraining factors that currently disable or have denied avenues of connecting with another *vernacular*.

Currently, there still is a lack of proper "translation," and, as can be seen in the discourses in popular science and among laymen, Asian medicine culture is usually either discarded as "esoterical non-sense" or embraced as if it was salvation in the absolute. Empirically, the problem is reduced to the question of evidence-based (reliability and large quantities) medicine versus anecdotal evidence.

---

[10] A *principle of charity* was first introduced by Neil Wilson in the late 1950s and made popular by W.V.O. Quine and Donald Davidson. It states that if a person A *wants* to understand a statement of a person B, then A will have to minimally assume that B is making a statement that is in some way rational by B's standards and thereby intended to be understood by A.

However, actual scientific evaluation and further investigation of anecdotal evidence has often been disabled by *constraining* factors in the history of western sciences (Daston, 1995, Daston/Galison, 2007). The goal, however, must be to find criteria of evaluating what is an actually effective treatment and isolating the factors that work for the patient, while at the same time unmasking ineffective practices that are employed on promises that cannot be kept. Human experience and empirical evidence show that there is a connection between brain/mind and body/mind targeted therapy (see: Harrington, 1987, 2008; Harrington/ Davidson, 2001).

c. Genealogy as Reopening

Genealogical analysis enables a reopening of possibilities by showing that categories and concepts like ADHD are generalizations that depend on long history of artificial conditions. We need these concepts to deal with reality as it presents itself. At the same time, we must be careful in their use not to confuse a general concept with an individual reality.[11]

Neo-marxist and other critics, on the other hand, often paint the picture of "unmasking" an evil pharmacological-industry that invents diseases and disorders and helps to keep people docile by putting them under the influence of tranquilizing drugs. This is, of course, an oversimplified account (Stingl, 2008; DeGrandpre, 2006). Indeed, there is an increase in the amount of diagnoses of behavioral disorders that exceeds the limit of those to be expected neuro-physiologically and/or genetically. Also, the amount of behavior-related drugs prescribed in the USA is by far higher than in some other countries. However, the reason for these developments owes itself to a multitude of factors, the least of which, if at all, is the—by the way, perfectly legitimate—interest in making a profit on a free and competitive market for medical/psychological treatment.[12] Closer to the truth is an account of the historic combination of many constraining social, political, economic, scientific and cultural factors that led to this development. Beyond that, there are cases of behavioral disorders where pharmacological treatment is a viable and rational (sometimes even the only) option. The question is whether the diagnosis or the treatment is always indicated or fitting the case. And, whether current research encompasses possible alternatives and what the actual goals of research, treatment and diagnoses are, that climax in the result as the semantics of the *vernacular* they are enunciated in.

A genealogical approach in Science Studies does not seek to destroy a concept from within or unmask an ideology. Instead, we need to understand where our concepts are coming from and how and why they came to be the way they are. In doing so, genealogy has the

---

[11] David Owen (2002) proposes this use of genealogy, which is superior to e.g., the *Frankfurter Schule* and *immanent criticism*. The latter kind of critical approach issues a general doubt toward every social or scientific fact of thought as part of a *total ideology*, that must be wrong and, therefore, blinding people to the true condition of society. *Immanent criticism* supposedly takes apart concepts from the inside, unmasking them "for the *fraud* they are" (Stingl, 2009a).

[12] The perquisites for navigating modern markets lie in knowledge and information, in itself ethical subjects also (see Stehr, 2007). The control of access to information for different interest groups is the crucial variable. This control is actually not in the hand of pharmaceutical corporations, but distributed and freely negotiated between governmental agencies, private consumer interest collectivities and industry interest collectivities. While there are naturally conflicts of interest, there are also dialogical solutions available and are usually the course of action (see Stingl, 2008b,d).

potential of opening up the conceptual frame to allow for new ideas or old ideas that have not been pursued. Genealogy can help "correct the picture" and enable constructive dialogue. In order to write the history of the present and the future, we need to return to the archives of science and re-assemble them. We need to address issues, such as "Complementary and Alternative Medicine" (CAM), and find proper translations in our own science history—for example in introspective psychology, which has recently received renewed interest in experimental neurophysiology (Kroker, 2003; Northoff/Heinzel, 2006)

II. ADHD Between Lay and Expert Discourses

a. Layman and Expert

There is a distinction between science and popular science in the distinction between laymen and experts. By experts, I refer to clinicians, researchers, academic teachers and practitioners. Their discourse is, in both *productive* and *constraining* factors, at a progressed stage than is laymen discourse, due to a time-lag in regard to concepts, expectations, and practices (Bordage, 2007; Savova, 2008; Cohen, 2008; Drolet, 2008). Even in an era of easy access to information via the internet, this has not really changed. There is a difference between absorbing information and understanding that information in context. In other words, the *regimes of truth* differ greatly between experts and laymen. It has been argued that laymen as patients can turn into experts. Indeed, patients suffering from chronic and progressive diseases such as *multiple sclerosis*, AIDS, or migraines, very often assimilate a wealth of knowledge regarding their conditions and plan their lives accordingly. But this does not necessarily make them experts. The crucial difference lies in the discursive practices they use, not the amount of information they have been exposed to. If a patient can indeed replicate the truth-regime of the clinical expert, he would be in the position to diagnose and treat others, i.e., make decisions for them. But this is rarely the case. A patient still perceives his individual disease and his/her goal is to get better individually, while an actual expert seeks to optimize an organism that he does not conceive to be an individual.[13] Additionally, many popular science publications today are produced by laymen and not experts. This widens the gap between laymen and experts. Parents with ADHD-children consequently participate in laymen discourses and not expert discourses. Their goals in a treatment differ in the definition and value of what is the presumed quality of life, what is self-realization, what is a happy family life, etc. Unfortunately, there is very little research available that has focused on the *semantics* that have emerged between the public and the popular science discourse in regard to the goals (or the teleology) of therapy.

[13] I do not want to belittle or negate the need for the participation of the patient in information-gathering. This is merely a neutral assessment of different modes of discourse that construct different kinds of truths. In each situation of debate, decision-making, etc., the interlocutors construct truth as a local phenomenon, very much like a micro-climate.

b. Laymen-Expert and Molar-Molecular Perceptions

Laymen/expert such as patient/parent-doctor discourses are shaped in the asymmetry between the *truth regimes* they are regulated by. In current bio-technology discourses, these can be described as an asymmetry between a molar/somatic and a molecular "style of thought" (Fleck, 1979) or gaze (Rose, 2007). A *molar/somatic gaze*, prevalent among laymen, imagines the body as an organism of limbs, tissues, organs, hormone flows, etc. Public discourses on genetics and neurology are also constructed from the viewpoint of the *molar gaze*. A *molecular* style of thought is shared by experts such as geneticists, neurologists, etc., and constructs their subject matter as nucleotides, coding sequences, ion channels, dopamine receptors, neuronal maps, and so forth. The dialogue between experts and laymen (such as patients and particularly parents in cases when the patient is a child) becomes problematic due to this asymmetry.

The expert-goal is the optimization of processes, constructed under the molecular regime, while parents seek to integrate the child as a "social organism" into a social environment—parents construct treatment thus as a social process aiming towards integration on a somatic level. All of these processes and entities on the parent's side are, therefore, social/cultural constructions, and such is the behavioral disorder they perceive in their child. For the expert, the truth is differently constructed, namely, a scientific fact perceived in the molecular gaze (Rose, 2007).

This would suggest that the dialogue between parents and doctors is a negotiation (Parsons, 1951) or a conflict between two different truth-regimes (Foucault, 1963; Rose, 2008), struggling for domination over which semantic—the molar or the molecular—will gain the power of defining (Foucault, 2007) the meaning of the diagnosis and the subsequent treatment (goal).

The increase in the number of behavioral disorders diagnosed in children and adolescents, and a subsequent rise in pharmaceutical treatments, has led to heated debates (Frazzetto et al., 2007), among laymen and experts, whether the phenomena observed are– at least in part—socially constructed and therefore subject to questionable validity. This argument has emerged in the discussion of ADHD, a group of symptoms clustered from the Hyperkinetic Disorders.[14] It must also be stressed that while diagnosis and treatment are generally on the rise, in some countries at least, the relative proportion is by far smaller than in others, as is the number of times that a pharmacological treatment is preferred over a psychological intervention or Cognitive Behavioral Therapy (CBT). The specific focus of these debates is in the U.S., where the density of diagnosis and pharmacological therapy is the highest.[15] First of all, ADHD has an existing neurophysiological basis. However, the number of cases reported by parents, teachers, and social workers is higher than even the expected amount of neuro-physiopathological cases. Factors accounting for this are a) culture specific and b) refer back to the concept of the human condition that shapes the semantic household of

[14] The definition of ADHD is produced by the American expert *arrangement*, while Hyperkinetic Disorders emerge from a European *arrangement*. American classifications are oriented more towards classroom performance, while European definitions are oriented towards interaction skills. It seems reasonable that it is the latter that open m ore room for the role of environment in diagnostic and therapeutic discourses.

[15] According to Frazzetto et al. (2007), the estimates for ADHD in schoolchildren in the US lie around 6.7%, though some estimates go up to 16% in the US. In Europe, estimates—with grave differences among nations—lie between 2% and 5%.

laymen and popular science discourses. As an example, we find in Frazzetto et al. the statement that "drug treatment is generally intended to grant and restore 'normal' functioning" (2007: 394). This semantic difference is crucial.

Typical American definitions of ADHD (in the standard literature, the DSM IV or online, e.g., mentalhealth.com) refer to problems with work and classroom activities, using metaphors such as "driven by a motor," etc. The "European" descriptions, such as the WHO classification, refer more to child development and social skills. I do not see that American nosology (classification of diseases) leaves very much room for the inclusion of the natural and the social environment in diagnostic and therapeutic schemes. If the interactions of the child with its different environments are accepted to be part of the development, it stands to reason that the European approach is potentially more inclusive. Diagnosis is problematic for it lies with the parents and teachers to report children/students and actively seek diagnosis and treatment, using different standardized (self-) report tests that have been made available to the public. But these tests are usually long categorical questions, much like a standardized multiple choice tests used in American High Schools and colleges. While reliability of these items is high, their validity is often questionable. If bombarded with a large set of questions of the nature "Does your child sleep tight (in comparison to other children)?" it is easy enough for parents to find several items where even the healthiest of children will be a "Yes!" Additionally, an often-used formula, "in comparison with other children," is a loaded aspect. With a declining number of children per family in Western countries, parents have less and less exposure to "other children" to draw comparisons. "What is a normal child?" is answered in dependence of a person's experience with children and their behavior in relation to accepted standards of "normal." Modern cities are not an environment suited for the development of children. In other words, the discursive practices, the assemblage ADHD, and the arrangements that currently handle children as patients promote underdetermined assemblages of the concepts "environment" and "health"—the latter is reduced to industrial functionality and physico-chemical optimum and reduces the influence of the prior, if it accounts for it at all.

A review of a wealth of literature allows the following statements

- ADHD is a compounded disorder itself. It consists of Attention[16] Deficit on the one hand and Hyperactivity on the other, which occur separately or in different kinds of combinations
- ADHD was defined from a cluster of often-unrelated symptoms[17]
- ADHD comes in different degrees
- ADHD has a high comorbidity[18] rate between 50% and 90% in observed cases, leading to misdiagnosis and problems in (pharmacological) treatment (Faraone/ Kunwar, 2007)

---

[16] It must also be pointed out that it is not neurologically clear what *attention* actually is (see Hunt, 2006a). The use of the concept *attention* in all branches of the cognitive sciences is frequent and glossed over. I know of no detailed semantic analysis. But it ranges from the ADHD discourse to visual attention and color perception, sensori-motor research to reading skills, etc.

[17] Adele Diamond has provided interesting cognitive development studies regarding this aspect (2005).

[18] The term "comorbid" currently has two definitions: 1) to indicate a medical condition existing simultaneously but independently with another condition in a patient (this is the older and more "correct" definition) and 2) to indicate a medical condition in a patient that causes, is caused by, or is otherwise related to another condition in the same patient (this is a newer, nonstandard definition and less well-accepted).

In the current literature, the role of non-physiological factors in causing, increasing or hastening ADHD, such as diet, environment, and social/family context is debated and controversial.

In summary of the condition, diagnosis, and treatment of ADHD (exemplary Hunt, 2006,; Frazzetto, 2007; Smoot et al., 2007):

- A rising number of children are reported to have behavioral problems.
- A large set of theses cases is subsumed under the label ADHD by parents and teachers but not necessarily by experts. This is a result of differences in the regimes of truth production.
- There *are* a neurophysiological aberrations in the brains of children suffering from severe ADHD, observed in studies using fMRI (Functional Magnetic Resonance Imaging), PET (Positron Emission Tomography), LDDMM (Large Deformation Diffeomorphic Mapping), and SPECT (Single Photon Emission Computed Tomography).
- The number of cases reported exceeds the number of cases that should be expected as neuro-physiopathological cases.
- In those cases where the condition was observed to be neuro-physiopathological, this condition seems related to the role of Dopamine, Serotonin and/or Norepinephrine in the neuro-chemistry system of the brain, and with inhibitory control (see: Oades, 2007; Hunt, 2006b; Diamond, 2005).
- There are reasons to assume that in actual neuro-physiopathological cases there is a relation to the genetic makeup. However, according to current literature, the neuro-physiopathological cases with a genetic predisposition account for between 2% and 5%, while there are between 2% to 16% children reported with ADHD.

A thorough review of the data presented by these and other studies reveals the following convergences that allow us to reconstruct an account of the following situation of a child that seems to provide a likely model or scenario:[19]

2% to 5% of children may have a genetic disposition of an ADHD career (!).[20]

Complex factors play into whether and in what fashion ADHD will actually occur and progress. That stands to reason since a) epigenetic factors[21] have to be accounted for, and b) the brain is continually developing through childhood and adolescence. One of the brain's central features is its plasticity, which is exemplified in its capacity to cope with even severe injuries.[22]

---

[19] There is a wealth of relevant literature listed in the bibliography below, which has served as basis for this synthetic model.

[20] Although this is debated. Researchers who assume the category is just "made up" have told this author that there is no such predisposition at all, while others insist that genetic disposition is as high as 25%.

[21] A recent study has suggested that obesity is contagious, if one accepts "social contagion" as a cause. Given the severity of the gene expression may vary epigenetically, within family or social settings where ADHD is already present, such a concept of social contagion may apply in the form of what I call developmental factors.

[22] As for example the famous Phineas Gage type injuries.

Therefore, factors like diet (see: Colter et al. 2008; Lakhan/Vieira, 2008; additionally Sagduyu et al., 2005), drinking water quality,[23] interaction with and stimulus by the environment, mass media exposure (see Spitzer, 2005), family relations and social interaction affect the development of the brain (see Smoot et al., 2007). The process of development is one of mutual interaction between social/cultural environment, natural environment and the brain/mind (Wexler, 2008).

To properly account for these different dimensions, we need to distinguish between causal and development factors. In cases without genetic predisposition for a disease or where the genetic predisposition merely increases the likelihood of a disorder emerging, but does not necessitate it, the developmental factors become causal factors. We can assume that in reported cases of ADHD, some are genetically caused, while others are cases that are not neuro-physiopathological cases, but either a psychopathological or even completely different type of behavioral problem, are caused by development factors. The same goes for the question of comorbidity. That said, in diagnosis and treatment, caution on the expert side is, therefore, imperative. However, the pressure existing in the social context (including parents, teachers, as well as health care administration) that is brought upon doctors is often high and interferes with a suitable diagnosis and solution for the actual pathology. This pressure is informed by concepts and world-views among laymen that gravely differ from the expert's understanding. But that does not mean that experts themselves are bias-free in general. While researchers and clinicians often have a keen understanding of the complexity of ADHD, many private practice therapists and doctors "in the field" find themselves in more difficult circumstances of running a business that must help them make a living, while they are faced with an increasing demand in various issues of health care administration, etc. Therapists, consequently, often succumb to the semantics of parents concerning the diagnosis of ADHD. The power structures in patient-doctor (and similar) interactions and semantics are a fairly new field, where more studies are highly necessary.

III. Factors, Environment and Systems

a. Causal, Developmental, Secondary Causality

The systems/environment[24] distinction is certainly crucial for modern science. For the "systems"-dimension, we have come to understand that there is a human organism that is set in interdependent, interpenetrating, dialectical and interactive relations with its environment on the one hand, and the social and cultural context on the other. The human organism is a dynamic and developing entity itself with various aspects that can be the subject of our focus. It can be understood as a physico-chemical system (Gomez-Pinilla, 2008) that incorporates social and cultural structures, which it reifies by acting in accordance or defiance to them; it develops a personality in a process of socialization and it has a conscious mind that has formed in development and is still open to change. Conscious mind is the name of the

---

[23] Drinking water in Western countries is often contaminated with chemical, medical and biological waste, including growth-hormones, antibiotics, lead, etc.. Even in trace amounts, these can effect the development of children.

interface where autonomous decisions are made (Wexler, 2008). At the same time, the human organism is not just a physico-chemical system with parts that can be studied in isolation.

The parts of the human organism are also interrelated in a whole system that adapts as a whole to its natural environment. Given evolutionary premises, it has limits and potentials to effect this adaptation. In the triadic relation between cultural/social environment, organism and natural environment, improper adaptations can occur either because of problems with the organism, changes in the natural environment, or demands or resources from the cultural environment that do not fit with the prerequisites of the organism or the natural environment. Usually, the result emerges in the interplay between all three, while a solution for emerging problems lies in the same interplay. The playground, however, is always the individual organism with its unique attributes with regard to limitations and potentials. The human organism is the result of a very long process of adaptation to a natural environment. The pace of cultural change helped a lifestyle to emerge that does not satisfy the outcome of this adaptation process. This was apparent in the late nineteenth century and adequately described in the classic sociological writings of Georg Simmel. The changes need, therefore, be consistent with the demands that the natural environment presents. Modern lifestyles need to be supplemented by physical activity, creativity and an adequate diet (see Langer, 1998; Ratey, 2008; Gomez-Pinilla, 2008). The analytical concepts that currently govern regimes of diagnosis and therapy have emerged in a long discourse about definitions of what it means to be human. Different scientific *assemblages and arrangements have used one or the other, tried to synthesize these concepts or negate one in favor of another*. Eventually, all of these aspects are merely conceptual perspectives that depend on the discourse that assembled them. Even if a genetic disposition exists in an individual, this does not necessitate that a disorder will occur. Even epigenetics suggests that there is more to disorder than "just DNA." The severity of the genetic expression is also not pre-defined. Neuro-pathological expression of ADHD depends on the developmental processes in brain-physiology and is subject to comorbidity, and the pathological career of comorbid disorders. It is influenced by parents/family, the social environment, nutritional diet, mass media exposure, physical engagement, intellectual stimulation, and so forth. Therefore, we can conclude that even with an existing genetic causal factor, a large number of cases of ADHD (or comorbid disorders) would be less severe, may sometimes even be prevented, if the array of developmental factors were optimized. Additionally, it seems reasonable to suggest that treatment should not only focus on the causal factor, but also on the developmental factor.

An integrated approach of

- Pharmacological treatment (PCT)
- Cognitive Behavioral Therapy (CBT)
- Physical Behavior Therapy (PBT)
- Environmental Interactive Factors (EIF)
- Nutritional Diet (ND)

would seem the most reasonable course of action, in order to account for the interplay of organism, cultural environment and natural environment.

---

[24] Problematized by Jakob von Uexkuell in 1909, for biosemiotics. His son, Thure von Uexkuell, championed research in psychosomatic medicine, another name for mind-body medicine.

There are differences in pharmacological strategies. Metylphendidates, such as the famous Ritalin are stimulants that have a "kind of a counter-effect" in neuro-pathological patients with ADHD, creating an ability to focus. But stimulants can make things worse in patients diagnosed with ADHD who have no severe neuropathological but only a mild psychopathological disorder. Not to mention side effects or unintended interaction with comorbid disorders, etc., and long-term effects in the development of brain-structures, leading to all kinds of disorders through malmedication, such as depression or substance abuse. There are new non-stimulant drugs on the market, like atomoxetine, which have proven useful, are sometimes less rigorous and seem to come with fewer side effects. But clinical trials that have distinguished among first/second/third line pharmacological treatments, still put methylphenidates up front and only upon their failure suggest atomoxetine as a third-line strategy. However, clinical trials are themselves not uncontested (Ionnanidis, 2008). This complexity suggests, against the grain of the majority of the existing literature, that each case must considered individually. Therefore, intervention options should vary the emphasis on the dimensions (PCT, CBT, PBT; EIF, ND). The best strategy is to opt for the healthiest possible lifestyle before resorting to any radical interventions in the physico-chemical system (Stingl, 2009b). This approach accounts for the fact that the human organism and the natural environment have evolved in mutual interaction before modern culture has created the modern lifestyle. The concepts "the normal" and "the pathological" are produced by discourses and not inherent in the natural environment or the organism. They are, at the same time, useful categories to deal with disruptions in the interplay between the parts and the whole of the organism or mal-adaptations between organism and natural environment.[25]

b. Negotiating Patient Careers between Organism, Natural and Cultural Environment

It stands to reason that within a patient-career,[26] treatments of any kind, even pharmacological regimes, may change while the patient is growing up. Side effects and comorbid disorders could be mitigated through a combined approach, even lesser dosages of drugs or non-stimulant medication could be considered in cases that responded so far only to strong medication. The long-term maintenance of latent drug-induced changes could be enabled through physical activity, mindfulness techniques, or meditation instead of promoting chronic drug use. Another aspect seems to be forgotten in the emphasis of pharmacological treatments—specifically in the U.S. with a focus in diagnosis and treatment on classroom performance: A child[27] suffering from ADHD is deprived of necessary learning processes

---

[25] To the knowledge of this author, there are no substantial studies available on the question of substituting treatment regimes with the goal of maintenance of therapy successes.

[26] There are interesting ethnographies available for depression careers. I am not aware of similar patient career-studies or life-course studies on ADHD. One must also bear in mind that patient career and life course are two different dimensions. Life course includes a comprehensive perspective that encompasses the quality of life as well as the environment. Patient career is a clinical perspective that reduces the patient's progress to clinical statistics. This distinction is a useful analytical tool itself, which is rarely used.

[27] In adolescents and young adults, ADHD careers are becoming rationalized or normalized differently. Often as legitimation for failings in youth, etc. The question could be whether these processes must be understood as the normalization of pathologies, or the pathologicalization of normalcy, provided ADHD is not viewed as a novel disorder, but rather merely as a previously normal occurrence that has only now become a pathology,

including interactions with its natural environment on the one hand, and, on the other, learning how to participate in social interaction, like learning group behavior, emotional budgeting, higher moral and logical reasoning,[28] etc. These have to be "relearned," once a treatment of the ADHD condition shows successes. It is here that developmental factors play a key role.

If these factors initially contributed to the ADHD condition in the patient, they will continue to work as boundaries for further cognitive, intellectual, and social development above all because the child may lack social and self-organizational skills appropriate for its age. This is certainly accounted for in several approaches in psychotherapy or CBT, however, not in all nor are all these successful.

It is in this regard that I propose to open up the conceptual frame of reference towards other medical cultures and ideas, however, subjecting these efforts to strict criteria of scientific method.

Donald Levine, a renowned sociologist from the University of Chicago, where he also served as Dean, and a long-time practitioner of Aikido, has made efforts in recent years to show that the conceptual frame of reference underlying Aikido is translatable into scientific and therapeutic concepts in the Western world (Levine, 1984, 1991) as well as into a curriculum for the Liberal Arts. Using concepts from Georg Simmel and Talcott Parsons, combining them with Western methods of psychology and of physical therapy (e.g., Feldenkrais), Levine included Aikido into a common conceptual frame of reference that can serve as a preliminary translation matrix. It has been demonstrated that the use of Aikido in psychotherapy, in working with children with Asperger's, or in treating cases of ADHD, has shown promising results. However, we are yet only at the beginning of developing and studying actual treatment strategies that integrate physical forms of therapy into ADHD therapies (Ratey, 2008; Mayes, 2009) on the one hand, as well as defining proper scientific criteria towards evaluating such efforts. Donald Levine, John Ratey, Peter Wayne[29] and an increasing number of scholars present translation matrices that can enable scientific criteria, as well as show that physical activity, and especially internal martial arts disciplines, have the potential to complement the Western medical canon in the prevention and the therapy of behavioral disorders[30].

Studies on the role of attention and postural stance (see: Schmid et al., 2005; Simoneau et al.,2006) have hinted at the fact that posture and movement are impaired by attention deficits. ADHD patient reports also indicate that self-location in spatial contexts, as well coordination capabilities, seem sub-optimal. Practicing a discipline like Aikido, which is focused on posture, self-location, and coordination, should be as beneficial for ADHD patients as is it with Asperger's patients (Levenson, 2003).[31]

---

either because developmental factors have changed to a point where the severity has increased or because of other reasons.

[28] While the rigidity of the scale suggested by Lawrence Kohlberg has been contested with good reason, the basic insight into a successive development of moral reasoning (and reasoning in general) must be accepted.

[29] Wayne teaches Tai Chi in Somerville, MA and works as a researcher at Harvard's Osher Center, focusing on hypertension and Life Course effects of internal martial arts.

[30] Hence, the term *alternative medicine*, seems an ill-fated choice, suggesting an either/or principle. It is more helpful to speak of *complementary healing methods* only.

[31] There is strong anecdotal evidence by ADHD patients' blogs that actually do practice Aikido that their postural and self-location capability has dramatically increased. There is a strong need for evidence-based, randomized, controlled trial studies to that regard to follow up on this anecdotal evidence.

It has also been demonstrated that simple meditative techniques (Linden, 2003) or mindfulness approaches (Langer, 1998) can help improve attention and other cognitive capabilities in children with behavioral disorders such as ADHD.[32] Additionally, a common comorbid disorder is found in ODD (oppositional defiance disorder), which necessitates a high rate of conflict with parents, teachers and peers. This, of course, leads to additional stress on family life, social interaction and, of course, the child's psychological make-up. Soft martial arts, such as Aikidio or Tai Chi are oriented towards avoidance or resolution of conflict. In practicing such an art, children and adolescents learn to integrate strategies to gravely reduce conflict and stress. If the practice of Aikido or Tai Chi is integrated into family life, social ties within the family, which may have been disrupted, could also be strengthened. Internal Martial Arts have proven to be very effective with regard to the integration of lifestyle and life course in therapy and prevention (Wayne et al.,2008).

c. Intervening in Whole-Systems: The Neuro-chemical Self, the Whole System, and the Importance of Environment

In sociology, the ideas of biological citizenship and the neuro-chemical self are continually developed from the seminal ideas of Michel Foucault in regard to the progress in biological technology and health care with the contributions Nikolas Rose, Susan Pickard or Paul Rabinow. We all acknowledge the fact that the technological progress, particularly in the life sciences, is at the very least a double-edged sword. It has brought new insights into different parts of physical-chemical mechanisms of the human body and broadened our spectrum of possible interventions in the form of diagnoses, therapies, and on occasion cures; these—let us call them—"opportunities" or "potentialities" arrive with ethical questions, structuring force, and paths of control dependencies. The human subject is understood in between those forces as a neuro-chemical system that can be changed, controlled, or optimized. And since the modern human condition is signified by the idea of our current modernity being a reflective modernity, we, as individual human beings, have absorbed this idea, been subjectified by it and understand ourselves each as a neuro-chemical self. As such, we participate in a society of other such neuro-chemical selves. We are subject to strategies and politics that, among other things, negotiate between conceptions of normal and pathological within life-courses and enable and constrain potentialities, opportunities, and decisions. The equation of life as a physico-chemical system that is identical with our very selves and the strategies that guide or discipline these selves or lives, we can understand with Nikolas Rose to be *the politics of life itself.* As actively engaged participants, we take on the citizenship of a collective political entity of lives that are neuro-chemical systems that reflect

[32] It could be suggested that more sophisticated techniques in meditation may be tried for treating ADHD, such as practiced in Zen-Buddhism. One the positive side, here stands the fact that Buddhist practices have received more attention in science (see exemplary Harrington, 2008; Harrington/Davidson, 2001); while Zen practitioners have included scientific methods, as well as discussion of Western psychological concepts and history into their own tradition (see exemplary the discussion of William James in: Sekida, 1975). On the downside, Zen meditation (Zazen) is very demanding and will not catch the interest of a child in a way Aikido may. However, the philosophy and practice of Zen can be also translated into Aikido concepts and may, therefore, be a viable alternative for adults suffering from ADHD symptoms, in particular adults that have prior martial arts experience. However, there is some excellent research on health and meditation gradually becoming available.

as selves on their being neuro-chemical systems and who can, in accordance with the politics governing these lives as biological lives, influence their life course (for good or ill). And here lies the actual point, for we must ask ourselves whether this kind of biological citizenship enables or constrains. Of course, we know that we cannot have one without the other: enablement comes with constraints and vice versa. We know that we cannot completely isolate the parts from the whole, nor view the whole without its parts, nor take them out of their environments. And yet policy, ethical, research decisions, the same as personal choices that we all have to make or participate in, will yield more freedom or more constraints under the new regime of bio-politics. Pessimists see and warn merely of the dangers, arguing that we will be subject to more control by anonymous forces or military-political-pharmaco-entrepreneurial complexes. Some optimists in the line of post-humanists like Nick Bostrom hail the power of technology as the salvation from humanity's pending doom, while a more sober approach, which I understand to be Nikolas Rose's, makes it clear that biological citizenship enables choices more than it constrains, if only we carefully understand what is enabled and what is constrained and how. Biological citizenship, in short, enables degrees of freedom over our life-courses. And while this author sympathizes with the sober stance, we ought to perceive a current crisis of this concept of biological citizenship and the promises of autonomy that it seems to make. And the disorder we know as ADHD is the very incident that represents this crisis in its conceptual aspects at the very best.

The neuro-chemical system (part) is integrated into a wider biological organism (the body, the whole); it has dialectical relationships with its natural, social and cultural environments (see Lewontin/Levins, 2007). The organism, culture and the natural environment effect changes of the neuro-chemical self just as much as the neuro-chemical system effects changes in the organism and its environment (see also Wexler, 2008). A change in either aspect cannot entirely go without repercussions in the others. This is the fundamental aspect that needs to be understood for a whole-systems perspective. Pragmatic or pluralist approaches to normalcy, pathology, and intervention must take this into account. We need to reassess our categories in order to create effective intervention regimes. This will necessarily go as far as re-assembling our conceptions of normalcy and pathology. It is not necessarily evident that a hyperactive child is suffering from pathology. Quite the contrary, medication can create obstacles in the development of the child when other factors are neglected. At the same time, in several cases, medication would support other treatment strategies, but is not prescribed for ideological reasons. The influence of the environment for the development of children has been downplayed. But is it not reasonable to ask the question, if a child grows up in a neighborhood without trees and grass, places to meet and play with friends and is instead put in front of a two-dimensional television or computer screen, will this not affect the development of the neuro-chemical self? We are faced with a paradigm shift of major proportions, and the discussion of the role of the environment in discourses on normalcy, pathology, and intervention is a key factor that could promote and shape this change. To succeed, we require a model that accounts for these aspects in a whole-systems approach in intervention on the one hand and a theoretical model of how discourses produce knowledge.

IV. CONCLUSION

The decision as to what form of treatment is best for a child remains case dependent. Physical activity, for example, is not a magic therapy but a tool within a combined treatment strategy. We need to integrate the natural, social and cultural environments perspective in our conceptual framework. The fallacy that specifically parents often fall for is to assume that there is a magic drug that solves all problems at once. In truth, it lies with experts, legislators, teachers and parents together to create environments and lifestyles that do justice to the needs of the child. This requires a combined treatment strategy that allows for an improvement of the quality of life.

Quality of life is a diverse and contingent concept itself, depending on the *regimes of knowledge* of discourse between laymen and experts. It is also a whole-systems concept that accounts for the adaptation of the organism to the natural environment. Quality of life can only be produced in the triadic interaction among organism, culture and nature. However, between parents and doctors, the choice for a child's integration is often reduced to the alternatives of serving as a functioning element in the labor market, which clashes with the idea of optimizing a neuro-physiological system. In between these alternatives, behavioral disorders emerge as a cultural construct *and* a scientific phenomenon and fact: ADHD, it seems, is the *assemblage* on which this, the current *episteme,* can be based. It is the history, the construction and practice of these phenomena, respectively the *assemblage* ADHD, within the scientific and public discourse that we need to uncover to open new avenues. Only a pragmatic and pluralist approach fits the demands of good science. The task of the *Science Studies* is to unravel the prerequisites for such. This must be viewed in light of the three questions the present chapter has set out to address:

- How does the discourse within science and between science and the public "construct" pathology and normalcy?
- What does this "construction" mean for the relation between health and environment in special regard to so-called lifestyle diseases, exemplified by case of attention-deficit and hyperactivity disorder (ADHD)?
- How need we effectively change current concepts of the human condition and our lifestyle?

Good Science is a kind of science that can question its basic concepts for their reliability *and* their validity. Good science has a clear understanding that its concepts are tools to be used, that they are not "natural kinds" (Davis, 2008; Daston, 1995) and that the same is true for its "objectivity" (Daston/Galison, 2007). Therefore, good science means science that clearly understands the limits of its explanatory power. That leads to a pluralist approach, namely the ability to *enable* the use of concepts *where they are appropriate* and *enable* critical relations between science, public and polity. Critical in the Kantian sense, meaning to ask questions about the potentials, possibilities and limits of science.

With regard to ADHD, we must understand that ADHD is an *assemblage*. It has emerged in a two-hundred-year-long discourse until it solidified in the current therapeutic and diagnostic practices.

In treating it as though it were a natural kind, the *constraints* of contemporary discourses are revealed. We understand that the *arrangements* health policies, pharmacological industries, parent organizations and health care bureaucracies, along with the modern lifestyle of lack of physical activity and unhealthy food, are confronted with demands of their natural, social and cultural environments, that they fail to meet. This failure or pathology re-*assembles* children into individual ADHD patients. Under the current *knowledge regime*, therapeutic practice is subsequently reduced in options. The current regime constructs each patient as a "neuro-chemical system," with a universal goal of optimization. A pluralist approach would try instead to integrate the individual needs of the child with the environmental and organic demands, before simply reacting with a preformed strategy.

Reconstructing the conceptual history, we can learn to integrate pathological concepts (such as ADHD) and diagnostic and therapeutic practices (such as neurophysiology) with a whole-systems approach. This would suggest to include the dynamic interactions between the developing neuro-chemical system, the organism, and the different types of environments. The *human condition*, which can serve as a goal or *regulative ideal* (Kant) for this development, can now be discussed under "quality of life" aspects. This perspective does not ask for optimization or social functionality, it instead poses the constant question: "What are the conditions of the possibility of living a good and healthy life?" *Lifestyle* should, therefore, not be viewed as predetermined *cultural* style, such as the classic "couch potato." Instead, *lifestyle* must be a reconciliation of adaptation processes among brain, organism and their environments: natural, social, and cultural. In other words, a predetermined *lifestyle* and artificial concepts of normalcy and pathology should not govern the development of a child. Instead, in becoming a person by adapting to its body and its environments, a child should create its own *lifestyle*.

In reviewing its own progress through the lens of *Science Studies*, science will have to understand what its prerequisites are—meaning: "How is science possible?." In short, the practices that govern the production of knowledge are similar to the practices that are governed by knowledge. In understanding their own development and the mechanisms that *enable* and *constrain* the development and progress of knowledge, the sciences that deal with knowledge about normalcies and pathologies should consider this analogy seriously and promote *enablement* of individual *lifestyles* over *constraints*.

This does include an active engagement of politicians and the public.

A whole-systems approach and respective "technologies of the self," (a.k.a. *lifestyles*) like Aikido, generate this kind of *enablement* towards a mindful and healthy development of children that can be sustained over life. *Science Studies* means in this case to promote a kind of pragmatism or pluralism that can integrate lessons from science history, theoretical analysis, and practical application.

REFERENCES

Anastassiou-Hadjicharalambous, David Warden (2008). Cognitive and affective perspective-taking in conduct-disordered children high and low on callous-unemotional traits In: *Child and Adolescent Psychiatry and Mental Health* vol. 2(16).

Anderson BJ, Rapp DN, Baek DH, McCloskey DP, Coburn-Litvak PS, Robinson JK (2000). Exercise influences spatial learning in the radialarm maze. *Physiol Behav*, 70:425-429.

Arenz, Stephan, et al. (2008). Intellectual Outcome, Motor Skills, and BMI of Children with Congenital Hypothyroidism: A Population based study: *Acta Pædiatrica/Acta Pædiatrica* 97: 447–450.

Benjamin, Sara E, Angie Cradock, Elizabeth M Walker, Meghan Slining, Matthew W Gillman (2008). Obesity prevention in child care: A review of U.S. state regulations. In: BMC Public Health, 8:188.

Blakemore, Colin (2002). From the "Public Understanding of Science" to the Scientist's Understanding of the Public In: *Neuroethics: Mapping the Field* Dana Press, Washington, DC: 212—221.

Brisswalter J, Collardeau M, Rene A. (2002). Effects of acute physical exercise characteristics on cognitive performance. *SportsMed*, 32:555-566.

Bordage, Georges (2007). Prototypes and semantic qualifiers: from past to present. In: *Medical Education* ; 41: 1117–1121.

Bloor, David (1997). Remember the Strong Program? In: *Science, Technology, and Human Values* 22: 373-385.

Carlson, Gabrielle A. et al. (2007). A pilot study for augmenting atomoxetine with methylphenidate: Safety and concomitant therapy in children with attention-deficit/hyperactivity disorder. In: *Child and Adolescent Psychiatry and Mental Health* vol. 1 (10).

Carson, John (1999). Minding Matter/Mattering Mind: Knowledge and the subject in Nineteenth Century Physiology. In: *Studies in History and Philosophy of Biological and Biomedical Science* vol. 30: 345—376.

Canguilhem, Georges (2007). *The Normal and the Pathological* New York, Zone Books.

Cavanaughm James T. Vicki S Mercer, Nicholas Stergiou (2007). Approximate entropy detects the effect of a secondary cognitive task on postural control in healthy young adults: A methodological report. In: *Journal of NeuroEngineering and Rehabilitation* vol. 4 (42).

Cohen, Trevor, Brett Blatter, Vimla Patel (2008). Simulating expert clinical comprehension: Adapting latent semantic analysis to accurately extract clinical concepts from psychiatric narrative. In: *Journal of Biomedical Informatics* 41: 1070–1087.

Coe DP, Pivarnik JM, Womack CJ, Reeves MJ, Malina RM: Effect of physical education and activity levels on academic achievement in children. *Medicine and Science in Sports and Exercise* 2006, 38:1515-1519.

Colcombe S, Kramer AF. (2003). Fitness effects on the cognitive function of older adults: a meta-analytic study. In: *Psychol Sci* 2003 (14):125–130.

Colter, Ashley L., Caroline Cutler, Kelly Anne Meckling (2008). Fatty acid status and behavioral symptoms of Attention Deficit Hyperactivity Disorder in adolescents: a case-control study. In: *Nutrition Journal* vol. 7 (8).

Cunningham, Charles E. (2007). Family-Centred Approach to Planning and Measuring the Outcome of Interventions for Children with Attention-Deficit/Hyperactivity Disorder. In: *Journal for Pediatric Psychology* Vol. 32 (6): 676—694.

Damasio, Antonio (2002). The Neural Basis of Social Behaviour. In: *Neuroethics: Mapping the Field.*

Dana Press, Washington, DC: 14-20.

Daston,. Lorraine (1995). The Moral Economy of Science. In: *Osiris* Vol. 10: 2—24.
Daston, Lorraine (1998). Fear and Loathing of the Imagination in Science. In: *Daedalus* 127/1: 73 - 95.
Daston, Lorraine; Peter Galison (2007). *Objectivity*. Zone Books, New York.
Davis, Andrew (2008). Ian Hacking, Learner Categories, and Human Taxonomies. In: *Journal of Philosophy of Education* 42/3-4.
DeGrandpre, Richard (2006). *The Cult of Pharmacology* Chapel Hill, NC, Duke UP.
Delisle, Richard G. (2000). The Biology/Culture Link in Human Evolution, 1750-1950: The Problem of Integration in Science. In: *Studies in History and Philosophy of Biological and Biomedical Science* vol. 31: 531-556.
Diamond, Adele (2005). Attention-deficit disorder (attention-deficit/hyperactivity disorder without hyperactivity): A neurobiologically and behaviorally distinct disorder from attention-deficit/hyperactivity disorder (with hyperactivity). In: *Development and Psychopathology* 7: 807—825.
Dilthey, Wilhelm (1886). *Dichterische Einblidungskraft und Wahnsinn* Leipzig, Duncker & Humboldt.
Drolet, Brian C., Kevin B. Johnson (2008). Categorizing the world of registries. In: *Journal of Biomedical Informatics* 41: 1009–1020.
Eisenberg, Dan, et al.(2008). Dopamine receptor genetic polymorphisms and body composition in undernourished pastoralists: An exploration of nutrition indices among nomadic and recently settled Ariaal men of northern Kenya. *BMC Evolutionary Biology* 2008, 8:173ff.
Elia, Josephine, Paul Ambrosini, Wade Berettini (2008). ADHD Characteristics I: Concurrent Comorbidity Patterns in Children and Adolescents. In: *Child and Adolescent Psychiatry and Mental Health* vol. 2(15).
Faggianelli, Patrick, David Lukoff (2006). Aikido and Psychotherapy: A Study of Psychotherapists who are Aikido Practitioners. In: *Journal of Transpersonal Psychology* vol. 38 (2): 159-178.
Farah, Martha J. (2002). Emerging Ethical Issues in Neuroscience. In: *Nature Neuroscience* 5: 1123 - 1129.
Faraone, Stephen V., Arun R. Kunwar (2007). ADHD in Children with Comorbid Conditions: Diagnosis, Misdiagnosis, and Keeping Tabs on Both. In: *Medscape Psychiatry and Mental Health: ADHD Expert Column Series* Retrieved (July 15th, 2008) at: www.medscape.com/viewarticle/555748.
Farr, Susan A, et al. Obesity and Hypertriglyceridemia Produce Cognitive Impairment. In: *Endocrinology* vol. 149, No. 5 2628-2636.
Filho, Gilberte Villal Figueiredo (2009). Medicine and the GERD of Immanuel Kant (1724-1804). In: *Otolaryngology—Head and Neck Surgery* 149: 9 - 12.
Fisher A, Reilly JJ, Kelly LA et al. (2005). Fundamental movement skills and habitual physical activity in young children. In: *Medicine and Science in Sports and Exercise* 2005;37:684–688.
Fleck, Ludwig (1979). *The Genesis and Development of a Scientific Fact*, Chicago, University of Chicago Press.
Foucault, Michel (1963). *Naissance de la Clinique*; Presses Universitaires de France.
Foucault, Michel (1966). Une historie restée muette. In: *La Quinzaine littéraire* 8: 3f.
Foucault, Michel (1966a). *Folie et déraison*, Paris: Gallimard.

Foucault, Michel (1966b). *Les mots et les choses*, Paris: Gallimard, 1966.

Foucault, Michel (1969). Qu'est-ce qu'un auteur? *Bulletin de la Société française de philosophie* 63/3: 73-104.

Foucault, Michel (1969a). *L'archéologie du savoir*, Paris: Gallimard, 1969.

Foucault, Michel (1975). *Surveiller et punir*, Paris: Gallimard, 1975.

Foucault, Michel (2007). *Psychiatric Power: Lectures* New York, Palgrave McMillan.

Frazzetto, Giovannua, Sinéad Keenan and Ilina Singh (2007). "Il Bamini e le Droghe": The Right to Ritalin vs. the Right to Childhood in Italy. In: *Biosocieties* Vol. 2 Part 4: 393-413.

Ghanizadeh, Ahmad (2008). Association of nail biting and psychiatric disorders in children and their parents in a psychiatrically referred sample of children. In: *Child and Adolescent Psychiatry and Mental Health* vol. 2 (13).

Gillet, Grant (2006). Medical Science, Culture, Truth. In: *Philosophy, Ethics, and Humanities in* Medicine vol. 1 (13).

Ginicola, Misty M. (2007). Children's unique experience if depression: Using a developmental approach to predict variation in symptomatology. In: *Child and Adolescent Psychiatry and Mental Health* vol. 1 (9).

Gomez-Pinilla, F. (2008). Brain foods: the effects of nutrients in brain function. In: *Nature Reviews Neuroscience* July 2008 | vol. 9: 568-578.

Harrington, Anne (1987). *Medicine, Mind and the Double Brain,* Princeton University Press.

Harrington, Anne (1997). *The Placebo Effect* Cambridge, MA, Harvard University Press.

Harrington, Anne (1999). *Reenchanted Science: Wholism in German Culture from Wilhelm II to Hitler* Princeton, Princeton University Press.

Harrington, Anne (2008). *The Cure Within: A History of Mind-Body Medicine*, New York, Norton.

Harrington, Anne, Richard Davdison (Eds.) (2001). *Visions of Compassion: Western Scientists and Tibetan Buddhists Examine Human Nature,* Oxford, Oxford University Press.

Henrich, Dieter (1991). *Konstellationen,* Suhrkamp, Frankfurt aM.

Hunt, Robert D. (2006a). The Neurobiology of ADHD. In: *Medscape Psychiatry and Mental Health* vol. 11 (2).

Hunt, Robert D. (2006b). Functional Roles of Norepinephrine and Dopamine in ADHD. In: *Medscape Psychiatry and Mental Health* vol. 11 (1).

Illes, Judy, Eric Racine (2005). Imaging or Imagining? A Neuroethics Challenge Informed by genetics. In: *American Journal of Bioethics* vol. 5 (2): 5-18.

Ionnanidis, John PA (2008). Effectiveness of antidepressants: A myth constructed from a thousand randomized trials? In: *Philosophy, Ethics, and Humanities in Medicine* vol. 3 (14).

James, William (1890). *The Principles of Psychology,* 2 vols. Dover Publications.

Judge S, Jahns L. (2007). Association of overweight with academic performance and social and behavioral problems: an update from the Early Childhood Longitudinal Study. In: *Journal of School Health*. 2007; 77: 672-678.

Kaitaro, Timo (2004). Brain-Mind Identities in Dualism and Materialism: A Historical Perspective. In: *Studies in History and Philosophy of Biological and Biomedical Science* vol. 35: 627-645.

Kaptchuk. Ted (2000). *The Web that has no Weaver* New York, McGraw-Hill.

Kay, Lily (2000). *Who wrote the Book of Life?* Palo Alto, Stanford University Press.

King S, Griffin S, Hodges Z, Weatherly H, Asseburg C, Richardson G, et al. (2006). A systematic review and economic model of the effectiveness and cost-effectiveness of methylphenidate, dexamfetamine and atomoxetine for the treatment of attention deficit hyperactivity disorder in children and adolescents. In: *Health Technol Assess*; 10(23).

Klassen, Anne F., Anton Miller and Stuart Fine (2004). Health-Related Quality of Life in Children and Adolescents Who Have a Diagnosis of Attention-Deficit/Hyperactivity Disorder. In: *Pediatrics* 2004;114;e541-e547.

Kristjansson et al. (2009). Adolescent Health Behavior, Contentment in School, and Academic Achievement. AM J Health Behav: 2009 22 (1): 69 - 79.

Kroker, Kenton (2003). The Progress of Introspection in America, 1896–1938. In: *Studies in History and Philosophy of Biological and Biomedical Science* Vol. 34: 77–108.

Lakhan, Shaheen E, Karen F. Vieira (2008). Nutritional Therapies for Mental Disorders. In: *Nutrition Journal* vol. 7 (2).

Lakoff, Andrew (2000). Adaptive Will: the Evolution of Attention Deficit Disorder. In: *Journal of the History of the Behavioral Sciences*, vol. 36(2), 149–169.

Lande, Marc et al. (2009). Parental Assessments of Internalizing and Externalizing Behavior and Executive Function in Children with Primary Hypertension. In: *The Journal of* Pediatrics February 2009: 207-212.

Landgraf, Jeanne M., Michael Rich, Leonard Rappaport (2002). Measuring Quality of Life in Children With Attention-deficit/Hyperactivity Disorder and Their Families Development and Evaluation of a New Tool. In: *Arch Pediatr Adolesc Med. 2002;156:384-391.*

Langer, Ellen (1998). *The Power of Mindful Learning* Da Capo Press.

Latour, Bruno (2006). *Reassembling The Social* Oxford: Oxford University Press.

Lauth, B. et al. (2008). Implementing the semi-structured interview Kiddie-SADS-PL into an in-patient adolescent clinical setting: impact on frequency of diagnoses. *Child and Adolescent Psychiatry and Mental Health* vol. 2 (14).

Levenson, Martha (2003). Choosing Aikido: An Opportunity for Children with Asperger's Syndrome. Retrieved from www.aiki-extensions.org July 13th, 2008.

Levine, Donald (1984). The Liberal Arts and the Martial Arts. In: Liberal Education. vol. 70 (3).

Levine, Donald (1991). Martial Arts as a Resource for Liberal Education: The Case of Aikido In: Featherstone, M. et al. (Eds.) *The Body: Social Process and Cultural Theory* Sage, London: 209-224.

Levinger, Itamar et al. (2008). BDNF, Metabolic Risk Factors, and Resistance Training in Middle-Aged Individuals. In: *Official Journal of the American College of Sports Medicine* 2008: 535-541.

Lewontin, Richard, Richard Levins (2007). *Biology under the Influence,* New York, Monthly Review Press.

Li, Yanfeng, Qi Dai, James C. Jackson, Jian Zhang (2008). Overweight Is Associated With Decreased Cognitive Functioning Among School-age Children and Adolescents. In: *Obesity* 16, 1809–1815.

Linde K, Allais G, Brinkhaus B, Manheimer E, Vickers A, White AR. (2008). Acupuncture for tension-type headache. *Cochrane Database of Systematic Reviews*, Issue 1.

Luhmann, Niklas (1984). *Soziale Systeme.* Frankfurt aM, Suhrkamp.

Matza, L.S., Swensen, A.R., Flood, E.M., Secnik, K., & Kline Leidy, N. (2004). Assessment of health-related quality of life in children: A review of conceptual, methodological and regulatory issues. In: *Value in Health, 7* (1), 79-92.

Mayes, Rick, et al. (2009). *Medicating Children* Cambridge, MA, Harvard University Press.

Muñoz-Solomando, Antonio, Tim Kendall, Craig J. Whittington (2008). Cognitive Behavioral Therapy for Children and Adolescents. In: *Current Opinion in Psychiatry* vol. 21 (4): 332-337.

Northoff, Georg, Alexander Heinzel (2006). First-Person Neuroscience: A New Methodological Approach for linking Mental and Neuronal States. In: *Philosophy, Ethics, and Humanities in Medicine* vol. 1 (3).

O'Shaughnessy, Roy J., Holly T. Andrade (2008). Forensic Psychiatry and Violent Adolescents" in: *Brief Treatment and Crisis Intervention* vol. 8 (1): 27-42.

Oades, Robert D. (2007). The Role of the Serontonin System in ADHD: Treatment Implications. In: *Expert Review of Neurotherapeutics* vol. 7 (10): 1357-1374.

Oberschall, Anthony (1992). *Social Movements* Edison, NJ, Transaction.

Owen, David (2001). Kritik und Gefangenschaft. Genealogie und kritische Theorie. In: Honneth/Saar (2002) *Foucault Konferenz,* Suhrkamp:122-144.

Parsons, Talcott (1949 (1937). *The Structure of Social Action* The Free Press, Glencoe, Ill.

Parsons, Talcott (1951). *The Social System* Free Press, New York.

Prasad, Suyash, et al. (2007). Baseline Values from electrocardiograms of children and adolescents with ADHD. In: *Child and Adolescent Psychiatry and Mental Health* vol. 1 (11).

Rabinbach, Anson (1990). *The Human Motor* California University Press.

Rabinow, Paul (2003). *Anthropos Today: Reflections on Modern Equipment,* Princeton University Press.

Ratey, John (2008). *Spark: The Revolutionary New Science of Exercise and the Brain* Little, Brown and Company.

Raviv S, Reches I, Hecht O (1994). Effects of activities in the motor-cognitive-learning center on academic achievements, psychomotor and emotional development of children (aged 5–7). In *Journal of Physical Education and Sports Science* (Israel) 1994, 2:50-84.

Reid, Lysette, Francoise Baylis (2005). Brains, Genes, and the Making of the Self. In: *American Journal of Bioethics* vol. 5 (2): 21-23.

Ribot, Theodule (1890). *The Psychology of Attention*, Chicago, Open Court.

Rose, Nikolas (2007a). *The Politics of Life Itself Biomedicine, Power, and Subjectivity in the Twenty-First Century* Princeton, Princeton University Press.

Rose, Nikolas (2007b). Molecular Politics. Somatic Ethics and the Spirit of Biocapital. In: *Social Theory & Health*, vol. 5 (1): 3-29.

Rossiter, Margaret W. (1985). Science and Public Policy since World War II. In: Osiris Vol.1: 273-294.

Rowland, Andrew S., Catherine A. Lesesne, and Ann J. Abramowitz (2002). The Epidemiology of Attention Deficit Hyperactivity Disorder ADHD): A Public Health View. In: *Mental Retardation and Developmental Disabilities Research Reviews 8: 162–170.*

Sadler, John Z. (2006). The rhetoricians craft, distinctions in science and political morality. (A Reply to Szasz). In: *Philosophy, Ethics, and Humanities in Medicine* vol. 1 (7).

Sagduyu, Kemal, et al. (2005). Omega-3 fatty acids decreased irritability of patients with bipolar disorder in an add-on, open label study. In: *Nutrition Journal* vol. 4 (6).

Sassen, Saskia (2007). *Territory, Authority, Right: From Medieval to Global Assemblages*, Princeton University Press.

Savova et al, Guergana K. (2008). Word sense disambiguation across two domains: Biomedical literature and clinical notes. In: *Journal of Biomedical Informatics* 41: 1088–1100.

Schlander, Michael (2008). The NICE ADHD health technology assessment: A review and critique. In: *Child and Adolescent Psychiatry and Mental Health* vol. 2 (1).

Schmid, Michael, et al. (2005). The Development of Postural Strategies in Children: A Factorial Design Study. In: *Journal of NeuroEngineering and Rehabilitation* vol. 2 (29).

Sekida, Katsuki (1975). *Zen Training* Weatherhill, New York, Tokyo.

Simoneau, Martin et al. (2006). The effects of moderate fatigue in dynamic balance control and attentional demands. In: *Journal of NeuroEngineering and Rehabilitation* vol.3 (22).

Sinzig, Judith et al. (2008). Inhibition, flexibility, working memory and planning in autism spectrum disorders with and without comorbid ADHD-symptoms. In: *Child and Adolescent Psychiatry and Mental Health* vol. 2 (4).

Smith, Roger (2005). The History of Psychological Categories. In: *Studies in History and Philosophy of Biological and Biomedical Science* vol. 36: 55-94.

Smoot, L.C., L.A. Boothby, R.C. Gillet (2007). Clinical Assessment and Treatment of ADHD in Children. In: *International Journal of Clinical Practice* vol. 61 (10): 1730-1738.

Spitzer, Manfred (2005). *Vorsicht, Bildschirm!* Stuttgart, Klett-Cotta.

Stehr, Nico (2007). *Die Moralisierung der Märkte*, Frankfurt aM, Suhrkamp.

Stein, Dan J. et al. (2006). Philosophy, Ethics and Humanities in Medicine: Expanding the open-access conversation on health care. In: *Philosophy, Ethics, and Humanities in Medicine* vol. 1 (1).

Steinhausen, Hans Cristoph (2008). Frequency, stability and differentiation of self-reported school fear and truancy in a community sample. In: *Child and Adolescent Psychiatry and Mental Health* vol. 2 (17).

Stingl, Alexander (2009 under contract). *The Biological Vernacular in Kant, James, Weber and Parsons* Lampeter, Edward Mellen Press.

Stingl, Alexander (2009a). *Aufklaerung als Flaschenpost oder Anthropologie der Gegenwart. Horkheimers und Adornos Immanente Kritik und Foucaults Interpretative Analytik* Saarbruecken, VDM.

Stingl, Alexander (2009b). Do we need a new *Wisdom of the Body?*—Obesity, Physical Activity, and Academic Performance. A review of recent studies. Manuscript.

Stingl, Alexander (2009c). Knowing the Normal from the Pathological: The Genealogical Reconstruction of *Attention* and its Pathologies under a Whole Systems Frame of Reference. Paper presented at the Annual Conference of the American Sociological Association 2009, San Francisco.

Stingl, Alexander (2009d). The Crisis of the Biological Citizen: ADHD as an Assemblage for the Early 21st century. Paper presented At the Annual Conference of the British Sociological Association 2009; Medical Health and Illness Panel.

Stingl, Alexander (2008a). Speaking Biology. From Kant to Parsons…. And beyond? Lecture presented at the Colloquium "Communication, Culture und Social Order, Friedrich-

Alexander University Erlangen-Nuremburg, Institute for Sociology; Philosophical Faculty.

Stingl, Alexander (2008b). Knowledge Based Economy. EBSCO *Sociology Notes,* EBSCO Academic Search Premier Online.

Stingl, Alexander (2008c). Posthumanism. EBSCO *Sociology Notes,* EBSCO Academic Premier Search Online.

Stingl, Alexander (2008d). Technology & Medicine. EBSCO *Sociology Notes,* EBSCO Academic Premier Search Online.

Stingl, Alexander (2008e). The Constitutive/Regulative difference in the History of German physiology and American Sociology and in Contemporary Debates. Presentation at the British Sociological Association's Annual Conference March 28th- 30th at Warwick University, UK.

Swanson, J.M. et al. (2007). Etiologic Subtypes of Attention-Deficit/Hyperactivity Disorder: Brain Imaging, Molecular Genetic and Environmental Factors and the Dopamine Hypothesis. In: *Neuropsychology Review* vol. 17 (1).

Vandenberghe, Frédéric (2001). From Structuralism to Culturalism. In: *European Journal of Social Theory* 4/4: 479-497.

Vaynman S, Ying Z, Gomez-Pinilla F. (2004). Hippocampal BDNF mediates the efficacy of exercise on synaptic plasticity and cognition. In: *Eur J Neurosci* 20:2580–2590.

Vaynman SS, Ying Z, Yin D, Gomez-Pinilla F. (2006a). Exercise differentially regulates synaptic proteins associated to the function of BDNF. In: *Brain Res* 1070:124–130.

Vaynman S, Ying Z, Wu A, Gomez-Pinilla F. (2006b). Coupling energy metabolism with a mechanism to support brain-derived neurotrophicfactor-mediated synaptic plasticity. In: *Neuroscience* 139:1221–1234.

Vaynman, Shoshanna, Fernando Gomez-Pinilla (2006c). Review "Revenge of the Sit": How Lifestyle Impacts Neuronal and Cognitive Health Through Molecular Systems That Interface Energy Metabolism With Neuronal Plasticity. In: *Journal of Neuroscience Research* 84:699–715 (2006).

Varni, James W. et al. (2007a). Parent proxy-report of their children's health-related quality of life: An analysis of 13,878 parents' reliability and validity across age subgroups using the PedsQLTM 4.0 Generic Core Scales. In: *Health and Quality of Life Outcomes* vol.5 (2).

Varni, James W. et al. (2007b). How young can children reliably and validly self-report their health-related Quality of Life?: An analysis of 8,591 children across age subgroups with PedsQLTM 4.0 Generic Core Scales. In: *Health and Quality of Life Outcomes* vol.5 (1).

Wayne, Peter, Ted Kaptchuk (2008a). Challenges Inherent to *T'ai Chi* Research: Part I— *T'ai Chi* as a Complex Multicomponent Intervention. In: *The Journal of Alternative and Complementary Medicine* vol. 14, No. 1: 95–102.

Wayne, Peter, Ted Kaptchuk (2008b). Challenges Inherent to *T'ai Chi* Research: Part II— Defining the Intervention and Optimal Study Design. In: *The Journal of Alternative and Complementary Medicine* vol. 14, No. 2: 191–197.

Wayne, Peter, Gloria Y. Yeh, Russell S. Phillips (2008). Tai Chi Exercise in Patients with Chronic Heart Failure. In: Hong Y (Ed.): Tai Chi Chuan. State of the Art in International Research. *Med Sport Sci.* Basel, Karger, 2008, vol. 52, pp 195–208.

Wayne, Peter, Kelley John, et al. (2008). Do placebo responders exist? In: *Contemporary Clinical Trials* 29 (2008) 587–595.

Wehmeier, Peter M., et al. (2008a). Emotional well-being in children and adolescents treated with atomoxetine for attention-deficit/hyperactivity disorder: Findings from a patient, parent and physician perspective using items from the Pediatric Adverse Event Rating Scale (PAERS). In: *Child and Adolescent Psychiatry and Mental Health* vol. 2 (11).

Wehmeier, Peter M., et al. (2008b). Global impression of perceived difficulties in children and adolescents with attention-deficit/hyperactivity disorder: Reliability and validity of a new instrument assessing perceived difficulties from a patient, parent, physician perspective over the day. In: *Child and Adolescent Psychiatry and Mental Health* vol. 2 (10).

Wexler, Bruce (2008). *Brain and Culture* Cambridge, MA, MIT Press.

In: Health and Environment: Social Science Perspectives ISBN 978-1-60876-216-3
Editors: Helen Kopnina and Hans Keune

Chapter 8

The Orientalist Lens and Traditional Medicine: Circumventing Western Biases in Studying Traditional Medicines

Barbara Potrata and Anzurat Akobirshoeva

Abstract

World Health Organisation's (WHO) strategy on traditional medicine (2002) reaffirmed the importance of non-biomedical medicine for the health of the world populations and recommended to include and promote traditional medicine into the national public health care systems to make public health care cheaper and more accessible. It also implied that such approach would benefit the environment, following the argument that if people use the naturally derived remedies, they have a particular interest in preserving them and their environment.

However, using our own ethnographic data from Central Asia, we argue that such integration of traditional modalities into public health care systems might be hindered due to several Orientalist assumptions implied in this influential document. Firstly, there is a tendency to study medical traditions as single entities rather than recognising vast differences in cultural and social traditions of Central Asian communities. Secondly, people in Central Asia who are frequently believed to have close links to their still apparently pristine environment are—following the decades of socialism as the ultimate modern project—often highly urban and as alienated from nature as their Western counterparts. For others, the economic needs might force them to overuse the plants and other natural resources, further contributing to the ecological decline in the region.

Thirdly, WHO Strategy is oriented towards the search for the evidence-based, effective and safe active ingredients of naturally derived remedies. This position reflects Orientalist intellectual imperialism, which disregards that "medicine," "effectiveness" and "safety" are themselves cultural constructs. Fourthly, studying traditional medicine(s) is often focused on what appears to Orientalist sensibilities as "exotic" religious and ritual practices, which might accompany traditional ways of healing, disregarding in this way those health-seeking practices, which might be more relevant to the Central Asian communities, and sustainable use of natural resources. In discussion, we, therefore, discuss the most important strategies to circumvent these Orientalist assumptions to benefit both the health of the populations and environment.

INTRODUCTION

At its summit on global health in 1978, the World Health Organization (WHO) recognised a potentially important role of non-biomedical systems for health care. In the following decades, the continued importance of traditional medicines in developing countries and the rise of complementary and alternative medicines in the developed world (CAM) resulted in the articulation of the WHO Strategy on traditional medicine (2002) for the following five years (in the continuation of this text referred to as WHO Strategy). There, "traditional medicine" is defined very broadly as: "including diverse health practices, approaches, knowledge and beliefs incorporating plant-, animal- and/or mineral-based medicines, spiritual therapies, manual techniques and exercises applied singularly or in combination to maintain well-being, as well as to treat, diagnose or prevent illness." (WHO, 2002, p. 7)

Picture 9. Grasses.

The WHO Strategy argues that traditional medicine, including the use of plants, might potentially offer cheaper and more accessible health care, based on traditional knowledge. As examples, it cites some African countries where up to 80% of the population uses traditional medicine for their health care needs, and the number of traditional healers is considerably higher than that of medical doctors. The WHO Strategy states that traditional medicine is substantially more accessible for many people living in developing countries. For example, herbal remedies are often much cheaper than pharmaceuticals, which makes such health care more affordable. The WHO Strategy lists many countries that are considering or have already included (some of the) traditional practices into some or all areas of their public health care systems—educational, clinical and pharmaceutical.

The WHO Strategy also acknowledges several challenges in implementation of this approach, such as development and promotion in national policies on traditional medicines and CAM; search for safe and effective remedies and their quality control; the challenge of equitable access to traditional remedies, which importantly depends on acceptance and co-operation of biomedicine of traditional medicine and its practitioners, and rational use of traditional remedies. The challenges of access to and rational use also encompass the issue of sustainable use and cultivation of naturally derived remedies. In this sense, environmental considerations impinge upon the health issues (WHO, 2002).

Other related documents that importantly complement the WHO Strategy (2002), such as the Agreement on Trade Related Aspects of Intellectual Property Rights (TRIPS), connect health, (traditional) medicine and naturally derived remedies in another way. It has been argued that since people are more likely to protect those plants and animals that they use, promotion of traditional medicine and natural remedies would, therefore, also contribute to environmental protection of the natural or naturally derived remedies (Gupta in Timmermans, 2003). The issues of traditional medicine are, therefore, strongly interwoven with access to protection, development, promotion and conservation of natural and genetic resources as well as biodiversity issues (Peria in Timmermans, 2003). The issues and challenges involved in the use of traditional remedies, therefore, importantly overlap with those involved in their rational and sustainable use.

ORIENTALISM

In this chapter we look at the WHO Strategy and its potential impact on both health and environmental issues in the geographical area of Central Asia where the WHO, at its summit on global health in Almaty in Kazakhstan in 1978, stressed the importance of traditional medical modalities for the health of the populations. This region is particularly suitable for studying issues of use of traditional medicines and their impact on environment since it is at the crossroads of many great medical traditions but presently experiences harsh socio-economic conditions. In the countries where up to 80 percent of population lives below the poverty line, such as Tajikistan (World Bank, 2009), there has been a marked interest in and use of traditional medicines and home remedies (K. Kuehnast, 1998)

We look at the issues of potential impact of WHO Strategy with help of Orientalist perspective because according to the influential Edward Said (1979), this is a long-standing Western approach to understand non-Western societies. Here, we use the term "Orientalist," as conceptualised by Said (2003) in both a general and specific sense. According to him, in general terms, it refers to the way Euro-American societies think about other, non-Western cultures. We will refer to this understanding in discussing the use of traditional medicines in developing countries other than Central Asian when such data for Central Asian countries does not exist. Furthermore, when discussing Central Asia, we will also use the term in its specific sense, which refers to Said's (2003) argument that the West has developed a particular understanding of the East as its inverse mirror image.

From Said's (2003) perspective, the main criteria to ponder and judge cultural and social considerations are Western beliefs and values, according to which the "Orient" has become everything that the West believes it is not. "Orient" stands for a vast, undifferentiated entity, plagued by despotic regimes and characterised by fatalistic and uncritical habitants and intellectuals. Those adhering to Orientalist perspectives do recognise that "Orient" might have been an area of splendid civilisations and amazing achievements, but they quickly point out that advanced civilisations are now a thing of the past with little possibility of revival (EW Said, 2003).

Said points out further differences, presented in antithetical terms. For example, whereas Euro-American societies stand for rationality and progress, epitomised by science, "Orient" is arguably backward and unable ever to change. The Western sense and rationality is juxtaposed against presumed Oriental irrationality of the senses; Western ratio against (exaggerated) Oriental emotions, and "male" versus "female." These Orientalist understandings might also include considering people as backward and ignorant and may include the Romantic idea of natives as "noble savages" living their lives in close proximity with nature (EW Said, 2003).

Orientalism, according, to Said is a part of Euro-American modernity and represents intellectual imperialism. Said himself believed that Orientalism is an old feature of Western discourse and even though the concept has often been criticised, Turner (1989) believes that most of these negative comments were only superficial. Moreover, Orientalism has only been strengthened by recent political events. For example, Nayak (2006) argues that this discourse is being (re-)employed as America struggles after 9/11 to understand why the "Orient" (Arabs/ Muslims) hates the Americans.

Orientalism, Health and Environment

The concept of Orientalism has been mostly partially employed to study phenomena in various medical traditions, including biomedicine, though rarely in connection with environmental issues. However, the connection among medicine, Orientalism and environment is most comprehensibly explored by those researchers studying manifold but often limited ways of imposition of European scientific and other standards, and indigenous resistance to them. They explored contradictory ways in which science and medicine have been employed to justify, consolidate and legitimise colonial and imperialist projects (for a general overview, see Harrison, 2005), especially in South Asia (D. Arnold, 1993; Harris & Ernst, 1999).

Scholars working within this area have been inspired both by Foucault and Said, who used many of Foucault's concepts on science, medicine and power to examine non-European, mostly "Oriental" societies. Since the 1980s, researchers influenced by Foucault and Said explored the ways in which science, especially medicine, supported, promoted and legitimised colonial domination (for South Asia, see A. Arnold, 1999; Guha, 2001; Robb, 1995). They explored how colonial medicine contributed to the articulation and promotion of Orientalist, racial stereotypes and scientific racism (for South Asia, see A. Arnold, 1999; Robb, 1995).

Racial stereotypes, in line with the Orientalist viewpoint, have also been used in relation to the physical and natural environment. Some researchers expanded their interests on how colonial authorities geographically organised the new territories, (Godlewska & Smith, 1994) and especially their attempts to attribute an important role in causation and transmission of various illnesses to environmental factors. Notions about the impact of transmission of illnesses and physical environment were frequently accompanied with fears that (tropical) climate, geographical and natural features of the colonised regions might have been detrimental for the health of the "Whites" and led to physical and moral "degeneration" of the "white race." (D. Arnold, 1993; Harrison, 1999)

A similar argument on the importance of factors of physical and natural environment for health has been explored by Michaels (2000; Michaels, 2003) in case of Central Asia in Soviet times, especially after the Second World War. Following Said (2003), she argues that Orientalist perceptions of Central Asia are not surprising since the countries in this region have been conceived as backward hinterlands of the Russian Empire. From the 19th century on, Central Asian countries were intentionally modelled as European colonial powers, and during much of the 20th century, seen as the "Orient" of the Soviet Union (Çeçen, 1999). Michaels writes that it was Marxist ideology according to which environmental factors were equally important in disease aetiology. Soviet authorities have consequently regularly portrayed Central Asia traditions, including medical ones, as contributing to backwardness, ignorance and ill health in societies (Michaels, 2000). Following Said (2003), she continues to argue how Soviet support for and promotion of biomedicine was control- and power-driven and similar to attempts undertaken by European colonial powers in other regions in the world.

Our present chapter follows a similar line of argument. The WHO Strategy is strongly influenced by the standards imposed by Western biomedicine, which implies neo-colonial tendencies. Based on our data from Central Asia, we will argue that the important and influential international document on traditional medicine—the WHO Strategy—clearly reflects Orientalist biases. The WHO Strategy and the Orientalist perspective are congruent with considering developing countries, including Central Asian, as vast, undifferentiated, backward areas. This influential document also clearly promotes Western scientific standards according to which all other medicines must be measured, and in this way represents continuation of Western intellectual imperialism, which is a feature of Orientalism. The WHO Strategy largely ignores vast socio-cultural differences in which traditional medicines are practised.

When social and cultural elements are taken into consideration, in the Orientalist tradition, the focus is mostly on so-called "exotic" and "irrational" elements, for example religious practices accompanying traditional medicine. In this way, the WHO Strategy leaves out less exotic, but more relevant practices for the health of the populations of Central Asia, for example home remedies. Since traditional remedies are mostly sourced from the physical and natural environment, they are linked to availability and the effects of the use of traditional remedies on biodiversity and conservation.

METHODOLOGY

This is a discussion chapter that (primarily) uses ethnographic data to illustrate main arguments. The data was collected within a research initiative *Documenting local knowledge*, undertaken by the University of Cambridge Central Asia Forum (Cambridge Central Asia Forum, 2008) and the local communities of academics and practitioners in Central Asia since 2006. The main research aims were to locate and map bio-diversity issues with special focus on Ferghana Valley as a region of great biodiversity and which is endangered because of the increasing encroachment of human interventions into the physical and natural environment. Keeping in mind that the standard methodologies might not satisfactorily capture the issues pertaining to conservation and biodiversity of a particular locale, another important aim of this research initiative was to develop new methodological approaches to complex environmental, conservation and biodiversity issues.

The researchers working within this initiative have assumed that new methodologies might better facilitate the collaboration between natural and social scientists, local and international researchers, academia and (lay) enthusiasts. Such collaboration would then promote more in-depth understanding and efficient management of environmental natural resources and related phenomena. As the researchers conceptualised "health" as an area of interactions between humans and their natural and physical environment, the issues of health and relations among humans and animals, and humans and plants were considered an important issue for closer examination.

The projects undertaken within this initiative were varied, but in the present chapter, we use findings from three particular research projects. First research was undertaken in November 2006, by a team of zoologists, biologists and an anthropologist. This was a week-long expedition to desert areas of Uzbekistan's part of Ferghana valley and riparian woodlands (areas along the riverbank) of Syr Darya, which were chosen for this particular project because they are considered by local and national scientists to be under increased environmental pressure. Natural scientists used established methods to assess ecological health and biodiversity status of these areas, while an anthropologist (who is simultaneously also an ornithologist) conducted a series of formal and informal interviews with the people (villagers, hunters, fishermen, herders) who live in these habitats in order to map practical, cognitive and symbolic meanings of their immediate environment.

The aim of the second research projects, undertaken in September 2006, was to assess environmental status of local plants used as herbal medicines, and/or dyeing, production of honey and tanning in the Issyk-Kul region in Kyrgiz Republic, a region known for its cultivation and use of multi-purpose plants. Plants used in local crafts were also foci of attention of two natural scientists, an environmental biologist and plant scientist who were established researchers from Kyrgiz educational and scientific institutions. The informal and formal, in-depth interviews with the people living in this particular area, and representatives of craft's associations to document local (mis)uses and knowledge about the plants were conducted by two peer researchers. The approach of peer research was chosen because it implies that the active researchers are those who are simultaneously also the members of the studied groups. In this way, the access to the studied groups and communities is greatly improved, and this approach gives greater power and ownership of the research to those who are not members of academic communities. The two peer researchers participating in this study both belonged to local NGOs respectively—one working on supporting local crafts, especially among the women, and second on monitoring biodiversity issues in the region.

The third research project was undertaken by the two authors of the chapter on repeated visits to Uzbekistan's capital Tashkent in August 2006, February 2007, and March 2008, and smaller Sogd in Tajikistan in March 2008. One of the researchers (Potrata) is a medical anthropologist, working within the sub discipline of social anthropology, which looks at how people and different groups understand the causes of illness (aetiology), define health and illness and what people believe needs to be done to regain health. Medical anthropologists are also interested in the perceptions of the body and its processes, its symbolic meanings and its relationships with other more individual phenomena, such as "psychology" or "mind."

Those working within the field of medical anthropology might study different medical systems, ranging from traditional medicines to contemporary clinical or bio-technology practices, and processes in which health care is taking place, such as health policies and institutions that shape healing and medical practices. As in this wider research initiative, "health" is considered as a site of many interactions between humans and their physical and natural environments, the aim of this particular project was to map places that urban populations might visit when in search to regain or improve their health. By concentrating on urban populations, the project complemented other projects within this research initiative that focused on rural areas. By concentrating on less exotic locations of marketplaces where naturally derived remedies are sold and pharmacies that sell over-the-counter medicines, this project addressed the under-researched area of home remedies and over-the-counter medicines (Qato et al., 2008).

For this purpose, both researchers conducted systematic observations in two major marketplaces in Tashkent in Uzbekistan and Sogd in Tajikistan, where people can buy home remedies for their ailments. The formal and informal interviews were conducted with the stall-holders selling and giving advice on the use of naturally derived remedies. We also interviewed the pharmacists working in the pharmacies in the vicinity of these markets on a range of topics regarding the use of over-the-counter medicines.

Many people also go to urban healers when they are unwell and in need of health treatment. Again, we interviewed several respected healers from urban Tashkent in Uzbekistan. As a particular case study, a traditional healer with a wide network of his apprentices reaching across the national borders was selected for the interview and observation. Since we wanted to observe his typical treatment, but it would be difficult to obtain a consent for observation from one of his clients, one of the members of our research team (who did not have any physical problems) underwent a diagnostic session and treatment (for a detected minor ailment), which was observed and recorded by the ethnographer.

Deconstructing Orientalist Assumption

In this chapter, we argue the Orientalist perspective to be one of the main determinants in the view of the WHO Strategy (2002) on the use and conservation of traditional naturally derived remedies. Firstly, Orientalist perspective is discernible in the very title of the WHO Strategy (2002) since it refers to "traditional medicine," and disregards a vast variety of different medical systems and their internal disparities. Even though the Strategy (2002) clearly recognises that there are many great medical traditions, it is not clearly acknowledged that in one country, there might be multiple, frequently very different, traditional systems and practices.

For example, the researchers encountered many different traditions by simply taking a stroll around the main marketplace in Tashkent. We encountered stall holders selling plants with a strong evidence base, such as calendula, well known from Russian scientific books on medicinal plants; minerals to be taken as calcium supplements are sold side by side with *mumio*, a resinous substance appearing like stalactites hanging down the caves and rocks of the high mountains of Central Asia, traditionally used to strengthen the bones, help in development and treatment of nervous system and believed to have antiseptic and anti-inflammatory properties.

On the stalls, various animal and plant oils, herbs and herbal teas known from Korean and Chinese Traditional Medicines (as also observed by Penkala-Gawetcka, 2002, in Kyrgiz Republic) are sold, while the vendors recommend spices to be used for medical purposes in accordance with Unani (Greco-Islamic medicine based on teachings of Hippocrates and Galen and Islamic scholars Ibn Sina and Al-Rhazi) medicine (Sheehan & Hussain, 2002). Many such remedies are used as home remedies and are frequently the first instance in health-seeking behaviour (as also observed by Waldstein & Adams, 2006). Some sell over-the-counter Ayurvedic remedies, one of the Indian traditional medicines (as also observed by Penkala-Gawetcka, 2002 in Kyrgiz Republic). In the literature, other researchers working in Central Asia also documented the presence of Turco-Mongolian traditional health practices, Persian traditional medicines, and various New Age spiritual practices (Rasanayagam, 2006)

Nevertheless, despite a wide variety of practices and traditional health modalities, the WHO Strategy discusses "traditional medicine" in its singular form and, therefore, considers countries and even whole regions as a vast, single, undifferentiated entity. Conflating whole regions has important methodological implications for research and practice of integration of traditional medicines into public health care systems and preservation of their natural and physical habitats. Keeping in mind the co-existence of many different traditional healing modalities and the fact that many of the hundred and twenty ethnic groups who live in Uzbekistan might have their own "medicine," it is therefore not clear which traditional medicine the government of Uzbekistan might attempt to promote and integrate into their public health care system.

This also has impacts on environmental issues. While the integration of traditional medicines might contribute to a greater respect for the environment from which they derive, a vast majority of those natural remedies that have not been integrated will not be bound by this logic. Worse, the natural remedies that might have stayed outside integration of traditional remedies into public health care systems might lose their perceived value and cause the loss of interest in their protection. Those plants or other similar remedies that are integrated might be cultivated as mono-cultures, reducing in this way biodiversity.

CHALLENGING NOTION OF BACKWARDNESS

Secondly, we argue that the WHO Strategy also implies the Orientalist assumption about the "unchangeable" nature of developing societies and Orientalist assumptions about backwardness and the consequent greater readiness to re-embrace "traditional" modalities and naturally derived remedies that are supposedly readily available for consumption. Indeed, the region of Central Asia does appear to fit into representations of "backwardness" since Central Asian countries are not among the developed ones. The situation became even more aggravated after the collapse of the Soviet Union, when the Central Asian region as a whole experienced profound socio-economic and political crises. Due to the lower standard of living, lower expenditures in health and since the state support is firstly channelled towards large bio-medical institutions, leaving primary care under-funded (Penkala-Gawetcka, 2002), the health of the populations suffered considerably. Many previously extinct or relatively well-controlled diseases, such as TB and malaria, have reappeared (Keshavjee, 2006).

Yet, the assumptions about backwardness and the consequent greater readiness to re-embrace "traditional" modalities does not take into consideration the dynamic, multiple trajectories of modernity, particularly that the Central Asian countries have already been subjected to "the" project of modernity—socialism (Nov., 1989). Due to the past robust processes of modernisation—urbanisation and industrialisation—many parts of Central Asia are no longer the unspoiled primordial natural oases they are often assumed to be. On the contrary, during the Soviet era, many agricultural areas (like Karakalpakstan, which is today a part of autonomous republic of Uzbekistan) were farmed to such intensity that agricultural pesticide and herbicides continue to be present to this day in many popular agricultural products and foods of animal origin to alarming degrees (Muntean et al., 2003). Environmental damage, underpinned by the increased technological interventions into the environment (for Kyrgiz Republic, see Penkala-Gawetcka, 2002; for Uzbekistan, see Sezik et al., 2004) continues, and affects health of the populations (Everest & Ozturk, 2005). Moreover, the plants traditionally used in many Central Asian medicines might no longer be easily available, because their number is dwindling due to increasing urbanisation and more intensive agricultural practices (Everest & Ozturk, 2005; Sezik et al., 2004).

THE LOSS OF TRADITIONAL KNOWLEDGE

The Soviet state established an extensive and comprehensive educational system and literacy levels continue to be high. Furthermore, because of widely accessible health care in the Soviet Union, people had become less inclined to use traditional modalities (Michaels, 2000) and thus were less likely to protect the environment where they might source naturally derived traditional remedies. Coupled with the attitude of the Soviet authorities, which tried to tie local populations to the state through biomedicine while simultaneously denigrating local traditions as backward (Michaels, 2000), local populations have lost a large degree of knowledge and respect for their traditions and knowledge of their environment. Consequently, this might no longer represent continuation of collective memory (Connerton, 1989; Gills, 1994) and thus further impinges on the use of medical remedies derived through traditional knowledge (Cambridge Central Asia Forum, 2008).

The WHO Strategy also implies the continuous and direct transmission of traditional knowledge and promotes greater respect for traditional modalities, developing cooperation between biomedical and traditional practitioners and giving greater visibility to the practitioners of traditional medicines. Yet, because of high literacy levels not only among the city dwellers, but also the rural population, today the citizens of Central Asian states are more likely to find out about herbal remedies from the internet (for Uzbekistan, see Sezik et al., 2004) and books and friends (Cambridge Central Asia Forum, 2008; Nelson, 2004) rather than from traditional holders of knowledge.

BUYING HEALTH IN UZBEKISTAN AND TAJIKISTAN

Our research also supports these findings. For example, the market stall holder selling herbal remedies in Sogd in Tajikistan did not learn about their traditional use from her relatives or somebody involved in its traditional applications, but from another person in the market selling agricultural products and herbal remedies. He was able to give advice on the use of the remedies. He had learned most from Russian books published after the collapse of Soviet Union by Russian authors with the highest academic degrees. He proudly showed his books to the researchers, but he also stressed his interest in traditional Chinese and Korean medicine, and how he tried to incorporate these insights into his knowledge of healing. Whereas the female stall holder in Sogd sold a limited number of remedies to those who already knew what they needed, and these were not her primary income, the male stall holder collected his own herbs and was able to offer a wider variety of herbal remedies. Due to a greater extent of his knowledge, he was also able to offer more complex types of herbal remedies, such as mixtures of raw herbs.

The situation in the main marketplace in Tashkent in Uzbekistan was somehow different. The researchers were directed to a stall holder who sold agricultural products and various naturally derived remedies. He was also able to give advice on the medicinal use of the plants, even though this was not his primary activity, as his main income came from selling agricultural products. He told the researchers that he learned most of what he knew from one of the tomes of Ibn Sina's (Avicenna's) *The Canon of Medicine*. He showed them the book and stressed how expensive the book had been.

The situation in Sogd and Tashkent, therefore, differed in that the vendor in Tashkent learned most of what he knew from the famous work by Ibn Sina (written in Arabic script), an Islamic scholar born near Bukhara in Uzbekistan, whereas references of the practitioner in Sogd were Russian academic books. Even though this might indicate greater cultural confidence of citizens of Uzbekistan, the fact remains that nobody learned their trade directly from traditional knowledge-keepers. Another such example is a well-known healer in Tashkent whose reputation was almost certainly helped by his high academic degree. Even though he had been strongly involved in informal exchanges of knowledge with other (herbal) practitioners in a wider region of Uzbekistan, Kyrgiz Republic and Tajikistan, he nevertheless learned much of what he knew through his work as a historian, and he freely consulted international popular literature. His work in traditional medicine was, therefore, not a continuation of tradition, but rather its "invention" (Hobsbawm & Ranger, 1983).

We argue that the loss of (direct) traditional knowledge (transfer) in the situation when there is a push towards greater use of traditional remedies has the potential of particularly negatively influencing the environment. For example, the reports (Cambridge Central Asia Forum, 2008) from the project on the use of edible and dyeing plants from Kyrgiz Republic, pointed out the extensive decline of local knowledge, which, in turn, affected environmental issues. The most widely accessible reading materials on medicinal plants in Kyrgiz Republic—Russian books—refer to the pink rodiola (*Rhodiola rosea*) as a medicinal plant. Yet this species does not grow in the Kyrgiz Republic. Instead, people collect a similar plant of entirely different species with no health benefits, believing it is pink rodiola (cf. Everest & Ozturk, 2005; Penkala-Gawetcka, 2002). By intensive collection, people might put the plant on the list of endangered species and because the plant has no medical property, they might loose trust in its health benefits, and interest in its environmental protection.

WESTERN SCIENCE AS ESSENTIALISED AND UNIVERSAL CRITERIA

Western studies of traditional medicine(s) are mostly focused on the search for the active components of the plants and remedies of animal origin, which is also a priority for the WHO Strategy (2002). Yet, this is only a very limited understanding of naturally derived remedies, and it is questionable whether this can be called "traditional medicine" at all. Sasaki (1999), therefore, points out that the belief that chemo-taxonomy and financial rewards must be the main motivators for everybody is Euro-centric, therefore belonging to a more general Orientalist tendency to impose Western science and medicine as universal and essentialist criteria, above all other systems of knowledge. Such an assumption disregards that people might not protect the plants (and animals) because of their user and instrumental value, but because they appreciate their aesthetics or symbolic meanings. This was evident in the case of the desert and riparian parts of Ferghana valley in Uzbekistan, where people protected certain birds simply because they found them beautiful or important for their community's identity, and not because they were useful to them. Well aware of environmental degradation, the local people protected both "useful" and "beautiful" plants and animals by taking initiative and creating a kind of micro-reservations where the natural habitat was left intact by human intervention (Cambridge Central Asia Forum, 2008).

One of the main foci for the WHO Strategy is the research into active components of traditional remedies and assessments of their effectiveness and safety. The WHO strategy therefore essentialises these Western standards as "objective" and thus "universal." We argue that such understanding of the issue is an example of Orientalist intellectual imperialism because it tries to hide that Western medicine is also a cultural system (Kleinman, 1978), and criteria, such as effectiveness and safety, are in fact social constructs and as such, cannot be essentialised. As in CAM therapies on the West, (Potrata, 2004) the practitioner of traditional medicine(s) and his or her client often negotiate what will be the outcome measure of a particular therapy. Whereas for some clients, the main outcome might be to cure the illness, for others it might be to simply alleviate particular symptoms or die in peace.

Various medical systems might also have outcome measures of different importance. Whereas for Western medicine, symptom alleviation is a priority, an Uzbekistani practitioner of Unani medicine might find alleviation of an overly "hot" organ to be equally, if not more, important. Traditional practitioners might prefer effects of traditional remedies to be long-term (Wachtel-Galor, Tomlinson, & Benzie, 2004) in order to avoid nasty side effects, while Western biomedical professionals prefer immediate effects, but sometimes accompanied by additional symptoms (Balick in Cardini et al., 2006). Safety is, therefore, also a social contract rather than an objective criterion.

Furthermore, in many traditional medical systems, like Traditional Chinese Medicine, which is popular in all the visited Central Asian countries (and natural remedies are not that rarely sourced in Central Asia), the practitioners rely on synergic effects of the herbs, rather than their active ingredients (Zhang et al., 2004). Traditional Chinese practitioners and those who use this traditional medicine are, therefore, interested in preservation of a much larger number of plants, minerals and animals, typically used in Traditional Chinese Medicine. Westerners, insisting on rigorous clinical evidence that is currently robust for a very small number of naturally derived medicines, might be interested in preservation of the environment of a much smaller number of naturally derived remedies. Again, the interest in environmental and conservation issues of those natural remedies that are not supported by clinical evidence might also decrease significantly.

Social and Cultural Lives of Naturally Derived Traditional Remedies

One of the main aims of the WHO Strategy (2002) is to look for scientific, "objective" evidence base for the use of traditional remedies. This goal does not take into consideration that efficacy of particular plants might not depend on their clinical effectiveness, but rather on social and cultural meanings that are ascribed to them. Moerman (2007) argues in case of medicinal plants that they, like placebos, are able to exert powerful physiological effects because of their cultural or symbolic powers. Because of their different socio-cultural meanings, the same plants might also be used for treatment of very different social and cultural ills.

This was the case in markets visited in Tajikistan's Sogd and Uzbekistan's Tashkent. Even though more plants of medicinal value were sold in the Tashkent's marketplace, a vast majority of plants and natural remedies sold in the cities visited in both countries were exactly the same. However, whereas most of the medicinal remedies in marketplace in Sogd were for the treatment of "depression," "nerves," and "pains in the heart," the very same plant species in Tashkent had a much wider and quite different spectre of uses, ranging from kidney, liver, and heart diseases to gynaecological problems, improvement of memory, sexual potency, sinusitis and for the treatment of diabetes.

For example, the stall holder in Sogd recommended the boiling of cloves (*Eugenia aromatica*) and using the liquid for "calming down the nerves," while the vendor in Tashkent recommended clove for bad breath, stronger gums and improving eyesight. Even though the same plants are frequently used in very different aims in other cultural contexts, it is worth noting that these traditional remedies were recommended for treatment of very different illnesses and aimed to alleviate very dissimilar historical and social circumstances—in Sogd for alleviation of anxiety of those who have experienced the trauma of recent civil war in Tajikistan, while "civilisation" diseases, like diabetes, were a sign of greater general standard of living in Uzbekistan. Perhaps the spice was efficacious both to calm down the nerves and in treatment of oral cavity and eyesight, only one of them, or neither. Yet, both those who believe in its efficacy as a sedative and in treatment of gum conditions and eyesight are interested in protecting the environment of this spice, even though it does not grow in this particular area.

Other socio-cultural factors to be taken into consideration when studying traditional medicines are the social meanings of health practices and systems themselves. Using Bourdieu's (2007) argument that in a consumer society class is no longer a sole function of socio-economic characteristics, but crucially depends on what is considered proper consumer choices and "tastes," certain (socio-economic) groups or those who aspire to them might base their preferences on the "distinction" that particular practices might bring them. By the same token, the use of traditional medicines might not depend on their objective effectiveness, safety and availability, but the users perhaps shun them because they convey the meanings of lower class and unrefined cultural tastes. For example, Linquist (2002) argues that in contemporary post-socialist Russia, the clients might be reluctant to consult a healer using rural healing traditions if they are perceived as being from low-status, uneducated social groups with no refined (cultural) interests. Middle-class Russian clients are, therefore, reluctant to use rural healing practices because these were associated with negative social meanings.

The situation in Central Asia was very similar because Soviet authorities ran intensive public health campaigns presenting Soviet biomedicine as progressive and prosperous, and traditional medicines as contributing to continuous backwardness and disease (Michaels, 2003). Local traditional remedies are therefore rated lower than foreign ones, despite their effectiveness. Indeed, both the pharmacists in Sogd and Tashkent said that given the same price, their clients would rather buy Western over-the-counter remedies than Russian and regional ones. Yet in Sogd, the pharmacists were convinced that any but Central Asian remedies were better than local ones, and they stocked Russian, Ukrainian, Pakistani and Indian herbal preparations. The pharmacists in Tashkent also thought that their clients preferred Western remedies and those coming from Russia or Ukraine, but they nevertheless showed greater cultural confidence and stocked a larger number of herbal remedies from Uzbekistan.

Tendency to Study the Exotic

The WHO Strategy (2002) explicitly encourages investigations into cultural and social factors involved in the use of traditional medicines. Even though we have already argued that one of the weaknesses of the WHO Strategy is the disregard for the social and cultural context in which traditional medicines are practiced and used, or socio-cultural meanings that are attached to remedies and medicines, the examinations into cultural and social factors might also mirror Orientalist tendency to study "exotic" and apparent "eccentric" religious and ritualistic practices that might accompany traditional healing modalities, if it leaves out an important research area that might be more relevant to the health of Central Asian communities.

We refer here to those (traditional) natural remedies that are used as home remedies, sold as nutritional supplements and over-the-counter-medicines. Several researchers have reported that especially home remedies have indeed become a valuable means for communities' health treatments in Central Asia (K Kuehnast, 1998; Nelson, 2004; Penkala-Gawetcka, 2002), whereas nutritional supplements and over-the-counter medicines are one of the fastest growing pharmaceutical areas in many other social contexts (Baran, Teul, & Ignys-O'Byrne,

2008).Yet these are not mentioned as the WHO Strategy's areas of interest, and systematic studies of pharmacies selling over-the-counter medicines, or the use of home remedies in general, are rather scarce (Qato et al., 2008).

As they appear mundane and perhaps even belonging more to the field of pharmacology, they easily escape attention of social scientists. Their role is important not only because home remedies, over-the-counter medicines and various nutritional supplements are one of the first strategies in health-seeking behaviour, but also because they also represent a very complex phenomenon. As data on Central Asia do not exist, we use findings from the study of Jones (2006), which was conducted in another challenging social context, South Africa. Jones found out that people use home remedies in up to 80% of their health complaints. Rather than just using traditional African herbs, he reported complex and sophisticated uses that included over-the-counter medicines, Dutch propriety medicines and Western pharmaceuticals.

Disregarding home remedies and over-the-counter medicines might thus leave out a vast health-seeking area where people use (traditional) naturally derived remedies. On one hand, the people might use the plants that they have close by, and they will have interest in preserving and protecting them. Onother hand, they might use remedies that are important for their health, for which clinical evidence of effectiveness and safety is available, but they have no influence of the environment from which they derive. This complex phenomenon, therefore, deserves much more attention than it usually receives.

Conclusion

The WHO global strategy on traditional medicine (2002) was articulated in consultation with all its member states, including the ones that would constitute the "Orient." Yet in this chapter, we argue that despite the input of many countries, it nevertheless implies a strong and consistent Orientalist bias. Since here we consider health as an intersection between humans and their natural and physical environment, this bias affects not only the area of health, but issues on conservation and biodiversity as well.

The WHO Strategy disregards that "medicine" does not exist in a social vacuum—even contemporary medicine underpinned by sophisticated science is a cultural system (Kleinman, 1978), and thus only one of the many. The WHO Strategy's term "traditional medicine" as a singular term is, therefore, Orientalist and disregards the vast differences existing outside the Western world, which are derived from natural resources that might be more relevant for the health of the populations. It is also based on potentially false assumptions, for example, that the people in developing world have full access to the naturally derived remedies and know well what to do with them.

Such Orientalist assumptions discount multiple trajectories of modernity and processes of globalisation. We argued that environment in many developing countries has degraded considerably, thus naturally derived remedies are often no longer easily available. The erosion of directly transmitted local knowledge is particularly detrimental to environmental issues. As people have lost, to various degrees, the knowledge of and respect for plants and other natural remedies in their vicinity (and also because of economic necessity), they over-pick the plants and over-source the minerals, collect wrong plants with little or no medicinal value, and, as these are not effective, further lose respect for such remedies and disregard the environment in which these are found.

We point out that it is therefore (directly) transmitted local knowledge that makes the connection between health, humans and their physical and natural environment meaningful. Rather than focusing all attention into chemo-taxonomy analyses and assessments of effectiveness and safety according to Western, apparently universalist (Orientalist) standards, documenting local knowledge and practices, combined with the search for effective (as assessed by the Western standards), biologically active plants appears to be a more fruitful avenue for both health and environment. It should not be forgotten that even those plants that might not pass scientific tests might be effective and safe for, and meaningful to, those who use them, and are therefore interested in protection of their environment.

Another crucial element is transmitting local knowledge through the educational system to the future generations and through informal knowledge systems to preserve traditional knowledge and keep the connection among health, humans and their environment meaningful. Not only respect and cooperation between biomedical and traditional practitioners, but respect for and a focus on direct transmission of local traditional knowledge should be one of the priorities of future WHO strategies on traditional medicines.

One of the challenges that might affect the use of traditional remedies is different social meanings, which might be ascribed to different naturally derived remedies and medicines as systems. Some medicines might be preferable not because people know them, have easy access to them, or are effective, but because they carry with them social meanings that people want to identify with. We argued that people might be interested in using Western herbal medicine because they want to identify with Western versions of modernity, not because their local herbal medicines are ineffective. This means that people do not have an interest in protecting the environment of naturally derived medicines over which they have very little influence, and they disregard their own remedies and lose interest in conservation of their environment.

"Traditional medicine" is therefore an empty term when not carefully examined among those who use it and in the context(s) in which it is used. Whereas such a holistic approach appears to require qualitative methodology, especially in an anthropological approach, it is necessary to counter-balance this research strategy and Orientalist tendency to look for the exotic and disregard in this way the most common use of naturally derived remedies with other research methods, including quantitative ones and natural sciences approach.

Another crucial element to circumvent many of the problems in studying traditional medicines is to bring into research those from the studied area as partners on an equal footing. This is because our local research partners will continue to remind us about the particularities of their research area and related issues and help us steer away from what might appear to us as a colourful phenomena, but which, according to their research agendas, is of little importance to the local people, and might not be relevant when in the transfer of good practices to our own social environments.

References

Arnold, A. (1999). "An Ancient Race Outworn": Malaria and Race in Colonial India, 1860–1930'. In W. Ernst & B. Harris (Eds.), *Race, Science and Medicine, 1700–1930* (pp. 123–143). London: Routledge.

Arnold, D. (1993). *Colonizing the Body: State Medicine and Epidemic Disease in Nineteenth-century India*. Berkeley: University of California Press.

Baran, S., Teul, I., & Ignys-O'Byrne, A. (2008). Use of Over-the-counter Medications in Prevention and Treatment of Upper Respiratory Tract Infections. *Journal of Physiology and Pharmacology, 59*(Suppl. 6), 135-143.

Bourdieu, P. (2007). *Distinction: A Social Critique of the Judgement of Taste*. Cambridge, MA: Harvard University Press.

Cambridge Central Asia Forum. (2008). Documenting Local Knowledge in Central Asia. *http://www.cambridge-centralasia.org/?page_id=59* Accessed 1.5. 2009.

Cardini, F., Wade, C., Regalia, A., Gui, S., Li, W., R, R., et al. (2006). Clinical Research in Traditional Medicine: Priorities and Methods. *Complementary Therapies in Medicine, 14*(4), 282-287.

Çeçen, A. (1999). Uzbekistan between Central Asia and the Middle East. In E. Ka (Ed.), *Rethinking Central Asia: Non-Eurocentric Studies in History, Social Structure and Identity*. Reading: Ithaca Press.

Connerton, P. (1989). *How Societies Remember*. Cambridge: Cambridge University Press.

Everest, A., & Ozturk, E. (2005). Focusing on the Ethnobotanical Uses of Plants in Mersin and Adana Provinces (Turkey). *Journal of Ethnobiology and Ethnomedicine, 1*(1), 6.

Gills, J. (Ed.). (1994). *The Politics of National Identity*. Princeton: Princeton University Press.

Godlewska, A., & Smith, N. (1994). *Geography and Empire* Oxford: Blackwell.

Guha, S. (2001). *Health and Population in South Asia from Earliest Times to the Present*. London: Hurst and Co.

Harris, B., & Ernst, W. (1999). *Race, Science and Medicine, 1700-1960*. London: Routledge.

Harrison, M. (1999). *Climates and Constitutions: Constitutions: Health, Race, Environment and British Imperialism in India 1600–1850* New Delhi: Oxford University Press.

Harrison, M. (2005). Science and British Empire. *Isis* (96), 56-63.

Hobsbawm, E., & Ranger, T. (1983). *The Invention of Tradition*. Cambridge: Cambridge University Press.

Jones, S. (2006). From Ancestors to Herbs: Innovation According to "Protestant Reformation" of African Medicine. *Cultural Transitions*(1), 177-197.

Keshavjee, S. (2006). Bleeding Babies in Badakhshan. Symbolism, Materialism and the Political Economy of Traditional Medicine in Post-Soviet Tajikistan. *Medical Anthropology Quarterly, 20*(1), 72- 93.

Kleinman, A. (1978). Concepts and a Model for the Comparison of Medical Systems and Cultural Systems. *Social Science and Medicine* (12), 85-93.

Kuehnast, K. (1998). From Pioneers to Entrepreneurs: Young Women, Consumerism, and the "World Picture" in Kyrgyzstan. *Central Asian Survey, 17*(1), 639-654.

Lindquist, G. (2002). Haling Efficacy and the Construction of Charisma: a Family's Journey through the Multiple Medical Field in Russia. *Anthropology and Medicine, 9*(3), 337-358.

Michaels, P. (2000). Medical propaganda and cultural revolution in Soviet Kazakhstan, 1928-41. *The Russian Review* (59), 159- 178.
Michaels, P. (2003). *Curative Powers: Medicine and Empire in Stalin's Central Asia*. Pittsburgh: University of Pittsburgh Press.
Moerman, D. (2007). Agreement and Meaning: Rethinking Consensus Analysis *Journal of Ethnopharmacology, 112*(3), 451-460
Muntean, N., Jermini, M., Small, I., Falzon, D., Fürst, P., Migliorati, G., et al. (2003). Assessment of Dietary Exposure to Some Persistent Organic Pollutants in the Republic of Karakalpakstan of Uzbekistan *Environmental Health Perspectives, 111*(10), 1306-1311.
Nayak, M. (2006). Orientalism and "Saving" U.S. State Identity after 9/11. *International Feminist Journal of Politics, 8*(1), 42 - 61.
Nelson, M. (2004). From Red Yurts to Capital Girls: Contextualizing the Use of Herbal Medicine and Re-emerging Medical Pluralism in Urban Kyrgyzstan. University of Kent.
Nove, A. (1989). *The Economics of Feasible Socialism*. London: George Allen & Unwin.
Penkala-Gawetcka, D. (2002). Korean Medicine in Kazakhstan: Ideas, Practices and Patients. *Anthropology and Medicine, 9*(3), 315-336.
Potrata, B. (2004). "Does It Work?" Holistic Healers and Assessing Breast Cancer Treatment. In S. Sered (Ed.), *Religious Healing in Boston: Body, Spirit, Community* (pp. 13- 16). Cambridge, MA: Harvard University Press.
Qato, D., Alexander, G., Conti, R., Johnson, M., Schumm, P., & Lindau, S. (2008). Use of Prescription and Over-the-counter Medications and Dietary Supplements Among Older Adults in the United States. *JAMA, 300*(24), 2867-2878.
Rasanayagam, J. (2006). Healing with Spirits and the Formation of Muslim Selfhood in Post-Soviet Uzbekistan. *Journal of Royal Anthropological Institute* (NS 12), 377-393.
Robb, P. (1995). *The Concept of Race in South Asia* Oxford: Oxford University Press.
Said, E. (1979). *Orientalism*. New York: Vintage.
Said, E. (2003). *Orientalism*. London: Penguin.
Sasaki, K. (1999). Beyond the Conventional Dichotomy/ Dualism: Occidental Value and Oriental Value *Eubios Journal of Asian and International Bioethics* (9), 74- 76.
Sezik, E., Yesilada, E., Shadidoyatov, H., Kulivey, Z., Nigmatullaev, A., Aripov, H., et al. (2004). Folk medicine in Uzbekistan: I. Toshkent, Djizzax, and Samarqand Provinces. *Journal of Ethnopharmacology, 92*(2-3), 197-207.
Sheehan, H., & Hussain, S. (2002). Unani Tibb: History, Theory, and Contemporary Practice in South Asia. *Annals of the American Academy of Political and Social Science, Global Perspectives on Complementary and Alternative Medicine* (583), 122-135.
Timmermans, K. (2003). Intellectual Property Rights and Traditional Medicine: Policy Dilemmas at the Interface *Social Science and Medicine, 57*(4), 745-756
Turner, B. (1989). From Orientalism to Global Sociology. *Sociology* (23), 629-638.
Wachtel-Galor, S., Tomlinson, B., & Benzie, I. (2004). Ganoderma Lucidum ("Lingzhi"), a Chinese Medicinal Mushroom: Biomarker Responses in a Controlled Human Supplementation Study. *British Journal of Nutrition, 91*(2), 263-269.
Waldstein, A., & Adams, C. (2006). The interface between medical anthropology and medical ethnobiology. *Journal of Royal Anthropological Institute* (NS), S95-S118.
WHO. (2002). Traditional Medicine Strategy 2002- 2005. *http://whqlibdoc.who.int/hq/2002/WHO_EDM_TRM_2002.1.pdf* Retrieved 5. 3. 2009.

World Bank. (2009). Tajikistan Country Brief. *http://lnweb90.worldbank.org/ECA/eca.nsf/Countries/Tajikistan/643F47CBD6B72A8A85256C2500671AF8?OpenDocument* Retrieved 1.5. 2009.

Zhang, B., Hu, P., Qiu, H., Qian, S., Chen, X., & Huang, G. (2004). Observation of Clinical Response of "Sheng Ban Recipe" for Platelets Decrease after Chemotherapy. *Ai Zheng, 23*((11 Suppl)), 1470-1472.

Part III. Policy and Society: In Search of Solutions

In: Health and Environment: Social Science Perspectives ISBN 978-1-60876-216-3
Editors: Helen Kopnina and Hans Keune

Chapter 9

SOCIAL SCIENTIFIC NAVIGATION IN ENVIRONMENT AND HEALTH COMPLEXITY

Hans Keune, Bert Morrens and Ilse Loots

ABSTRACT

Two cases of complexity in relation to environment and health are discussed in order to illustrate what complexity is and how we may deal with complexity. Climate change is discussed by focusing on the work of the Intergovernmental Panel on Climate Change and examples of practical experience as social scientists in the Centre of expertise for Environment and Health in Flanders (Belgium) are introduced. The authors explain how a complexity approach is helpful to the field of environment and health research and how social science can make a relevant contribution. The approach is characterized by modesty, respect and transparency with regard to complexity of reality, and openness to different relevant forms of knowledge and opinion. The theoretical issues of objectivity and the inter-subjectivist nature of science are discussed as well as difficulties of applying the complexity approach in practice. The approach will not make work easier in all respects, but it will enhance the quality by respecting the complexity of real world problems.

INTRODUCTION

In this chapter, two cases of complexity in relation to environment and health will be introduced in order to illustrate what complexity is and how we may deal with complexity. We do this from a social scientific perspective. We introduce the case of climate change by focussing on the work of the Intergovernmental Panel on Climate Change (http://www.ipcc.ch) and our practical experience as social scientists in Centre of expertise for Environment and Health (CEH; http://www.milieu-en-gezondheid.be/English)[1] in Flanders (Belgium). We want to illustrate how a complexity approach is helpful to the field of environment and health research and that social science can make a relevant contribution to solving some of the problems.

Whereas scientific knowledge has evolved enormously in the modern era, also insight about its limitations has grown. These are limitations of understanding and predictive power and limitations on foreseeing possible negative impacts of some applications of scientific knowledge. Within science, more and more attention is being raised regarding the complexity of phenomena that are of interest to science, policy and society, such as environmental issues. Critical questions are being raised on traditional scientific approaches: how does traditional reductionist[2] science relate to a complex reality? How can science deal with its ever-growing amount of increasingly rich data, its growing capacity of extensive data collection and data analysis? How sharp is the contrast sometimes to getting a grip on complex reality: more information about a complex system does not necessarily result in better understanding. The social and policy relevance of the traditional scientific approach (both in natural and social science) is questioned ever more, in line with earlier critics/discourse of forerunners in interdisciplinary research and boundary work of environmental studies in de '70s and '80s (Leroy, 2004).

How can science, policy and society deal with complexity, especially when complex issues are of great importance to society? We will focus on the question of how we can deal with the inherent complexity of environmental and health issues in practice.[3] How can we do this in a relevant and transparent manner? Does recognition of limited knowledge compel us (except for further research) to passivity with regard to action? How can we communicate about this in society and policy debate? After this introduction, we will focus on the meaning of complexity. We do this by briefly introducing some theoretical notions and presenting two cases: one on climate change and one on environment and health research. We will then introduce some characteristics of complexity science and deal with issues of problem-framing, objectivity and reduction. After that, we will focus on the question of how we can deal with complexity. We discuss how social scientists and policymaking can cooperate and how social involvement can contribute to dealing with environment and health complexity. We discuss the approach of post normal[4] science (and policymaking) and compare an open attitude with a closed expert attitude. We introduce a precautionary perspective and compare how the right to know may be thwarted by fear of panic. Finally, we present some practical reflections based on our discussion of complexity.

---

[1] The Centre of expertise for Environment and Health was funded and steered by the Flemish government. In this Centre, environmental health experts from all Flemish universities and two research institutes jointly investigate the complex relationship between environment and health. In addition, a social scientific expert unit is part of this Centre. After a first programme period, the Centre was granted continuation for the next five years (2007-2011), again including the social scientific unit. The studies of the Flemish Center of Expertise on Environment and Health were commissioned, financed and steered by the Ministry of the Flemish Community (Department of Economics, Science and Innovation; Flemish Agency for Care and Health; and Department of Environment, Nature and Energy).

[2] Traditional reductionist science tries to understand complex phenomena by investigating parts of this complexity. By understanding its parts, it is believed that one can understand the whole as a causal sum of its parts. Compelxity theory opposes this view.

[3] By practice, we refer both to scientific and policy practice, especially with regard to policy relevant or action research.

1. What Is Complexity?

Complexity Theory

Different definitions of complexity circulate, both in natural and social science. Some definitions are similar to each other; some are rather different (Cilliers, 1998, 2005a, 2005b; Byrne, 1998; Funtowicz et al., 1999; Urry, 2005; Richardson, 2001; Richardson et al., 2005; Nowotny, 2005; Netherlands Organisation for Scientific Research, 2008). An important common feature seems to be that the complexity of a complex phenomenon by definition cannot be fully described or understood. The concept of "full," of course, can be debated. But is this discussion relevant? The complicatedness of something perhaps implies that describing and understanding is limited, but need not exclude the possibility of a useful framing. Another characteristic of complexity is the presence of a great number of (often simple) system components that interact in a manner that cannot be explained by the characteristics of the individual components is often mentioned. This is called emergence. To make things even more complicated, Cilliers (2005a) states: "*More than one description of a complex system is possible. Different descriptions will decompose the system in different ways. Different descriptions may also have different degrees of complexity.*" This, of course, also may be true for some simpler systems. We should thus especially remember that complexity is largely irreducible due to emergence. Interesting examples are to be found in the natural sciences (e.g., chaos theory) and social sciences (the chaos of the current financial crisis). It is, by the way, a misunderstanding to interpret these two examples of chaos as having equal meaning in relation to complexity. The natural scientific meaning of chaos in chaos theory[5] (see e.g., Aula, 1999) is an example of complexity and order, and not of chaos as it is normally understood. The apparent absence of understanding and control[6] thus should not be attributed to the complex system, but to the lack of understanding and control of human beings or society as a whole. The chaos of the present financial crisis thus need not be contrary to the self-organizing capacity of a complex system, of which the financial crisis is a symptom. The complex non-linear interaction between system components is a more influential explanation for the powerlessness felt world wide with regard to the global economy.

We now introduce two examples of complexity: climate change and environment and health.

The Complexity of Climate Change

Climate change science and policy deal with the question of whether the modern global climate is changing due to human activities like energy production/use, traffic and industry. Important issues that cause concern are, for example, global warming and sea level rise. An important feature of scientific debate on climate change is scientific uncertainty. This is

---

[4] Post-normal science (developed by Funtowicz and Ravetz) refers to an alternative scientific approach rather than traditional (normal) science; we will elaborate on this later on in this chapter

[5] The academic meaning of chaos is different to as how it is commonly understood; academically in relation to complex systems: "chaos refers to the complex behavior of a system that has the appearance of randomness but is nevertheless governed by its deterministic properties" (Aula, 1999)

[6] Control in the sense of knowing how to and to be able to steer (the development of) a system to a preferred state.

mainly caused by the complexity of the phenomena involved (Van Asselt and Rotmans, 2002). Important characteristics of complex systems (e.g., emergence, non linearity) are applicable to climate change and imply that (certain forms of) scientific uncertainty and incomplete knowledge of the lack of capacity of predicting or understand complex developments, are inherent to climate science. Van Asselt and Rotmans (2002) mention different forms of uncertainty and present examples in relation to climate change. As an example of uncertainty due to variability (things being variable), they mention the non-linear and unpredictable character of natural phenomena like clouds and oceans, the diversity of norms and values and perceptions with regard to issues relevant to climate change, and the non-rational behaviour of human beings in relation to, for example, energy use. The debate on uncertainty has become a debate in itself and receives special attention e.g., in the framework of the *Intergovernmental Panel on Climate Change* (IPCC) (IPCC, 2005; Moss and Schneider, 2000; Risbey and Kandlikar, 2007; Swart et al., 2008). The IPCC (http://www.ipcc.ch) was established in 1988, by the World Meteorological Organization (WMO) and the United Nations Environment Programme (UNEP) in order to evaluate the risks of climate change. The panel exists of hundreds of scientists from all over the world, collecting scientific, technical and socio-economic information relevant to the interpretation of the risks of climate change. Next to scientists, also the governments of the member states of the United Nations are involved. Because of its intergovernmental character, important decisions and reports of the IPCC have to be accorded by both scientists and governments.

In the first and second Assessment Report of the IPCC, uncertainty assessment was not centrally or systematically dealt with (Swart et al., 2008). It was up to individual scientists as to how to deal with these issues. The third Assessment Report (IPCC 2001) had a "*guidance paper*" attached with regard to uncertainty with some general guiding principles for consistent use of terminology on uncertainty proposed (Moss and Schneider, 2000). Allen et al. (2001) mention three important causes for uncertainty with regard to the third Assessment Report (IPCC 2001). 1. Socio-economic trends (and related emissions) are hardly or unpredictable. 2. The difficulty of reaching consensus on quantification of e.g., climate sensitivity. 3. The possibility of a nonlinear response in the carbon cycle or ocean circulation to very high late-21st-century greenhouse gas concentrations.

According to Giles (2007), in the IPCC, confidence in climate science has grown due to the fact that several earlier predictions proved to be correct. Giles mentions the example of global warming. Schiermeier (2007) mentions the examples of frequency and heath of hot days and the increased change of heavy rainfall. According to Schiermeier, there are still several uncertainties that remain in science on climate change. This concerns the carbon cycle, the ocean's ability to absorb more carbon dioxide and sea level rise.

The Complexity of Environment and Health

The relation between environment and health is very complex and is, to a large extent, unknown (as was already illustrated in the chapter by Keune and Van Larebeke: Appendix). In the European Union, tens of thousands of chemical substances are on the market. For a number of individual toxic substances, the health effects of high doses are well known. Unknown are the effects of small doses of different substances over a longer period of time. Also unknown are the combined effects of different substances. There are clues, though, about DNA-damage, hormone disruptions, loss of sperm quality and risk of cancer (McCally, 2002, Harremoës et al., 2002).

It is not always possible to prove unambiguously that there is a relation between environmental pollution and specific health effects (Shriver et al., 2008). Scientific assessments of environmental health risks are being faced with large (partly irreducible) uncertainties, knowledge gaps, and imperfect understanding, and often go hand in hand with conflicting claims and scientific controversy (Funtowicz and Ravetz, 1990; Ravetz, 2002; McCally, 2002; Harremoës et al., 2002). We will briefly discuss these concepts. Scientific uncertainty may be understood as a container for different types of uncertainty in our knowledge about reality. Van Asselt and Rotmans (2002) differentiate on the basis of variability and limited knowledge. Variability is to be understood as a collection of aspects of complex reality that are difficult to grasp or understand, such as system elements and complex interaction between those elements. We can think of natural and social phenomena. Van Asselt and Rotmans call this objective uncertainty. The complexity of these issues largely will remain in the dark, precisely because of its complex character. Limited knowledge, according to Van Asselt and Rotmans (2002), is mainly connected to either the scientific researcher or to the epistemological debate about knowledge. On the one hand, we can think of limited measurability due to measurement errors, lack of data or immeasurable data. On the other hand, we can think of unknowns. Or, as Rumsfeld (2002) put it: *"There are known knowns. There are things we know that we know. There are known unknowns. That is to say, there are things that we now know we don't know. But there are also unknown unknowns. There are things we do not know we don't know."* When we moreover realize there to be a difference between measuring and understanding, we introduce a next nuance of limited knowledge: limited understanding. Depending on which question of meaning we project on the problem data, we can roughly differentiate between on the one hand knowledge about how reality works, e.g., the mechanism explaining how environmental pollution instigates health effects, and on the other hand, the social meaning of conclusions about problems, such as the weighing of the importance of a public health risk and the economic benefit of a risk producing activity. Harremoës et al. (2002) distinguish uncertainty and ignorance from a precautionary perspective. They qualify these concepts in terms of knowledge about the kind of effects and the probability of the actual occurrence of such effects. In case of risks, we have known effects with known probabilities; in case of uncertainty, we have known effects with unknown probabilities; and in case of ignorance, both the effects and the probabilities are unknown. Apart from being natural scientifically complex, risks are also socially complex because they are interwoven with our way of life, our perceptions and interests, our norms and values (Covello, 1991, Slovic, 1998, Renn and Rohrmann, 2000).

In 2003, HK (the first author of this chapter) organized semi-structured interviews with both scientists and policy representatives involved in the work of the CEH. The main topic was to discuss the genetic research that was part of the research of the CEH. Next to this specific topic, more general topics with regard to environment and health science and policy were also discussed. In the discussion, complexity, social relevance, boundary work between science and policy and the role of social science were also discussed. Our natural scientific colleagues of the CEH presented some very interesting practical examples to illustrate the complexity of policy relevant research on environment and health (Keune & Goorden, 2004).

First, we present an example one of the interviewed scientists mentioned with regard to complexity: fish. Eating fish is good for your health because of some specific ingredients fish contains. It helps to prevent heart and coronary diseases. In Belgium, the amount fish that is eaten is nevertheless quite low. Therefore, government should probably stimulate people to eat two to three times more fish. At the same time though, eating fish is probably unhealthy, because pollutants, such as PCB's, contaminate fish these days more and more. We probably have to say, because the scientific basis for this assessment is still not very stable: the relation between exposure to PCB's through eating fish and health problems such as cancer, is scientifically not yet very clear. So, what should the government decide? Apart from being a dilemma for public health policy, this will also be a dilemma for environmental and economic policy.

A second example with regard to complexity concerns the ozone problem. Here, scientific complexity seems less of a problem according to another of the interviewed scientists. It is well known that certain groups of people are extra vulnerable to ozone exposure. In the past, according to this scientist, governments stated that the ozone policy should be tuned to these most vulnerable groups. Consequently, a severe reduction of car traffic might be a wise policy measure from this perspective. In Belgium, this has never happened before though. Apparently, scientific knowledge and praiseworthy ideals are not sufficient to solve the problem; it remains complex, socially complex in this case.

The complexity of issues concerning environment and health can be characterized both as a matter of scientific research, "measuring the issue," and social choice, "what is important when dealing with the issue." What was striking in interviews with our colleagues (both scientists and policy representatives; Keune and Goorden, 2004) was the use of complexity as an argument: both pro and contra (some forms of) environment and health research. Proponents seem to look at complexity as a necessary challenge that will have to be dealt with in order to make the work policy relevant. Critics look at complexity as a weakness. Whether the biomonitoring research will result in hard evidence is questionable to some. Some argue that it will only generate hypotheses for further research. Proponents, though, think it will contribute pieces of evidence: the more pieces, the more proof.

2. Complexity Science

A clear view on the characteristic properties of complex issues can be hindered by isolating elements from complex reality from their context and studying them without taking into account the interaction of those elements with other components of complex reality. This, of course, poses the problem of, on the one hand, the incapacity to fully grasp complexity of real world problems and on the other hand, the need for a reduction of real world complexity to make research practically performable. Practice from the CEH clearly is an example. Not all aspects of environment and health can be investigated in one research project. By designing the research, one also frames the problem. Not only practical and scientific considerations play a role in the case of the CEH, the political ones also interfere. At the start of the CEH, scientists had a different research design in mind than the one that eventually came into being. Due to discussions with policy representatives, the design changed e.g., with regard to research areas. One of the researchers: "*It also has disadvantages doesn't it? (HK:*

cooperation with policy representatives) Actually, we have made concessions to our client in designing the research. (...) Of course, one is responsible for the concessions one makes. But afterwards, one may have scientific doubts, did we do the right thing, e.g., in the case of so many research areas? Originally, it was our idea to take Flanders and two problem areas.[7] *(...) And of course, in those discussions, the voice of the client is strong, as they pay for the research*" (Keune and Goorden, 2004).

Also within the IPCC, science and policymakers cooperate closely. The involvement of politics in the IPCC, according to Giles, behind the scenes results in intense debate between scientists and policy representatives. On the basis of a draft summary document with the most significant findings, produced by the scientists, policy representatives enter debate. Representatives of the scientists are present as to, according to Giles: *"(...) make sure vested interests didn't distort the science."* In the most recent debate on the last IPCC report (2007), according to Giles (2007), China and Saudi Arabia put political pressure to discard the assertion that global warming induced by human activities is at least is five times higher than that by the sun. Nevertheless, this statement survived in the IPCC report.

Complexity and Objectivity

The issue of framing is quite important with regard to complexity and science in general. In order to deal with complexity, by definition we have to reduce complexity, since we can never fully encompass complexity. This problem of choice has consequences concerning the traditional concept of objectivity. Traditionally, scientists are portrayed as being objective neutral professionals active in the quest for objective knowledge: the truth is out there. The "*Gatekeepers of Truth*" as Richardson (2005) calls them (the natural scientists) or "robots" without personal opinion (Smaling, 2008). This idea of truth-dust being dug up from the mines of reality by objective miners is criticized. Reality is too complex to justify the claim by some to have sole access to the truth, and thus denying any critical perspective. They *"(...) keep us in a kind of false consciousness by not restoring the world to its original difficulty. It is only by acknowledging that we are in trouble that we can start grappling with the complexities around us*" (Cilliers, 2005a). This is especially true for complex issues such as environmental problems (Norgaard and Baer, 2005), where truth may be considered "*a luxury, indeed an irrelevance*" (Funtowitcz et al., 1999). "Truth" or "reality" is framed: "*(...) knowledge is provisional. We cannot make purely objective and final claims about our complex world. We have to make choices and thus we cannot escape the normative or ethical domain*" (Cilliers, 2005a). Do we hereby have to leave the idea of "objectivity" altogether? This is a matter of definition.

Smaling (2008) refers to Neurath (1933) and Popper (1934, 1945) as early birds in characterizing the quest for objective knowledge to be inter-subjectivist by nature: objectivity is the product of (scientific) discussion. Latour (2000) even takes us back to "good old-fashioned laboratory conditions" to make a similar statement: "*Objectivity does not refer to a special quality of the mind, an inner state of justice and fairness, but to the presence of*

---

[7] In the end, it was decided on to investigate eight problem areas: two urban areas (Ghent and Antwerp), four areas with different types of industry (harbour, non-ferrous smelter, chemical industry, waste incinerator), a rural area (the countryside), and area near fruit orchards.

objects which have been rendered 'able' (...) to object to what is told about them (...). A laboratory experiment is a rare, costly, local, artificial set up in which it becomes possible for objects to become relevant for statements made by scientists (...)." We need not limit this view on objectivity to the study of "things": "*The paradox is that when quantitative social scientists imitate the natural sciences they avoid precisely those features that would render their discipline really objective*" (Ibid).

When objectivity is open for discussion, does this imply that *anything goes*? Need we splash ourselves in a post-modern bath of relativism and vagueness? Allen and Torrens (2005) rightfully state that "*(...) we would almost certainly prefer to fly in an aeroplane designed by engineers rather than post-modernists (...).*" The question is: does this hold for complex issues, in which experts are blinded by irreducible uncertainties and unknowns? Complex issues, moreover, in which the normative dimension is important, such as in environmental issues? Relativism implies that there are no grounds for difference in quality to be attributed to different forms of knowledge. That is, no other grounds than personal opinion. Such radical social constructionist or post-modern perspective is criticized by Cilliers (2005a): "*'Limited' knowledge is not equivalent to 'any' knowledge. If this were so, any modest claim, i.e., any claim with some provisionality or qualification attached to it, would be relativistic. (...) Modest claims are not relativistic and, therefore, weak. They become an invitation to continue the process of generating understanding.*" On the issue of vagueness, Cilliers (2005a) also proposes a modest perspective: "*This does not imply that we can know nothing about complex systems, or that the knowledge claims we make about them have to be vague, insipid or weak. We can make strong claims, but since these claims are limited, we have to be modest about them.*"

Complexity and Reduction

In 2008, a working document of an interdisciplinary group of Dutch scientists, collaborating in complexity research was published (Netherlands Organisation for Scientific Research, 2008). In one of the lectures held during the workshop in which the document was presented, a biologist stipulated that the clue for dealing with complexity is not building ever more complex and data-rich models and scenarios. Mapping complexity is not the equivalent of understanding complexity, he stated. He pleaded for the use of simple models, to be tested in reality. "*On what grounds then do you design your simple model?*" was asked from the audience, since there are a lot of options to choose from. The questioner, moreover, mentioned several problems, such as the phenomenon that different models might result in the same outcomes. The biologist agreed to these complications and replied that his scientific intuition is the basis for his simple models. When open for discussion and in an endless ocean of options, how then can we tackle (to some extent), *and* at the same time respect, complexity?

"*Objection will take us a step further*" a Dutch scientist/writer stated in 1974 (Koopmans, 1974). An important feature of the complexity approach is the inter-subjectivist nature of the process of knowledge production: science cannot tackle complexity without dialogue. Interdisciplinary and trans-disciplinary cooperation are melodious bells in this respect (Nowotny, 2005, Maasen and Lieven, 2006). In the CEH, we experimented with the input of a diversity of actors with regard to problem/research framing. The aim was selection of hotspots: in the period 2009-2011, a limited number of local environment and health issues will be investigated by human biomonitoring.[8] We designed a procedure to make an inventory of possibly relevant issues so that we can select a limited number for research. We opened up our discussion and invited (other) scientists, policy representatives and stakeholders to propose research options. Moreover, this diversity of actors is involved in the selection procedure. We will not go into details here, but we will merely use this procedure as a practical example from a complexity point of view. We both tried to respect and reduce complexity by opening up debate and by means of a selection procedure. Our natural scientific colleagues mostly agree with the goals and principles of such an approach but tend to hesitate once practical consequences are discussed. They then tend to question the relevance of procedures that involve a diversity of actors, visions and other disciplinary (e.g., social scientific) perspectives than their own. Their main argument is practical complexity. They fear that the research schedule will be in danger, that things will become too complex. This we may call the PR-problem of the complexity approach in practice: embracing complexity will initially not necessarily make life easier (Nowotny, 2005; Funtowicz et al., 1999). The PR-problem especially concerns the social scientific contribution to the complexity approach: a focus on process aspects, uncertainties and diversity of opinion. Whereas our natural scientific colleagues are used to dealing with rather complicated and nuanced routines and quality standards within the natural scientific domain, there seems to be some allergy for "other complicatedness" such as opening up debate to other perspectives. They are not to be blamed for this; they are trained that way (Funtowicz et al., 1999). The perspective of non-scientists or non-experts (stakeholders) is easily disqualified by natural scientists because of the subjectivist nature. This supposes their perspective to be solely of the objective kind. From the inter-subjectivist point of view on objectivity, we may question whether this labelling of objectivity is correct. This traditional *closed* attitude is criticized by the complexity approach (Funtowicz et al., 1999; Stirling, 2008). We will further discuss this later on in relation to policymaking, where this seems even more relevant.

Non-traditional problem-framing thus is suspect to the *creation* of complexity (opening up to a diversity of viewpoints) in accommodating its promise to *tackle* complexity (reduction). The promise of traditional (*modern*) science also seems to be of dual nature with respect to complexity: science *creates* complexity by building ever more complex technologies that produce all kinds of benefits to mankind. At the same time, science seeks to *reduce* complexity when it comes to undesirable consequences of these technologies (Nowotny, 2005). The complexities of the non-traditional complexity approach may be part of the explanation why the comfort of seeming clarity and control of the traditional approach is sometimes more welcome to policymakers, companies and so on. At the workshop of

[8] Human bio-monitoring is the measurement of pollutants and health effects in human beings, using biomarkers. Exposure markers measure the amount of chemicals present in tissue fluids such as blood. Markers for effect measure the possible biological and health impacts of these exposures. Apart from the environment, other factors may influence these markers, such as lifestyle, health status, working conditions, and food intake. If we want to investigate the contribution of environmental exposure to the biomarkers, we need to correct for the contribution of other factors, the confounding factors and co-variables. Many of these co-variables are quantified through questionnaires. In the 2001-2006 project of the CEH, chlorinated hydrocarbons such as PCBs, dioxin-like compounds, persistent pesticides such as HCB, and pp'-DDE (a key metabolite of pp'-DDT), heavy metals (such as lead and cadmium) and metabolites of PAHs and benzene were measured. To determine the possible biological and health impacts of these pollutants, biomarkers of effect, such as cardiovascular markers, markers of DNA damage and of cancer risks, were used. Markers of respiratory problems (such as asthma) were collected via an extensive questionnaire.

Dutch scientists where the working document on complexity (Netherlands Organisation for Scientific Research, 2008) was presented, a research advisor to a large multinational company stated that company management does not like complexity, and thus does not like a complexity approach. The PR-problem of the complexity approach perhaps also explains why the outdated and refuted rational actor model (Ormerod, 2006) in economics is often still favoured over a complexity approach. In the rational actor-model, people or organisations are considered to be perfectly informed and capable all the time to make rational choices based on setting priorities and benefit or profit maximisation.

3. How Can we Deal with Complexity?

The modern promise of clarity and control more and more is loosing its feathers, resulting, amongst other things, in attention for the risks involved in climate change and the rise of the research and policy domain of environment and health. This means that policymakers are confronted with different challenges: how to deal with limited knowledge on socially important but complex issues? Science no longer is a safe haven of certainty; the problems are not easily measured, interpreted, understood or tackled with traditional approaches. The questions poses itself whether society should invest mainly in more science, or to also partly reconcile with the fact that there will never be clear-cut, unambiguous and fully satisfying information on complex issues such as climate change and health risks due to environmental pollution. But, more importantly, the question poses itself as how to deal with this in policy uptake and communication to the general public.

Science and Policy

"*A picture of reality which reduces complex phenomena to their simple, atomic elements can be very effective to controlled experimentation and abstract theory-building. However, it is not best suited for the tasks of environmental policy today. The scientific mind-set fosters expectations of regularity, simplicity and certainty in the phenomena and in our interventions. But these can inhibit the growth of our understanding of the problems and of appropriate methods to their solution.*" (Funtowicz et al., 1999) Collecting data does not automatically generate suitable scientific knowledge: we should not confuse data with knowledge (Cilliers, 2005b). "*Knowledge is interpreted data. This leads us to the next big question: what is involved in interpretation, and who (or what) can do it?*" (Ibid). Nor do data automatically become policy relevant. From a policymaking point of view, different interpretative questions will be put to data about problems than from a purely scientific perspective. Scientists generally will focus more on further description (data collection) and understanding (why does this happen like this?) of phenomena. Actors with policy responsibility, when confronted with socially relevant problems, will be mainly interested in the burden of the problem (e.g., the seriousness of health risks), problem-solving perspectives (how can we deal with a problem?) and the importance of the problem and its solving compared to other issues of social interest (e.g., health impact compared to economic impact).

Hopkin (2007) typifies science in the framework of the *Intergovernmental Panel on Climate Change* (IPCC) as *"consensus science."* The consensus though often masks quite differences of opinion. This is especially the case when it concerns politically sensitive topics. The scientific uncertainties make way for quite some room for manoeuvre for different points of view. More and more, though, in recent years, the consensus on important climate change issues has gained broader support within the IPCC, and the scientific basis for this has become more robust. The most recent report from the IPCC (2007) leaves less room for manoeuvre for sceptics than before, both within and outside the IPCC. Hopkin cites a scientist: "*Their argument continues to shift" (...) "That makes it clear that the issue for them is not the science. Whatever the science is, they will try to find ways to question it..*" According to Hopkin, climate sceptics shift their attention to the economic consequences of policy adaptations with regard to climate problems. The article of Hopkin's clearly demonstrates the social and political aspects of climate change science. Swart et al. (2008) focus on the interdisciplinary character of the scientific exercise and the political context of it. They characterize the assessment of data and knowledge by pointing at both objective and subjective forms of information. The subjective is not only situated in the political issues: the social scientific contribution to the work of the IPCC mainly deals with freedom of choice and the choices made. It is also situated in the qualitative approach of uncertainty: uncertainty within the IPCC is interpreted by assessing the confidence in science. This is done by means of confidence levels; see Figure 1.

Similar to Hopkin (2007), Giles (2007) describes how the scientific foundation for climate change has been strengthened over the years. The next step according to Giles is translation into policy: how can scientists help policymakers to formulate policy measures based on the scientific assessment of problems? Giles cites a scientist pointing at the need for model builders to generate information that can be used by policymakers directly: "*Some of what is in the research domain needs to become operational.*" An important condition for this, according to another scientist, one of the co-authors of the IPCC, is the cooperation between social and natural scientists (Giles, 2007). Also Swart et al. (2008) stress the importance of interdisciplinarity. They criticize the opinion that discovery of "the truth" is only hampered by temporary uncertainties, mainly caused by limited scientific resources. They point at the subjective aspects in the definition of uncertainty.

Confidence terminology	Degree of confidence in being correct
Very high confidence	At least 9 out of 10 chance
High confidence	About 8 out of 10 chance
Medium confidence	About 5 out of 10 chance
Low confidence	About 2 out of 10 chance
Very low confidence	Less than 1 out of 10 chance
IPCC (2007)	

Figure 1.

Social Involvement

In the framework of the Centre of expertise for Environment and Health (CEH), together with medical and environmental scientific experts and policymakers, we developed an action-plan for interpretation and policy measures with regard to the biomonitoring results (Koppen et al., 2005; Keune et al., 2007, 2008a, 2009c). At first, the discussions in the working group mainly focussed on environmental and medical scientific interpretation of the monitoring data. It was thought of as a merely scientific quest: with the right group of experts, the interpretation with regard to policy priorities will follow automatically. While trying to build bridges towards policy interpretation, though, the limitations of an exclusively scientific endeavour clearly showed: no scientist or group of scientists dared to claim to possess the necessary and overarching knowledge that answers difficult questions. Questions e.g., on policy priorities when also other than medical and environmental scientific factors had to be taken into account (economics, social preferences, feasibility of policy measures; issues introduced by the social scientists). The social scientists, therefore, proposed the formation of a stakeholder jury that will judge relevant data and knowledge in order to give advice to the government. This approach can be characterized as an example of the analytical-deliberative approach (Stern and Fineberg, 1996; Stern, 2005) or of the extended peer review (Funtowicz et al., 1999; Pereira and Funtowicz, 2006): next to experts also stakeholders are consulted.

Several approaches on dealing with complexity (though not fully alike) show some similar features. The opening up of debate (Stirling, 2008; Funtowicz et al., 1999) to different fields of expertise, both scientific and non-scientific is one of the key features. We will focus on one of the approaches on clarification: post-normal science (Funtowicz and Ravetz, 1990). Post-normal refers to an approach different from the normal approach of traditional modernistic science and its promise for a sound base for policymaking. With respect to complex issues (such as environmental problems) ''where facts are uncertain, values in dispute, stakes high and decisions urgent'' (Funtowicz and Ravetz, 1991, 1994), the normal (traditional) approach is no longer sufficient.

In the case of the CEH (with regard to risk communication; Keune et al., 2009a), the following was agreed about the involvement of relevant stakeholders: "*A question that needs to be answered is who relevant stakeholders may be. We, for example, think of (local) governmental institutions, citizens, scientists, social sectors and social groups. Apart from this, also the question needs to be answered within which geographical area we look for these groups. Defining and choosing relevant stakeholders to be involved will be dependent on the way the actors responsible for the work of the CEH assess the issues.*" Selecting stakeholders thus also is subject to discussion and negotiation. Next to the geographical marking, also the definition of timeframe in principal is relevant: in many issues also the interests of future generations may be important for considerations.

Post-Normal Science

Normal science is characterized as puzzle solving: "*(...) particular puzzles to be solved within an accepted paradigm or framework of assumptions and methods*" (Funtowicz et al., 1999). In our experience in the CEH, such a traditional "truth-seeking" approach is recognizable in the occasional appearance and disappearance of problems under study. On the basis e.g., of part of the biomonitoring results, further, deeper research is proposed by scientists in order to find out what the causes of problems are and how they can be solved. After diving into the details of some of these very complex issues concerning the impact of environmental pollution on public health, scientists and experts, cannot easily come up with crystal clear scientific conclusions or practical advice to policymakers. Conclusions such as *we cannot proof a causal relation between A and B* or *the problem is in need of further research* are not exceptional. Practically, this may lead to further research but not to policy action, and as such, the problem more or less (temporarily) disappears from the agenda. We interpret it as a symptom both of unpredictable complexities and of the problematic relation between the traditional scientific approach and current complex policy problems. Moreover, they are not solely responsible for this *paralysis by analysis*, since also policymakers have a responsibility in coping with problems that are socially important but inherently or currently show lack of scientific proof or understanding and also can also use research projects for postponing political decision-making. Normal policymaking can be qualified as dependent on the idea that: "*(...) routine puzzle solving by experts provides an adequate knowledge base for policy decisions*" (Funtowicz et al., 1999).

The post-normal approach stresses the importance of the quality of the process. "The criteria for selection of data, truncation of models, and formation of theoretical constructs are value-laden, and the values are those embodied in the societal or institutional system in which the science is being done. This is not a proclamation of relativism or anarchy. Rather, it is a reminder that the decision process on environmental policies must include dialogue among those who have an interest in the issue and a commitment to its solution. It also suggests that the process towards a decision may be as important as the details of the decision that is finally achieved. (…) For these new problems, quality depends on open dialogue between all those affected. This we call an 'extended peer community,' consisting not merely of persons with some form or other of institutional accreditation, but rather of all those with a desire to participate in the resolution of the issue" (Funtowicz et al., 1999).

Opening up versus Expert Closure

The secret of success of dealing with complexity, e.g., by translating scientific knowledge into adequate policy actions, not only is warranted by open processes alone. Stirling (2008) describes how the increasing openness of politics at national and European Union levels towards public participation in science and technology assessment, is accompanied at the same time by rather closed, deterministic and linear notions of innovation and technological progress. Maasen and Lieven (2006) characterize some open participatory processes to be merely participatory for PR reasons. The quality of such processes is, therefore, not only dependent on the openness itself, but also on use of and respect for content.

Policy representatives[9] involved in the CEH agree with a lot of the features of extended, open approaches, but nevertheless in daily practice some "old" reflexes show up. Similar to

[9] We work together with three types of (what we call) policy representatives. First, the politically appointed members of staff working directly for the Ministers: cabinet staff. We mainly discuss our work with them in the steering group attached to the CEH, rather than directly cooperate on a more regular basis. When reaching decisions, they are often at the "end of the pipeline." Second, the civil servants working for the administrations. We cooperate and discuss with some of them rather directly. Others play a role similar to the

our natural scientific colleagues, in our experience, they sometimes seem to be allergic to effectively opening up debate. In the practice of the action plan (described above), they show uneasiness mainly about two things. One, instead of presenting a diversity of options to the minister, they tend to focus on "the best option" in their advice. This probably relates more to their (often natural scientific) training than to one of the essential responsibilities of a politician: making a well-informed decision about relevant options. In order to be well informed, a minister may be helped better with clear view on a diversity of relevant options, accompanied by documentation and argumentation, rather than deciding on the one black box option. The second uneasiness is an allergy to difference of opinion or to the involvement of different opinions, from experts, or from other actors such as stakeholders. When this is looked upon as a threat to the autonomy of policy, they overlook the value of relevant information that will enrich and improve the decision-making process. Perception research among participants of the biomonitoring (Keune 2008b) shows that the vast majority agrees with this division of responsibility: a vast majority agrees on the importance of public participation in problem-solving and policy and state that their opinions should be dealt with seriously, but that in the end, it's up to politicians to decide. In the case of the action plan, this is the deal: at the end of the process, the minister(s) decides, based on all input gathered during the process. No loss of autonomy for politics, but well-informed transparent and well-argued decision-making. Thus, an open approach does not necessarily mean referendum, revolution or anarchy; constructively, it can be a well-informed process, indeed, one that is rendered to be a more objective basis for decision-making.

Precautionary Principle

Apart from opening up the process, also the precautionary principle is important with regard to both science and policy concerning complex but socially important environmental problems. Also this concept is a container of several meanings. We present one of the most famous definitions, the so-called 15th principle of the Rio declaration (United Nations Conference on Environment and Development, 1992): "*In order to protect the environment, the precautionary approach shall be widely applied by States according to their capabilities. Where there are threats of serious or irreversible damage, lack of full scientific certainty shall not be used as a reason for postponing cost-effective measures to prevent environmental degradation.*" Also in the EU, this principle has gained a position in the diversity of policy considerations (Commission of the European Communities, 2000). The European Environment Agency (EEA) focused attention for this principle with an interesting report (Harremoës et al., 2002), in which several examples from practice are presented. These examples show how "*harmful and costly misuse or neglect of the precautionary principle can be*" according to Poul Harremoës, chair of the report's editorial team, at the presentation of the report (Harremoës, 2002). Harremoës continues by pointing out that too much precaution may be harmful because opportunities for innovation as well as scientific opportunities may be lost. At the same time, he stresses the importance of increased diversity of information, form various sources, to guarantee that a better balance is established between innovation and

third group: experts from governmental expert institutions like monitoring agencies. They are mainly involved as experts in steering groups and expert consultation.

potential future harm. The report draws some lessons on the basis of the case studies and proposes some guidelines. Harremoës added: "*None of the lessons would themselves remove the dilemmas of decision-making under situations of uncertainty and high stakes. They cannot eradicate uncertainties or avoid the consequences of ignorance. But they would at least increase the chances of anticipating costly impacts, of achieving a better balance between the pros and cons of technological innovations and of minimising the costs of unpleasant surprises.*"

The lessons are applicable to subjects discussed in this chapter. They also plea for the inclusion of a relevant diversity of expertise and interests, in order to guarantee a broad and sufficient information basis for risk assessment. Barriers to exchange of information among relevant groups should be diminished, both within science and policy. At the same time, the independence for the organisation responsible for the coordination and steering of the assessment should be warranted. The report stresses the importance of looking at "*real world conditions*" as to take into account as much as possible the complexity of reality and to not reduce it in a way that more or less blinds sight on the real issue. Finally, the report warns not to invest too much time and effort in the process of investigating and interpreting the problems whilst the actual tackling of the problems may be justified already.

Right to Know versus Fear of Panic

The complexity of communication is often underestimated (Keune et al., 2008c). It takes time and effort and also a lot of reflection: external communication puts scientific results in a format different from a scientific or policy perspective. What is discussed inside is not automatically communicated externally. Part of the anxieties concerns the fear of panic, misunderstanding and misuse. Especially when scientific conclusions are not straightforward or crystal clear (because of uncertainties, nuances, lack of easy interpretations especially with regard to policy and individual health effects or lifestyles), such considerations pop up. An interesting paradox in this respect is the apparent contradiction between, on the one hand, caution, and to some extent reluctance to extensively communicate nuanced and complex problems and results, especially when it concerns politically sensitive issues, and, on the other hand, the drive to make people more aware of such problems and of potential prevention measures. It seems as if problems disappear because of absence of clear-cut scientific interpretation. At the same time, though, both scientists and policymakers are sometimes wondering why people stick to unhealthy lifestyles or show lack of concern for environment and health. The *allergy* to the *opening up-approach* thus also seems to work in the other direction.

The deficit model (Wynne, 1996), the model stating that one-way communication of objective sound scientific information from experts to the ignorant public will be sufficient in order to make them aware of problems and adapt behaviour (a rather traditional expert approach to communication), does not tell us how to treat important but uncertain ambiguous information. Internationally, public disclosure of environmental information is high on the political agendas, for example the Aarhus Convention (UNECE, 1998). There has been quite some experimentation with "right-to-know initiatives" that tried to put this into practice. One of the most interesting examples, rather unique because of its attention for health effects (cancer risk) of pollutants, is the "right-to–know website" in the Netherlands (Ragas et al.,

2006). Part of the website was a forum, on which an interesting diversity of opinion showed on the release of such information to the broader public: fear of panic and misuse, on the one hand; on the other, it is stated that withholding information may cause more unrest than presenting uncertain information. Ragas et al. (2006) criticize the reluctance to communicate because of uncertainties. It is used by regulators and risk assessors; why not let the public know about it? The public, moreover, is capable of dealing with complex issues (e.g., Marris et al., 2001). "*The challenge is to communicate the risk estimates so that they are understandable and that the risks and associated uncertainty can be put into a personal perspective*" (Ragas et al., 2006), refer to Slovic (2001)). Nevertheless, the dominant framework with regard to communicating science to the public is still a traditional one-way communication (the deficit model) and is rather negative (Davies, 2008). Communication is predominantly seen as difficult and dangerous (panic, misuse, etc.) whereas a more positive approach acknowledges the complexity of communication but stresses the positive value: it makes science accountable, and it may empower the public.

4. Conclusion

First of all, we may conclude that in the past decennia, a lot of critical food for thought has been developed and was published concerning the complexity of environmental problems. Priority should be given to actually implementing these ideas in practice and "*learn by doing*." This requires a context and problem-specific approach. Considering the important choices to be made during the assessment and policy design, the negotiation process between actors involved is an important element. During the process, the involvement of stakeholders (from outside the group of actors involved in coordinating and steering the process) is an important asset.

Cooperation between social and natural scientists is essential. Practice shows that on the level of ambitions, this is easily agreed, but practical implementation is far from easy. This is also true for the involvement of stakeholders, especially when this concerns groups that are not professionally involved in these issues. Our experience shows that specialized expertise is called specialized for a reason. For example: social scientific expertise relevant to such processes is not easily transferred to natural scientific experts. Moreover, each specialist has a full agenda of activities and very limited time, energy and resources for incorporating complementary elements from another expertise. Practice also shows that phenomena that appear to be only complex from a natural scientific point of view also show to be socially complex, especially when they are to be looked at from a social and policy perspective. In this broader perspective, different challenges are posed than within a purely (natural) scientific context.

A problem-solving perspective for complex problems demands more than a purely scientific approach. Moreover, the scientific approach itself is to be questioned. We may put question marks behind the ambition to fully know complex phenomena. More importantly, we should think in terms of quality and weight of knowledge. Which qualitative knowledge is relevant and what does this mean for interpreting problems and what may be the policy consequences? Moreover, we have to stress the importance of the intersubjective nature of science: *the proof of the science is in the discussion.*

Transparency about the production of knowledge is very important for better and realistic understanding. This concerns all choices involved: scientific framing and approach, actors involved and knowledge limitations. Practice learns that presenting a kind of "how to deal with complex issues" cookery book is not realistic. The secret of success is in close cooperation and deliberation of all relevant scientific disciplines and stakeholders.

ACKNOWLEDGMENTS

The studies of the Flemish Centre of Expertise on Environment and Health were commissioned, financed and steered by the Ministry of the Flemish Community (Department of Economics, Science and Innovation; Flemish Agency for Care and Health; and Department of Environment, Nature and Energy). We thank the colleagues from the Centre of Expertise for Environment and Health for their cooperation: Prof. Willy Baeyens, Prof. Nik Van Larebeke, Prof. Greet Schoeters, Prof. Stefaan De Henauw, Prof. Geert Molenberghs, Dr. Vera Nelen and Els Van de Mieroop, Maaike Bilau, Liesbeth Bruckers, Gudrun Koppen, Elly Denhond, Greet Van Kersschaver and Karen Goeyens. We gratefully acknowledge the collaboration of the participants in the human biomonitoring research.

REFERENCES

Allen M., Raper S. and Mitchell J. (2001). Climate change: Uncertainty in the IPCC's Third Assessment Report. In: *Science* 20 July 2001: vol. 293. no. 5529, pp. 430-433, DOI: 10.1126/science.1062823.

Allen P. M., Torrens P. M. (2005). Knowledge and complexity. In: *Futures* 37 (2005) 581-584.

Aula P. (1999). Chaos, Communication, and Cultural Change: Beyond the Management of Organization. In: Lissack M.R. and Gunz H.P. (1999), *Managing Complexity in Organizations. A View in Many Directions*, Quorum Books, Westport Connecticut, London; pp180-196; ISBN: 1-56720-285-3.

Bal R., Bijker W. E. en Hendriks R. (2002). *Paradox of Scientific Authority—About the societal influence of advice of the Health Council, Health Council*, Den Haag, (in Dutch).

Byrne D. (1998). *Complexity and the Social Sciences*, Routledge, London 1998.

Centre of Expertise for Environment and Health, http://www.milieu-en-gezondheid.be/

Cilliers P. (1998). *Complexity and Postmodernism. Understanding Complex Systems*, Routledge, London, 1998.

Cilliers P. (2005a). Complexity, Deconstruction and Relativism, in: *Theory, Culture &* Society 2005; 22; 255-267.

Cilliers P. (2005b). Knowledge, limits and boundaries, in: *Futures* 37 (2005) 605–613.

Covello V. (1991). Risk comparisons and risk communication: Issues and problems in communicating health and environmental risks, in Kasperson R.E. and Stallen P.J.M. (Eds.), *Communicating Risks to the Public: international perspectives*, Dordrecht.

Davies S. R. (2008). Constructing Communication: Talking to Scientists About Talking to the Public. In: *Science Communication* 2008; 29; 413-434.

Fisher F. (2000). *Citizens, Experts, and the Environment, The politics of local knowledge*, Duke University Press, Durham/London.

Funtowicz S. O., Martinez-Aler J., Munda G. and Ravetz J. R. (1999). *Information tools for environmental policy under conditions of complexity*, Environmental Issues Series 9, EEA, Copenhagen.

Funtowicz S. O. and Ravetz J. R. (1990). *Uncertainty and quality in science for policy*, Dordrecht, Kluwer.

Funtowicz, S.O. and Ravetz, J.R. (1991). A new scientific methodology for global environmental issues, in: Costanza, R. (Ed.), *Ecological Economics*. Columbia, New York, pp. 137–152.

Funtowicz, S.O. and Ravetz, J.R. (1994). The worth of a songbird: Ecological economics as a post-normal science, in: *Ecological Economics* 10, 197–207.

Giles J. (2007). Special Report: From words to action, in: *Nature* 445, 578-579 (8 February 2007) | doi:10.1038/445578a

Harremoës P. (2002). News Release European Environment Agency10 January 2002.

Harremoës P., Gee D., MacGarvin M., Stirling A., Keys J., Wynne B., Guedes Vaz S. (Eds.) (2002). *The precautionary principle in the 20th century, Late lessons from early warnings*, Earthscan Publications Ltd, London England.

Hopkin M. (2007). Climate sceptics switch focus to economics. In: *Nature* 445, 582-583 (8 February 2007) | doi:10.1038/445582a.

IPCC (2001). *Climate change 2001*, Cambridge University Press, Cambridge.

IPCC (2005). *Guidance Notes for Lead Authors of the IPCC Fourth Assessment Report on Addressing Uncertainties*, July 2005.

IPCC (2007). *Climate change 2007*, Cambridge University Press, Cambridge.

IPCC, http://www.ipcc.ch

Keune H. and Goorden L. (2004). *Accidental complexity, Societal embedding of Science on Health & Environment, case study Genetic Research*, Centre of Expertise for Environment and Health (in Dutch).

Keune H., Koppen G., Van Campenhout K. (2007). Knowledge for action: joint reflection on environment & health-data, In: *Sociale interventie*, 16:1(2007), p. 33-40.

Keune H., Koppen G., Van Campenhout K. (2008a). From environment & health data to policymaking: the case of DDE in Belgium. In: *International journal of environment and health*, 2:2(2008), p. 209-224.

Keune H., Loots I., Bruckers L., Bilau M., Koppen G., van Larebeke N., Schoeters G., Nelen V., Baeyens W. (2008b). Monitoring environment, health and perception: an experimental survey on health and environment in Flanders, Belgium, In: *International journal of global environmental issues*, Special issue on "Risk Perception and Social Trust," 8:1/2(2008), p. 90-111.

Keune H, Loots I., Morrens B. (2009a). *Guidelines risk communication*, Centre of Expertise for Environment and Health (in Dutch).

Keune H., Morrens B., Loots I. (2008c). Risk communication and human biomonitoring: which practical lessons from the Belgian experience are of use for the EU perspective? In: *Environmental health*, 7:S1(2008), p. S11.

Keune H., Morrens B., Springael J., Koppen G., Colles A., Loots I., Van Campenhout K., Chovanova H., Bilau M., Bruckers L., Nelen V., Baeyens W., Van Larebeke N. (2009c). Policy interpretation of human biomonitoring research results in Belgium; priorities and complexity, politics and science., in: *Env. Pol. Gov.* 19, 115–129 (2009), DOI: 10.1002/eet.500.

Koopmans R. (1974). *Tegenspraak brengt ons verder: Over de strategie van de oppositionele beroepspraktijk*, De Bezige Bij (in Dutch).

Koppen G., Keune H., Casteleyn L., (2005). *Action plan Biomonitoring results*; Centre of Expertise for Environment and Health (in Dutch).

Latour B. (2000). When things strike back: a possible contribution of "science studies" to the social sciences, in: *British Journal of Sociology*, vol. 51, No. 1, 1 January 2000, pp. 107-123.

Law J. and Urry J. (2004). Enacting the social, in: *Economy and Society*, vol. 33, No. 3, August, 2004, pp. 390-410.

Leroy P. (2004). Sciences environnementales et interdisciplinarité: une réflexion partant des débats aux Pays-Bas, in: *Natures, Sciences, Sociétés*, 2004, 12, 3: 274-284.

Maasen, S and Lieven O (2006). Transdisciplinarity: a new mode of governing science? In: *Science and Public Policy*, vol. 33, No. 6, 1 July 2006, Special issue on discipline and research: practices of inter-/trans-disciplinary cooperation in science, pp. 399-410.

Marris C., Wynne B., Simmons P., Weldon S., Cárceres J., De Marchi B., Klinke A., Lemkov L., Pellizzoni, Pfennig U., Renn O. and Sentmartí R. (2001). *Public Perceptions of Agricutural Biotechnologies in Europe*, Final Report of the PABE research project, European Communities.

McCally M. (Ed.) (2002). *Life Support, The Environment and Human Health*, The MIT Press, Cambridge, Massachusetts, London England.

McLennan G. (2003). Sociology's Complexity, in: *Sociology* 2003 37: 547-564.

Moss, R.H. and Schneider, S.H. (2000). Uncertainties in the IPCC TAR: Recommendations to lead authors for more consistent assessment and reporting. In: *Guidance Papers on the Cross Cutting Issues of the Third Assessment Report of the IPCC* [eds. R. Pachauri, T. Taniguchi and K. Tanaka], World Meteorological Organization, Geneva, pp. 33-51.

Netherlands Organisation for Scientific Research (2008). *Complexity—Outline of the NWO strategic theme Dynamics of complex systems*, The Hague.

Neurath O. (1933). Protokollsätze, in: *Erkenntnis*, 1932-33, III, pg. 204-214.

Norgaard R. B. and Baer P. (2005). Collectively Seeing Complex Systems: The Nature of the Problem, In: *BioScience* November 2005 / vol. 55 no. 11, 953-960.

Nowotny H. (2005). The Increase of Complexity and its Reduction: Emergent Interfaces between the Natural Sciences, Humanities and Social Sciences, in: *Theory, Culture & Society* 2005 22: 15-31.

Ormerod P. (2006). Shun the rational agent to rebuild economics, *Post-autistic economics review*, issue no. 40, 1 December 2006, article 7, pp. 59-60, http://www.paecon.net/PAEReview/issue40/Ormerod40.pdf

Pereira G. and Funtowicz S. O. (2006). Knowledge representation and mediation for transdisciplinary frameworks: tools to inform debates, dialogues & deliberations, in: *International Journal of Transdisciplinary Research* vol. 1, no. 1, 2006, Pages 34-50.

Popper K. R. (1935). *Logik der Forschung*, Springer, Wien.

Popper K. R. (1945). *The open society and its enemies*, vol. II: The high tide of prophecy: Hegel, Marx and the aftermath.

Ragas A., Huijbregts M., van Kaathoven E., Wolsink J., and Wemmenhove J. (2006). Development and Implementation of a Right-to-Know Web Site That Presents Estimated

Cancer Risks for Air Emissions of Large Industrial Facilities. In: *Integrated Environmental Assessment and Management*, Vol. 2, Issue 4 (October 2006), 365-374.

Ravetz J., (2002). *The Post-Normal Science of Precaution*, Paper for the thematic workshop of the Joint Research Centre "Managing Uncertainty in science for sustainability: future research challenges for Europe," at the Launch Conference of the 6th Framework Programme: "European Research 2002, The European Research Area and the Framework Programme," Brussels, 11-12-13 November 2002.

Renn O. and Rohrmann B. (Eds.) (2000). *Cross-cultural risk perception: a survey of empirical studies*, Dordrecht, Kluwer.

Richardson K. (2005). The hegemony of the physical sciences: an exploration in complexity thinking, in: *Futures* 37 (2005) 615–653.

Richardson K., Cilliers P., and Michael Lissack M. (2001). Complexity Science: A 'Grey' Science for the 'Stuff in Between', in: *Emergence*, 3(2), 6–18.

Risbey J.S. and Kandlikar M. (2007). Expressions of likelihood and confidence in the IPCC uncertainty assessment process, in: *Climatic Change* (2007) 85:19–31, DOI 10.1007/s10584-007-9315-7.

Rumsfeld D. (2002). *There are known knowns. There are things we know that we know. There are known unknowns. That is to say, there are things that we now know we don't know. But there are also unknown unknowns. There are things we do not know we don't know.* Press conference (12-02-2002) as Secretary of Defence of the United States of America.

Schiermeier Q. (2007). What we don't know about climate change. In: *Nature* 445: 580-581.

Shriver T. E., Cable S. and Kennedy D. (2008), Mining for Conflict and Staking Claims: Contested Illness at the Tar Creek Superfund Site, in: *Sociological Inquiry*, vol. 78, no. 4, November 2008, 558–579.

Smaling A. (2008). *Dialogue and empathy in methodology*, SWP, Amsterdam (in Dutch).

Slovic P. (1998). Perceived Risk, Trust and Democracy. In Löfstedt R. and Frewer L. (Eds.), *Risk and Modern Society*, pg 181-192, London.

Slovic P. (2001). The risk game. In: *Journal of Hazardous Materials* 86:17–24.

Stern, P. C. (2005). Deliberative Methods for Understanding Environmental Systems. In: *BioScience*, vol. 55, no. 11, November 2005, pp. 976-982(7).

Stern P. C. and Fineberg H. V. (Eds.) (1996). *Understanding risk: information decisions in a democratic society*, National Research Council, National Academy Press, Washington D. C.

Stirling A. (2008). "Opening Up" and "Closing Down": Power, Participation, and Pluralism in the Social Appraisal of Technology. In: *Science, Technology & Human Values* 2008 33: 262-294.

Swart R., Bernstein L., Ha-Duong M. and Petersen A. (2008). Agreeing to disagree: uncertainty management in assessing climate change, impacts and responses by the IPCC, in: *Climatic Change* DOI 10.1007/s10584-008-9444-7.

UNECE (1998). *Convention on Acces to Information, Public Participation in Decision-making and Acces to Justice in Environmental Matters.*

United Nations Conference on Environment and Development (1992). *Rio Declaration on Environment and Development,* Rio de Janeiro from 3 to 14 June 1992, http://www.unep.org/Documents.multilingual/Default.asp?DocumentID=78enArticleID=1163

Urry J. (2005). The Complexity Turn. In: *Theory, Culture & Society* 2005 22: 1-14.

Van Asselt M.B.A., Rotmans J. (2002). Uncertainty in Integrated Assessment Modelling, in: *Climatic Change*, 54, 1-2: 75-105.

Wynne B. (1996). Misunderstood Misunderstandings: social identities and public uptake of science. In: A. Irwin & B. Wynne (eds.), *Misunderstanding Science? The Public Reconstruction of Science and Technology* (pp.19-46), University Press, Cambridge, 1996.

In: Health and Environment: Social Science Perspectives ISBN 978-1-60876-216-3
Editors: Helen Kopnina and Hans Keune

Chapter 10

HEALTH AND ENVIRONMENT POLICIES IN THE EUROPEAN UNION

Helen Kopnina

ABSTRACT

Pendulum shifts from the more "humanitarian" focus on public health and welfare to greater attention to environmental issues continued throughout the twenty-first century in Europe. Concerns about economical progress (with the perceived risks thereof to both health and environment) as well as technological innovation inherent in this process (with possible advantages to health and environment) are reflected in European policies. These shifts of public interest and policy priorities are still very much present, as European governments debate priorities in spending. This chapter examines the evolution of European health and environmental policies, and the role of European institutions as well as lobby and interest groups in shaping these policies.

This chapter outlines the necessity of social scientific research into the effect of the European policies at the national, regional and local policies. The author argues that social scientists are in unique position to contribute to the expansion of knowledge on health and environment by analyzing the differences in interpretation, adaptation and implementation of health and environmental policies in cross-national perspective. Two cases of European policies promoting public health and environmental protection in the Dutch context are addressed. European policy, targeted at the improvement of air quality in one case, and at the reduction of illegal timber trade in another case, is met with only partial compliance and even resistance at the Dutch national level.

INTRODUCTION

We shall first examine the principles of European policymaking and review some of the policies and institutions related to health and environment. European and national stakeholder groups such as policymakers, lobbyists and interest groups have actively engaged in

European health and environmental initiatives. We examine the role of those groups in the shaping of the European policy agendas, as well as address the current priority issues in the areas of health and/or environment. Consequently, we turn to the Europeans' concerns regarding policy agendas as indicated by the Public Opinion Monitor, Eurobarometer, and relate these concerns to the current research agendas.

Picture 10. Europe.

As contributors to our book demonstrate, there are many ways in which social scientists can become more actively engaged in areas traditionally relegated to natural, medical and environmental scientists as well as policymakers. At the European level, comparative case studies of the effects of the health and environmental policies at the national actions form a significant base for the contribution of social scientists. European citizens' perceptions of generically formulated policies and actions, as well as their perceptions of local health and environment issues and their views on effectiveness of policies in regard to these issues, form a rich area of qualitative inquiry.

Development of the European Union brought forth rapid changes in economic, political, cultural and social landscapes of European nation-states. Due to rapid expansion and intensification of trade and migration, the differences between European countries became blurred, the conceptual boundaries shifted. New alliances are being formed, new disparaging lines are being drawn between nations, and new legal mechanisms and policies come into place (Miller and Hashmi, 2001). While the idea of the bounded nature of the nation-state used to rest on the assumption that it is a territorial entity with its own borders encompassing a national "culture," nowadays Europe presents us with some new and unexplored issues: it represents an entity in which borders are removed as a result of the neo-liberal idea of the single market, and where certain categorical distinctions (West/East, North/South, Left/Right) that have so far been taken for granted are subsequently blurred (Stacul et al, 2005).

The crisis of structural-functionalism in the 1960s-70s, the political unrest that characterized that period and the emergence of Marxism in social theory entailed a shift in theoretical focus: studies on rural communities became less popular, and sociologists and bureaucratization anthropologists paid increasing attention to the interrelationship between the local and the national, most notably to macro-social processes of state formation and bureaucratization, even though they continued to focus on the Mediterranean area (Stacul et al., 2005). In the 1980s, by contrast, the theoretical framework around which most of Europeanist social science was built (i.e. that associated with "positivism") came under attack from the postmodernist approach, which gave primacy to deconstruction, textuality and the politics of identity (Goddard, Llobera and Shore, 1994: 19). It is within such a context that social scientists started focusing on European identities rather than on "communities" (McDonald, 1993) and, more recently, on European integration and its paradoxes (Holmes 2000; Shore, 2000; 2002). In regard to health and environment issues, social scientists could further engage with the topics discussed in the Introduction to this volume. The concept of the risk society (Beck, 1992; 1996; 2006; Cohen, 1997; Lash et al., 1996) may be particularly useful for social scientists attempting to tackle political involvement of European citizens in the areas of public health and environment. According to Beck (1992, 2006), we could distinguish between ecological and financial dangers, which can be conceptualized as side effects, and the threat from terrorist networks as intentional catastrophes. The principle of deliberately exploiting the vulnerability of modern civil society replaces the principle of chance and accident (Beck, 2006:329). We need to note, however, that Beck does not include the notions of "nature," "destruction of nature" or "environmental problems," choosing instead to focus on the sociological analysis of ecological questions, a conceptual framework allowing grasping "ecological problems" not as problems of environment or surrounding world, but of the inner world of society (Beck, 2006:1). Social scientists (among whom are contributors to this volume) may further ponder how this society-based approach may be compared to the more environment-centred approaches pioneered by Catton and Dunlap (1978) in the European context.

Another area of engagement of social scientists in relation to health and environment issues in Europe in their positioning within the larger context of capitalism and industrialisation. We may recall the discussion of "natural capitalism" model (Speth, 2008) or "cradle-to-cradle" principle (McDonough and Braungart, 2002) or sustainable business model (Hawkes et al., 1999; Kolk and Pinske, 2008). Social scientists expand cross-national studies to provide evidence as to how different European nations deal with health and environmental issues in relation to these models and principles. Europeanization could be compared to globalization process, and European health and environment policies could be examined through the prism of global relations and interdependencies. The removal of either physical or conceptual boundaries between nation-states represents the site of a complex, overlapping, disjunctive order to follow Appadurai (1990: 296). In this order, the significance of health and environment issues is debated, contested, and re-asserted by different European member states and their citizens. Once the cradle of nationalism, which conveys ideas of divisions, much of Europe now represents a political and territorial entity in which frontiers can be crossed freely without going through checkpoints. Although it would be contentious to postulate that a European identity has the potential to replace national and regional identities, it seems clear that concepts such as "nation," "region," "locality" as well as ethnicity, identity, class and community (to name a few) have now to be redefined in relation to this expanding entity (Stakul et al., 2005). As the two Dutch examples of health- or environment-related topics provided at the end of the chapter demonstrate, not only generically formulated European policy, but also its interaction with the national, regional and local regulations, behavioral patterns and perceptions, play a large role in how health- or environment-related issues are being resolved at the grass-roots level.

At the empirical level, social scientists could engage in comparative studies focused on local perceptions of the problems by local stakeholder, interest and lobbyist groups involved with the issues of public health and environment (Giddens, 2009). While the scope of this chapter does not allow us to delve deeper into such cross-national cases studies, we need to note that most of them point to the large differences in the compliance or resistance to common European policies, as well as in great varieties of interpretation, implementation and integration of such policies at the national level (McCormick, 2008).

LEGISLATION WITHIN EU

The primary legislation of the European Union (EU) consists of Treaties (such as the Treaty of Lisbon, which is largely reflective of what pro-European leaders conceived of as the "European constitution"). These treaties are the basis for a large body of "secondary" legislation that has a direct impact on the daily lives of EU citizens and consists mainly of regulations, directives and recommendations adopted by the EU institutions. The most significant European institutions involved in legislation are the Council (including both the Council of the European Union, also known as the Council of Ministers and the European Council), The Commission and The European Parliament.

The Parliament is the only institution within the EU that is directly elected by European citizens. Members of Parliament come from different political groups, such as European People's Party and Socialist Group. Between these political groups, they represent all views on European integration, from the strongly pro-federalist to Eurosceptic (European Communities, europa/eu.int/comm/publications).

The Parliament shares the decision-making power with other European institutions under the "cooperation," "assent" and co-decision" procedures. According to the Treaties, it is normally the European Commission that proposes legislature to The Council and The Parliament. Members of The European Commission are appointed and can be dismissed by the Parliament. Together with the Council, The Parliament makes decisions of Commission's proposals. The Council has to take its decisions either by a simple majority vote (especially in cases when common European policy is proposed), a "qualified majority" vote or unanimously, depending on the subject to be decided (Europa: Gateway to the European Union, 2009).

Other significant European institutions include the Court of Justice, the Court of Auditors and various regional committees, such as the European Economic and Social Committee

(EESC) and the Committee of the Regions (CoR), which is usually consulted by the Council and Commission on important legislative decisions. Increasingly, legislature related to public health and environment address policies formulated at the common European level, thus affecting national policies of all European states.

PRINCIPLES OF EUROPEAN POLICIES IN RELATION TO HEALTH AND ENVIRONMENT

The principles of European policy include integration, prevention, subsidiary and derogation. Integration refers not only to integration of existing and new member states into the European Union, but also to integration of all EU policies including consumer protection, culture, human health and natural resources. Prevention refers to action to prevent emergence of problems, rather then responding to them as they arrive. Subsidiary refers to EU restricting itself to those problems that could not be tackled at the national levels only. Derogation means that those member states unable to comply with regulations on time are given deadline extensions or lower targets or financial assistance (McCormick, 2008:186).

An important moral and political principle in relation to health and environment is the precaution(ary) principle, which states that if policy might cause severe or irreversible harm to the public or to the environment, in the absence of a scientific consensus that harm would not ensue, the burden of proof falls on those who would advocate taking the action (Raffensperger and Tickner,1999). In the European Union law, the precautionary principle is also a general and compulsory principle of law (Commission of the European Communities, 2000; Recuerda, 2006). As formulated by the European Environmental Bureau (1999), the precautionary principle justifies early action in the case of uncertainty and ignorance in order to prevent potential harm. Uncertainty refers to the situation when there is knowledge about the parameters of a system, but there is no information as to the quantitative significance of the relevant factors. We may distinguish between different types of "uncertainty" including risk (where boundaries of the system in question are deemed to be known), ignorance (that which is not known and that which we are not aware that we don't know) and indeterminacy (referring to that which cannot be known, either because it is produced through unforeseeable actions of the respective actors, or because of certain ontological considerations, recognizing the conditional nature of all knowledge, which precludes any absolute certainty) (Stirling, 1999).

Another principle of European law is the substitution principle allowing for technology-driven changes instead of waiting for the proof of harm. The principle is normally applied in a wide sense including the consideration of alternative products or services to serve the same function in addition to alternative materials for the same product (European Union Biocides Directive 98/8/EC). This principle is recognized by a number of environmentally advanced member states.

Integration, precaution and substitution principles were recognised in environmental policies since the 1970s, underpinning ideals of the Environment Action Plan of the European Union. Sustainable development became the magic bullet linking different policies together. Sustainable development refers to natural resources such as air, water and forests, which should be used in a way that ensures their availability for future generations.

Health policies, however, have been slower to respond. Only with the adoption of the Amsterdam Treaty in 1997, did the EU make a formal commitment to ensure human health protection. World Health Organization (WHO) established national environmental health action plans in 2003. The EU is currently implementing the European Environment and Health Action Plan 2004-2010, with the ultimate goal to "fill the knowledge gap by strengthening research and addressing the emerging issues on environment and health, and to make available relevant information to the public" (European Environment and Health Action Plan, 2009).

INTEGRATING HEALTH AND ENVIRONMENT AT THE POLICY LEVEL

Since the forging of the EU, health and environment were closely linked to both public and political agendas. At the initial stages of European integration, pressure to harmonize health and environmental policies came from the desire to remove the barriers to the single market created by different health and environmental standards (McCormick, 2008:102).

Environmental policy, concerns for which were all but unknown in the 1950s, grew in the 1960s, as the air and water pollution created by economic growth promoted a reaction from increasingly affluent Europeans. Differences in environmental standards and regulations also posed a barrier to the single market, so that by the 1980s, there was new political support for addressing the problem (McCormick, 2008:166).

While emergent from the economic and political imperatives, health and environment policies have gained their own momentum.

The European Environment Health Action Plan led to the expansion of the EU-funded projects aimed at developing and applying new, integrated methods for assessment to support environmental health policies. The medical and bio-physical projects upon which The European Environmental Health Action Plan depends stress the need for an environment and health cause-effect framework that will provide the necessary information for the development of community policy dealing with sources and the impact pathway of health stressors. Social scientists could add their knowledge in order to address this cause-effect framework in economic, societal, political and cultural contexts.

European "risk society" has been keen on examining the underlying causes of illness, linking it to environmental factors. In the opening statement of Director-General for Research at the European Commission Achelleas Mitros of the EU Research on Environment and Health (2004) stated:

> "Everyone now accepts that the world around us has an important influence on our health."

Many conditions—from asthma and allergies to new infectious and emerging diseases—have been linked in some way to the environment. Studies have shown that environmental pollutants, food contaminants, noise, and issues like climate change, can all affect our well-being.

As the European Commission report (2008) states, "Current consumption and production trends have significant environmental impacts, including greenhouse gases, pollution and the running down of the Earth's natural resources. But there are some easy ways of reducing environmental damage without affecting our well-being". The link between health and environment is clearly made in the statement of European Environmental Agency (2008):

> "Human health has always been threatened by natural hazards such as storms, floods, fires, landslides and droughts. Their consequences are being worsened by a lack of preparedness and by human actions such as deforestation, climate change and biodiversity loss."

These natural hazards are being translated into practical problems that need to be addressed both at the social and political levels. We may recall from the Introduction of this volume, however, that the issue of environmental illness is highly contested (Brown, 2007; Brown et al, 2003; Kroll-Smith, 2000). As some contributors to this volume argue (Baralt, Dépelteau, Keune et al., Potrata and Akobirshoeva, Zavestovski), "scientific truths" about environmental illness need to be carefully examined as part of social processes, before any policy action is to take place. While the very cause of disease is widely disputed by different stakeholders (the illness victims, doctors, policymakers, environmental and health "experts," industrial developers, etc.), European policies tend to dictate generic, one-size-fits-all (or at least aspires to fit all) prescriptions. It is the interpretation of such policies at the national level that should warrant particular attention of social scientists in examining the various economic, social, political and cultural factors that lead national governments and the public to accept, reject, negotiate, or adopt European environmental health policies. As contributors to this volume argue, social scientists could do well by addressing the social processes involved in interpreting European environmental health policy initiatives and translating them into national and local contexts, as well as debating the truth and objectivity of certain claims through the process of multi-disciplinary (social) scientific research.

The traditional domain of sociologists, social phychologists and anthropologists, the society as a whole and sub-groups within the society, are rich areas of study in effectiveness of generically formulated policies. The European societies are actively involved in the process of interpretation, integration, adaptation or contestation of certain generically formulated policies, as well as in the sense-making process underlying both health and environmental agendas. The knowledge and beliefs about health and environment present within the nation states has large implications for how the public, or interest or lobbying groups perceive European policies. For example, some Europeans (particularly the recently joined East European nations) may camp with the heritage of the protectionist state and the "tragedy of the commons" (see Chapter 1 for discussion). Their perception of health risk or environmental damage might be different from those of West European states, where health and environmental protection have been popular topics of political and media discussions for decades.

It is thus not only the European Union institutions that determine policy agendas concerned with health or environmental protection of European nation-states and their public. The public, in turn, has also been influencing European health and environment agendas and have, at the grass-roots level, determined how useful, effective or indeed desired certain policies are for their own countries.

Social pressure groups, NGO's, charities, and interest groups, have been increasingly active in influencing EU policy in the past three decades. More importantly, such stakeholder groups have influenced the way European citizens view both health and environment.

European Interest and Lobbying Groups

According to Carl Smith, a sociologist concerned with human rights and political movements, at the time of the publication of the Universal Declaration of Human Rights (1948) "the darkness hovering over nations was not pollution, but the residue of war and political oppression" (Smith, 2003:1). Soon it became apparent, however, that peacetime presented Europeans with new dangers. *The Silent Spring* (1962), the influential book by American marine biologist Rachel Carson (1907–1964) about the use of chemical pesticides and their influence on health and environment was soon discovered by Europeans. The book raised awareness of the dangers of chemicals and lead to the formation of environmental protection movements in many European countries.

The "toxic fog" otherwise known as the "Killer Fog" of '52 was probably the first large-scale environmental health disaster recognized by Europeans. According a recent study in the journal *Environmental Health Perspectives* (2009), 12,000 may have been killed by the great smog, particularly in the London area.

The so-called killer fog is not an especially well-remembered event, even though it changed the way the world looks at pollution. Before the incident, people in cities tended to accept pollution as a part of life. Afterward, more and more, they fought to limit the poisonous side effects of the industrial age (Nielsen, 2009).

By the time the Declaration of Human Rights was ratified by 35 states and entered into force in 1976, thousands of Londoners had died in toxic fog. DDT had suffered a fall from modern miracle to emblem of dangerous science. The publication of *Silent Spring* had inspired an international environmental movement...(Smith, 2003:1).

The European public was stirred by these events, placing environmental health and other issues associated with either health or environment at the top of their agendas. The driving force behind European politics, as well as the work of charities and NGO's from the late nineteen sixties, became interest groups, "covering not only political pressure groups but also the great variety of formal and informal organizations in areas such as welfare, sports, education and religion" (Rucht, 2006:118). The notion of corporate social responsibility (CSR) has entered the European discourse involving not only governments but multinational and national businesses into the debates (Carrol, 1999).

In Eastern Europe, controlled at the time by the Communist governments, health care and environmental protection were mostly under exclusive government control. Patients' rights groups or environmentalists were largely marginalized or silenced and practically non-existent. Large-scale disasters, such as nuclear reactor accident in Chernobyl, Ukraine, in 1986, were hushed up by the Soviet government. The International Atomic Energy Agency (IAEA)'s report 1986 analysis attributed the main cause of the accident to the operators' actions. In 1993, the IAEA issued a revised analysis, attributing the main cause to the reactor's design (IAEA, 1986; 1993). An estimated 600,000 mostly Belorussian citizens suffered radiation exposure, of which 4,000 suffered cancer deaths. Approximately 100,000 fatal cancers can be still expected (IAEA, 1993; 2009; Osborn, 2006).

Investigation commissions in Ukraine itself were secretive, despite the efforts of emerging lobby groups, which included the reactor's designers, power-plant personnel, and Belo-Russian and Ukrainian residents most affected by the disaster. "The mere fact that the operators were carrying out an experiment that had not been approved by higher officials indicates that something was wrong with the chain of command. The State Committee on Safety in the Atomic Power Industry is permanently represented at the Chernobyl station. Yet the engineers and experts in that office were not informed about the programme. In part, the tragedy was the product of administrative anarchy or the attempt to keep everything secret (Medvedev, 1991)." However, the Chernobyl disaster created—perhaps for the first time in Soviet history—a degree of public awareness and mobilization of different stockholders' groups sharing a common hope on finding the causes and trying to prevent similar disasters.

Only after the fall of the Soviet regimes did the issues affecting health and environment start to be openly discussed on the public stage or in the media. This public awareness and media coverage in many East European countries are still very much lagging behind West European countries. In many countries of former Soviet Union, construction workers may be seen working with dangerous substances such as asbestos without necessary protection. Oil spillage, water, air and soil contamination are commonplace in the Baltic states, countries of Central Asia and the Caucuses. While some countries permit more open access to information to its own citizens, international public and media are habitually are not allowed to "intrude" (Kopnina, 2005; 2007). The scope of this chapter does not allow us to delve deeper into the differences between Eastern and Western, as well as Northern and Southern Europe. Suffice is to say that much more detailed historical investigation of differences in the treatment of health and environment issues in these countries opens a wide stage for further research.

Returning to Western Europe, we may note that until the late nineteen seventies, public interest groups representing patients and environmental organizations focused their attention on national governments since most of the policy was still done at national level. Interest group politics, including the so-called associational sectors such as nature conservation, experienced significant growth in the past three decades (Rucht, 2006:122). Many interest groups started forming more powerful alliances with governmental and non-governmental organizations.

As the Community became more active on the environment in the mid-1980s, it became a more profitable target for interest group pressure. The new emphasis given to the EU-level activities was reflected in the opening of offices in Brussels by such groups as Friends of the Earth, Greenpeace, and the World Wide Fund for Nature, while many other groups employed full-time lobbyists (McCormick, 2008: 139).

While the industrial lobbyists have been lobbying with the EU since the nineteen fifties, they now had to compete with the demands of environmental and patients' groups also lobbying at the EU level. The view of the groups representing the industrial perspective (such as the European Chemical Industry Council, Eurelectic and the European Crop Protection Association) on environmental issues was quite different from that of environmentalist groups (such as Epha Environmental Network, Friends of the Earth Europe, and Friends of Nature). A group of "Green 10" laid down common rules for lobbying from the turn of this millennium (http://www.eeb.org/activities/transparency/g-10-position-220206.pdf).

Groups like European Federation for Transport and Environment and Bankwatch network often position themselves between industrialist and environmentalist platforms. Another example of the effort to reconcile commercial and environmental interests is the European Principles for the Environment (EPE) initiative. EPE is an initiative launched by the five signatory European-based Multilateral Financing Institutions (MFIs)—mostly banks—in response to the drive for increased harmonisation of environmental principles, practices and standards associated with the financing of projects. The initiative is founded on the commitment to ensuring environmental protection and promoting sustainable development globally and across all sectors of their activities (European Investment Bank, 2009).

Patients' rights groups have also been more vocally present at the EU level since the late nineteen eighties. Seen as the extension of human rights, the rights of the patients became recognized in Europe in the past two decades. The European Consultation on the Rights of Patients, held in Amsterdam 1994, under the auspices of the WHO Regional Office for Europe (WHO-EURO), defined principles and strategies for promoting the rights of patients, within the context of the health care reform process in Europe (http://aitel.hist.no/~walterk/wkeim/patients.htm). WHO remains one of the most significant organizations involved in the logistical process, issuing a declaration on the promotion of patients' rights in Europe (Amsterdam, 2004). The famous example of the clash between industrialist and patients' groups' interests are asthma patients' organizations and the tobacco industry. Group platforms were drawn out and new laws came into place, including Patients' Rights laws and Charters of the Rights of Patients.

We may recall the discussion of the Charity Paradox in the Introduction to this book. We may note that interest groups concerned with environment and/or health do not necessarily share the same objectives. The basic differences in ideologies—be it broadly defined by their human or conservation interests—underlie interest groups' agendas and reflect on the European policy agendas. Most of the lobbing groups concerned with environmental health refer to environment strictly in the sense of (human) surroundings and their effect on (human) health. Conservationist groups consider environment outside of human sphere as having an intrinsic right to exist independent of human interests. Lobbying groups dealing with both patients' rights and the environment have emerged, particularly united by issues concerning exposure to toxic substances, pollution, water, air and soil quality. The Public Health Portal of the European Union lists groups and research projects involved in tackling both health and environmental risks associated with air pollution, noise, chemicals, UV, endocrine disrupters, and electromagnetic fields.

Increasingly, economists, political scientists, sociologists, anthropologists and psychologists get involved in the study of European interest groups. More of interdisciplinary projects are needed to appreciate the complexity of economic, political, social and cultural factors influencing not only the policy processes, but policy's embeddedness in the Euopean societies and stakeholder groups within it.

Environmental Health Issues on the EU Agenda

European Environment and Health Committee (EEHC) supports countries as they try to reduce environmental hazards affecting human health. The EEHC is a unique coalition that brings together representatives from health ministries, environment ministries, intergovernmental organizations, and civil-society organizations (http://www.euro.who.int/eehc). WHO/Europe, specializing in health and environment, covers a wide range of issues, both specific and cross-cutting, ranging from air quality, water and sanitation, etc.

In 2008, European Commission published Environment, Consumers and Health reports, which distinguished among the following categories: Consumers (consumer protection rights and soil, civil protection, chemicals, noise); Environment (Climate, Global warming, nature and biodiversity, waste, water, air); food safety; health (human diseases, tobacco, medicine, drugs); maritime policy; and sustainable development.

The health section contains a section on "my environment," which includes subtopics such as social environment; environmental health; consumer, road and food safety; chemical, physical and biological risks; as well as bioterrorism (http://ec.europa.eu/health-eu/my_environment/index_en.htm). We may note that the topics classified under "environment" with the exception of "nature and biodiversity" overlap with consumer and health topics.

Generally, policy initiatives targeting environmental health attempt to reconcile environmental and health interests and focus on common solutions that may be beneficial both to human health and natural environment. One such EU initiative is the "green capital award" based on the broad idea that curbing urban pollution has benefits for both people and nature. The green capital award is supposed to "encourage improvements in quality of life for millions of city-dwellers" and depends on the subjective ratings of European citizens of their city as "green and pleasant." Credit is given for green spaces, clean public transport and creative thinking on waste management and noise pollution. The panel of experts and the jury will also assess candidates' environmental track record and forward-planning (European Commission 2008 http://ec.europa.eu/news/environment/080522_1_en.htm).

European policy related to health and sickness may be divided in disability policy, equality policy, public health policy and employment policy. While the recent shift in European policy has been towards prevention, most of these policies do not necessarily address people at risk of becoming chronically ill or disabled (Grammenos, 2005). Employment equity policies employ mediating mechanisms (such as employment services and vocational rehabilitation), disability management focuses on reintegration (early intervention and mediation) and job retention (human resources/equal opportunities). Incentives for both mediating and reintegration mechanisms include anti-discrimination levies such as quotas and levies (Wynne, 2008).

The record on implementation of these policies is not always good, having to do with some common (for both health and environment policies) problems, such as lack of financial and technical resources; organizational problems within EU institutions; EU's focus on policy development rather than monitoring implementation in nation-states (McCormick, 2008:186).

According to European Environment and Health Information System (EEHIS), effective policies can lead to significant improvements in health and environmental conditions as "clean environment is clearly one of the prerequisites for good health." The trans-boundary nature of environment and health issues has resulted in international pan-European policy actions, also affecting national policymaking. The key players at the pan-European level are the European Commission and WHO/Europe. At the national level, the policy approach to environmental health issues varies by country. National policies benefit individual countries and contribute to improve the global situation

According to a WHO report issued in 2006, in the WHO European Region as much as 14% to 19% of disease is caused by environmental exposures that can be averted. The report further estimates that more than 24% of deaths and 22% of disease in children under the age of 14 are caused by environmental exposures

According to the European office of the WHO, the following are among the major diseases and conditions associated with environmental determinants: cancers, gastrointestinal and cardiovascular and respiratory diseases, being overweight, injury and developmental disorders (http://www.euro.who.int/childhealthenv/Monitoring/20040519_1). The Public Health portal for the European Union lists environmental factors, such as exposure to pollutants through water, food or air, as important determinants of health. For example, it has been estimated that up to one-sixth of all child fatalities and disease can be attributed to environmental factors. Individuals can make certain choices that affect their lifestyle and health on their own, but they also rely on public authorities for protection from health threats.

Environnemental Contaminants, Chemicals and Pollution

In Europe, the major environment-related health concerns are related to outdoor and indoor air pollution, poor water quality, poor sanitation and hazardous chemicals. Health risks resulting from air pollution include respiratory and cardiovascular diseases, cancer, asthma and allergies, as well as reproductive and neurodevelopment disorders. There is growing concern about the effects of exposure to mixtures of chemicals at low levels and for long periods over our life-time, in particular during early childhood and pregnancy. According to The EU's Clean Air for Europe (CAFE) report, fine particulate matter and ground-level ozone are the main threats to human health from air pollution. The EU's Clean Air for Europe (CAFE) programme estimated a total of 348,000 premature deaths per year due to exposure to fine particles (PM2.5).

Registration, Evaluation, Authorisation and Restriction of Chemical substances (REACH) is a new European Community Regulation on chemicals and their safe use (EC, 1907/2006), which entered into force in 2007. The aim of REACH is "to improve the protection of human health and the environment through the better and earlier identification of the intrinsic properties of chemical substances. At the same time, innovative capability and competitiveness of the EU chemicals industry should be enhanced. The benefits of the REACH system will come gradually, as more and more substances are phased into REACH" (http://ec.europa.eu/environment/chemicals/reach/reach_intro.htm).

Many pollutants known to affect human health are gradually coming under regulatory control. However, there are emerging issues for which environmental pathways and effects on health are as yet poorly understood. Examples are electromagnetic fields (EMF), pharmaceuticals in the environment and some infectious diseases (the spreading of which may be affected by climate change). The development of "early warning" systems should be encouraged to shorten the time between detection of a potential hazard and a policy action or intervention (European Environmental Agency).

Environmental contaminants are substances that, when accidentally or deliberately introduced into the environment, may have the potential to harm people, wildlife and plants (http://www.hc-sc.gc.ca/ewh-semt/contaminants/index-eng.php). These include dioxins and furans, drugs and personal care products, polychlorinated biphenyls (PCBs), lead, radiation and many others. The European Environmental agency warns that persistent chemicals with long-term effects, such as and chlorofluorocarbons CFCs, and those used in long-life structures—for example construction materials—may present risks even after their production has been phased out.

Low birth weight was proven to be associated with industrial air pollution when the cases of births in polluted urban area were compared to those in rural areas in the neighboring territories in Norway (Hansteen et al., 1998). In the Italian study of air pollutants, researchers discovered that results add evidence for an association between air pollution and early mortality or morbidity and support the hypothesis of a synergism between meteorological variables and air pollution (Biggeri et al., 2005). Another Italian study (in the framework of the International Study of Asthma and Allergies in Childhood study) demonstrated the link between the prevalence of asthma and allergies in childhood, and environmental factors including exposure to chemical compounds in the air (Galassi et al., 2005). Other European studies of asthma showed strong correlations between air pollution by traffic exhaust and asthma occurrence (Andersen et al., 2008; Braback and Forsberg, 2009).

At the Fourth Ministerial Conference on Environment and Health (Budapest), member states made commitments to reduce exposure to environmental health hazards (European Environmental and Health Committee, 2004). Final evaluation of results stemming from such commitments is still on the way, but it is already safe to assume that there is a huge difference in the actual implementation of policies by different European countries (http://www.euro.who.int/eehc/ctryinfo/ctryinfo). Intergovernmental Mid-term Review Reporting on the Declaration of the Budapest Conference (Vienna, 2007) pointed out discrepancies between outlined goals and the actual results of taken measures (European Environment and Health Committee). The recent meeting organized by the WHO Regional Office for Europe, entitled Quantification, Comparison and Ranking of Environmental Stressors in selected European countries (Bonn, February 2009), established the project plan, agreeing on methodological issues for the development of national assessments in participating countries (Belgium, Finland, France, Germany, Italy, the Netherlands, Norway, and the United Kingdom).

RESEARCH IN EUROPE

Since 1973, the European Commission has been monitoring the evolution of public opinion in the Member States, thus helping the preparation of texts, decision-making and the evaluation of its work. Eurobarometer, a public opinion portal of the European Commission, reflects on the effects of European level policies on national policies (European Commission, Public Opinion Eurobarometer, 2009). Qualitative studies of Eurobarometer presently involve "in-depth the motivations, the feelings, the reactions of selected social groups towards a given subject or concept, by listening and analysing their way of expressing themselves in discussion groups or with non-directive interviews" (http://ec.europa.eu/public_opinion /quali_en.htm).

The areas that most Europeans raise issues that give rise to interest or hope—or, on the contrary to reservations and concerns—are 1). On the plus side, the medical-pharmaceudical field followed by research into solutions to energy, environment and climate problems; 2). On the minus side, risks of genetic manipulations, GMO, other concerns related to health, preoccupations linked to the environment, and those that involve the science for the destructive purposes (European Comission, Eurobarometer, 2008:8).

The recently published study by Eurobarometer *Qualitative Study on the Image of Science and the Research Policy of the European Union* (http://ec.europa.eu/ public_opinion/quali/ql_science_en.pdf) examines the role (social) scientists are perceived to have by European citizens. Particularly significant for the readers of this volume is the fact that the role of social scientists in addressing health and environment related issues is barely recognized. Social sciences are often perceived as very different to natural or medical science (European Comission, Eurobarometer, 2008:31-32). It is interesting to note that while one of the top priorities almost all European citizens perceive as highly important (the issue of environment and health) is seen completetly outside the domain of social scientists.

CORDIS, interdisciplinary research programs and projects sponsored by EU, have been largely focussing on health and environment as purely natural scientific, medical or technical domains, leaving little room for social scientific expertise (http://cordis.europa.eu/home_en.html). As a result, very few social scientists worked on these multidisciplinary projects, leaving the research terrain to natural and medical scientists, technical experts, policymakers and interest groups. Hopefully, some changes in the involvement of social scientists in the European health and environment research are currently under way. Social scientists attempting to analyze European policies increasingly address the results of Eurobarometer surveys, complementing them by cross-national case studies.

Conflicting Interests

Issues associated with urban air, water and soil pollution remain high on European agendas. Unfortunately, little is known about the health impacts of pollutants, and the true scale of environmental degradation is also hard to access. While patients' rights and environmental groups produce research reports pointing to the dangers of chemicals and other pollutants, other European industries that often potentially counteract such efforts continue to prosper. We can think of a few examples as to how these conflicts of interest express themselves at the European level.

Health and environmental effects related to *climate change* in Europe are yet to be explored in much greater detail (Kovats et al., 1997; Watson et al., 2005; Brenner and Parker, 2009). The already existing evidence points to the fact that effects of increasing heat have particularly devastating effects on health of Europeans (Kosatsky, 2005, 148-149). The fear of flooding is also very much felt in the home countries of the editors of this volume, Holland and Belgium. Another, perhaps unintended, aspect of climate change influencing Europe is the increase in migration from environmental disasters prone areas from post-Communist and developing countries into Western Europe (Myers, 2001; Stacul et al., 2005; Kopnina, 2007). Measures to combat climate change are taken at the national, European and international level. As many critics have pointed out, however, political efforts at combating climate change are futile if industrial processes will continue at the present pace.

Dutch Example 1: Air Quality

One such industrial processes is that of constant expansion within European manufacturing industries, notably the automobile industry.

To take a concrete example mentioned in the Introduction of this volume, the car driving can be said to have a negative effect on *both* health and environment. As we may recall from the Introduction of this book, cars were discussed as seen in a paradoxical relationship between human and nature interests. Europe today is the largest producer of cars, with a 33% share of global production with about 16.900 thousands per year as of 2006 (European Automobile Manufacturers Association http://www.acea.be/). In 2006, there were 49,886,549 passenger cars produced in the world, with an increase of 6.45% over the previous year. For 2009, global vehicle sales were affected by the financial crisis leading to declines in the mature markets of the United States, Western Europe and Japan. Nonetheless, to date (April 2009), there are about 51,000,000 produced. Vehicle penetration is 749 vehicles per 1,000 people in the mature markets of the G7 (International Organization of Motor Vehicle Manufacturers).

Directive 2001/81/EC on national emission ceilings for certain atmospheric pollutants that cause acidification and the formation of ground-level ozone: sulphur dioxide (SO2), nitrogen oxides (NOx), volatile organic compounds (VOCs), and ammonia (NH3). After two years of negotiation, it was adopted by the Council of Ministers and the European Parliament in July 2001 (http://www.airclim.org/factsheets/factsheet16.htm). New Air Quality Directive, 2008/50/EC of the European Parliament and of the Council of 21 May 2008, on ambient air quality and cleaner air for Europe entered into force on 11 June 2008 (http://ec.europa.eu/environment/air/quality/legislation/existing_leg.htm).

The European Commission (EC) is checking up on action taken by European Union (EU) member states to comply with the EU's ambient air quality standard for dangerous airborne particles known as PM10., Nitrogen Dioxide and Benzene. The EC's action follows the entry into force on June 11, of a new EU air quality directive, which allows member states to request limited extra time to meet the PM10 standard under certain conditions. Particularly the particles linked in large part to car emissions, PM10 and NO2, were demonstrated to bring serious health risks by independent researchers. The EU has set the target limit of 40 microgram per cubic meter. In the case of the Netherlands, where the author resides, concerns about the fine particle matter have been expressed by both environmental organizations and different citizen groups.

Since the target of reducing fine particle matter has not been met and Dutch government gained deadline extension from the EU, the national Cooperation Air Quality Program (Nationaal Samenwerkingsprogramma Luchtkwaliteit) (NSL) has been put into place by the Dutch government in response to European initiative. The Royal Institute for Public Health and Environment (Rijksinstituut voor Volksgezondheid en Milieu) noticed a slight decline in the mass of fine particles in the air. However, this decline is not stable, and local concentration of such particles in certain urban areas has actually increased (Raad van Verkeer en Waterstaat, VROM-raad, Algemene Energieraad, 2008). Marga Jacobs, the chairwoman of another environmental organization with strong health focus, Vereniging Leefmilieu, commented that the fine particle concentration is still much higher than targeted standards and very unevenly distributed, endangering certain segments of the population disproportionably more than others (Oussoren, 2009:27; http://www.leefmilieu.nl/).

On the basis of Aarhus Treaty, environmental organizations can appeal against such decisions. Appeal against the decision to postpone the deadline has been made by one of the largest environmental organizations, Milieudefensie in 2008. Meta Meijer, responsible for Transport branch of this organization, has issued a statement: "With this [European] decision, Dutch government gets awarded for many years of ignoring the issue" (Milieudefensie,

2009). Other Dutch citizens' and activists groups have been protesting against this extension of the deadline.

While "green" or "eco" cars are more popular and some European states give tax breaks or priority parking privileges' to "sustainable" car owners, ministries of transport keep sponsoring building and expansion of the roads, car makers continue producing cars and European citizens, despite the recent financial crisis, keep on buying them. This situation largely applies to the Netherlands. Paradoxically, while the Dutch media abounds in examples of how poor air quality affects public health, there are few articles addressing the possible connection between increasing traffic and the air quality. The author did not attempt the detailed content analysis of the Dutch media on this subject, nor have any qualitative studies of the subject been made. While the scope of this chapter does not allow for detailed analysis of the Dutch situation, a few directions for future research could be outlined.

Studies of the effect of poor air quality on health and environment from societal perspective could help gain more insight into social and policy processes lying behind this question. Social scientists could examine economic, political social and cultural configurations connected with issues such as the health effects of the poor air quality. They could also examine the perceptions of the possible connection between poor air quality and car production as well as ideas about possible solutions by the Dutch citizens.

Dutch Case 2: Illegal Timber Trade

Another example of how certain EU policies have not yet led to positive results because of the conflicts with commercial organizations is illegal logging and import and export of timber into and out of EU (Falconer, 2008). The question of illegal logging is not only salient to the question of environmental protection, it has direct relevance to climate change and, indirectly, to European health. In 2003, European Union decided on the action plan for Forest Law Enforcement, Governance and Trade (FLEGT) to combat illegal logging (http://www.dfid.gov.uk/eupresidency2005/flegt.asp). So far, this plan has not shown many practical results, as the Joint NGO statement has shown (http://www.fern.org/media/documents/document_3384_3385.pdf).

The recent report of World Fund for Nature on Illegal Logging and the EU (WWF, 2008).

"Illegal timber is still a major problem in the EU, with an estimated 16% to 19% of EU imports being illegal or from suspicious sources," said Anke Schulmeister, Forest Policy Officer at WWF. "We need a specific and effective timber law in the EU, for which it is clear, there is wide public support." This law

"will hopefully be debated at EU level, is urgently required to ensure that only wood and paper products from legal sources are traded—with an obligation to provide proof for processors as well as traders. This is the only way to exclude illegal sources following processing and import through transit countries and to stop commercial illegal logging in EU Member States.

> A legal origin is only a first step towards sustainable provenance from ecologically and socially responsible forestry management. Consumers and companies already have the option

> today of using Forest Stewardship Council (FSC) certificate, which guarantees both the legality and sustainability of products" (WWF, 2008).

To date, political debates continue, and no such law has been issued yet. Despite the efforts, particularly by lobbying environmentalist groups, to curb illegal exports for over a decade, import and trade in timber, many European agencies, industries, and, last but not least, consumers continue buying illegal timber products. A recent report by the Greens, the European Free Alliance in the European Parliament, celebrates modest progress in adopting measures in combating illegal timber trade. In the paper entitled *Illegal timber trade: EU Parliament backs Green report to axe illegal deforestation and timber trade*, UK Green Member of European Parliament Caroline Lucas commented on the illegal trade of timber and wood products:

> "I am delighted that MEPs have backed my proposals to transform the Commission's vitally important—but distressingly weak—proposal on illegally-sourced timber into meaningful and effective legislation. For years, the EU has preached against illegal timber but continued to provide one of the biggest markets for it. Between 20% and 40% of global industrial wood production is estimated to come from illegal sources, and up to 20% finds its way into the EU—with biodiversity and forest-dependent peoples paying the price. The longer term effects are potentially even more serious, since deforestation accounts for nearly a fifth of the global total emissions causing climate change" (Lucas, 2009).

The Greens, as well as WWF reports, demonstrate large differences in adherence to European policies related to import and export of timber at the national level. Generally, recent Eastern European members of the EU (although not all) seem to practically ignore both European and national efforts at curbing illegal trade in timber.

Considering the author's own country of residence, the Netherlands, one could trace the historical developments in the timber trade industry, concerning both the local forests and import from other countries. The Dutch forests (which diminished greatly already during the Middle Ages making way for extensive agriculture) have been replenished in the recent decades. Since 1899, the Dutch government instated Staatsbosbeheer (State Forest Management) to manage Dutch forests and other nature areas. Most protected forest areas are designated as recreation areas, others as "natural monuments," yet others as parks. Around 25.000 hectares of forest are presently protected and managed by Staatsbosbeheer.

Picture 11. Forest.

As far as import of timber in the Netherlands is concerned, about half of it is imported from illegal timber trading countries. Director Andre de Boer of the Dutch Timber Trade Association VVNH admitted that the Netherlands imports a large quantity of illegal wood every year, and large scale improvements can be made in relation to imports from tropical countries (VVNH, 2003). These countries include Malaysia, Indonesia, Cameroon and Brazil (EU Forest Watch, 2001). This confession came in the midst of a Greenpeace protest campaign against illegal hardwood (2003), claiming that about 50 percent of the annual Dutch import of several hundred thousands tons of hardwood is illegally felled.

Historically, however, the Dutch seemed to have a special fascination with the tropics, not the least through the East Indian Company (the VOC) and other colonial connections. Many botanical gardens in Dutch cities such as Amsterdam and Leiden were laid out in the 17th century, serving mostly as medicinal herb gardens. At that time, herbs were of vital importance as the basis of medicines, and the city had just experienced a plague epidemic. Doctors and pharmacists trained in the preparation of prescriptions at the botanical Gardens (Hortus Botanicus). Since that time, the VOC ships brought exotic ornamental plants, which were greatly admired by the public (Hortus Botanicus).

While the scope of this chapter does not allow us to delve deeper into the history and attitudes of the Dutch citizens towards plants and nature, the briefly outlined case of timber in

The Netherlands may be instructive in outlining broad areas within which social scientists could contribute to the study of timber trade regulating policy at the national level. Social scientists could examine historical, social and cultural configurations present in the Netherlands in relation to distribution, use and consumption of timber, as well as attitudes towards the conceptions of "plants," "nature" and "forest" in relation to national values. Attitudes survey results could be instructive in indicating what values Dutch citizens attribute to their own country's or export country's forests or plants, and their beliefs about relation of forests to health and environment.

At the more theoretical level, social scientists may place the issue of illegal timber trade in the context of globalization, where the removal of certain trade restrictions in Europe creates a context for the observation of the effects of a process that is occurring across the globe. Trade in timber is a global process, yet national governments can work together or against generically formulated regulations. European governments may choose to largely ignore European regulations, and European institutions may fail to convince the national governments of their authority and more importantly of the necessity to follow the regulations. National citizens, represented by the members of European Parliament, as well as by lobbying and interest groups, may aid the process of regulation or work against it. While they need to be examined in much more depth, the Dutch cases demonstrate that just examining European policies does not reveal much about the role and the power of national government, institutions, agencies, lobby groups and organizations and, last and not least, the people themselves.

CONCLUSION

We may recall the discussion at the beginning of this chapter about the precautionary and substitution principles, which implies that the European Union has a responsibility to protect the public and the environment from exposure to harm where scientific investigation discovers a plausible risk in the course of having screened for other suspected causes (Schwarzenberger, 1999). The measures taken to mitigate suspected risks to health or environment can be relaxed only if further scientific findings emerge that point to an alternative diagnosis of the problem (Santillo et al., 1999). Considering these principles in relation to examples of air quality and illegal timber control, we may note that European member states respond quite differently to European policies and directives. The impact on national governments by EU policies can be characterized by (partial) resistance or (partial) compliance, considering how advantageous (or not) European countries view such policies for themselves and how much they truly aspire to act in accordance to common interests.

This chapter attempted to reveal some of the on-going trends and agendas related to health and environment as well as outlined a few areas in which social scientists could contribute to the development of effective European and national level policies. As mentioned in the Introduction of this chapter, social scientists could greatly contribute to the expansion of knowledge about the effects of public health and environmental policies at the grass-roots national level. It is argued that social scientists could contribute to understanding of this complex interaction between European policy initiative and its interpretation and adaptation at the national and regional levels. While political scientists could examine the institutions involved in both European and national policymaking, sociologists could examine the underlying social structures and activist lobbyist groups and anthropologists could focus on cultural factors playing a role in the interpretation and adaptation of policies at the grass-roots level.

As other chapters of this volume indicate, close cooperation between social scientists and policymakers, natural scientists and other stakeholders could lead not only to fruitful discussions but to effective means to assess, evaluate and monitor effectiveness of policies. The expertise of social scientists involved in either health- or environment-related studies, in combination with knowledge of other stakeholders' groups (such as health or environment specialists), may lead to greater problem-solving capacities in the areas of both health and environment at the European and potentially, indeed, at the global level.

Another area that could be addressed by social scientists is that of public opinions, interest and lobbying groups, expressing themselves both at European and national levels. By examining Eurobarometer, we may note that public opinion on matters concerning health and environment vary greatly from country to country. Reasons behind this variation, as well as shifts of opinions within one country over time, need to be better understood and theorized.

Social scientists could also examine the efficiency of European level policies at the national level and thus contribute to the on-going dialogue with other stakeholders, potentially leading to the problem-solving strategies. Social scientists, by the virtue of their distinct position as societal observers, examine the underlying social, political and cultural roots of environmental health policies, as well as examine the role of other stakeholder groups in shaping policy agenda at both European and national levels. Cross-national longtitudanal studies of policy initiatives could lead to greater understanding of interaction between European and local-level policies. To give a few simplified examples, economists could look into financial incentives, sociologists into social processes and interest groups, political scientists into the role of European and national institutions, anthropologists into national cultures and identities in relation to health and environmental policies. Both macro- and micro-level studies, as well as qualitative, quantitative and mixed methodologies are needed to tackle very complex issues associated with health and environment policies in Europe.

REFERENCES

Andersen, Z.J., Loft S, Ketzel M, Stage M, Scheike T, Hermansen MN, Bisgaard H (2008). Ambient air pollution triggers wheezing symptoms in infants. *Thorax,* 63:710-716.

Appadurai, A. (1990). Disjuncture and difference in the global cultural economy. In *Global Culture*, ed. M. Featherstone. London, SAGE.

Atkinson RW, Anderson HR, Sunyer J, Ayres J, Baccini M, Vonk JM, Boumghar A, Forastiere F, Forsberg B, Touloumi G, Schwartz J, Katsouyanni K (2001). Acute effects of particulate air pollution on respiratory admissions: results from APHEA 2 project. Air Pollution and Health: a European Approach. *American Journal of Respiratory and Critical Care Medicine*, 164:1860-1866.

Beck, U. (1992). *Risk Society: Towards a new modernity*, Londen, SAGE.

Beck, U. (1996). World Risk Society as Cosmopolitan Society. Ecological Questions in the Framework of Manufactured Uncertainties. *Theory, Culture & Society,* vol. 13, no. 4, 1-32.

Beck, U. (2006). Living in the world risk society. A Hobhouse Memorial Public Lecture given on Wednesday 15 February 2006 at the London School of Economics. In *Economy and Society,* vol. 35, Issue 3 August 2006, pages 329-345.

Biggeri, A., Baccini, M., Bellini, P. and Terracini, B. (2005). Meta-analysis of the Italian Studies of Short-term Effects of Air Pollution (MISA), 1990–1999. In *International Journal of Occupational and Environmental Health,* vol. 11, no. 1. http://www.ijoeh.com/index.php/ijoeh/article/view/128

Braback L, Forsberg B (2009). Does traffic exhaust contribute to the development of asthma and allergic sensitization in children: findings from recent cohort studies? *Environmental Health* 2009, 8:17. 16 April.

Brenner, S. A. and Parker, C. L. (2009). Health Effects of Global Climate Change: How Health Professionals Can Be Part of the Solution CME/CE. MedScape CME http://cme.medscape.com/viewarticle/589058

Brown, P. (2007). Toxic Exposures: Contested Illnesses and the Environmental Health Movement, Columbia University Press. New York.

Brown, P. Zavestoski, S. Luebke, T., Mandelbaum, J., McCormick, S. and Mayer, B. (2003). The Health Politics of Asthma: Environmental Justice and Collective Illness Experience in the United States. Social Science and Medicine 57:453-464. Reprinted in Where We Live, Work, and Play: A Critical Appraisal of the Environmental Justice Movement, David Pellow and Robert Brulle, eds. Cambridge: MIT Press.

Carrol, A. (1999). Corporate Social Responsibility Business & Society, vol. 38, no. 3, 268-295 http://bas.sagepub.com/cgi/content/abstract/38/3/268

Catton, W.R., & Dunlap, R.E. (1978). Environmental Sociology: A New Paradigm. *The American Sociologist,* 13, 41-49.

Cohen, M. J. (1997). Risk society and ecological modernisation alternative visions for post-industrial nations. *Elsevier.* vol. 29, Issue 2, March 1997, Pages 105-119.

Commission of the European Communities (2000). Communication from the Commission.
on the precautionary principle (COM(2000) 1), (Brussels, Belgium, 2000).

CORDIS Community Research and Development Information Services. European Commission http://cordis.europa.eu/home_en.html

Galassi, C., Biggeri, A., Ciccone, G. and Forastiere, F. (2005). Environment and Respiratory Diseases in Childhood: The Italian Experience. In *International Journal of Occupational and Environmental Health,* vol. 11, no. 1.

Giddens, A. (2009) *The Politics of Climate Change.* Polity Press.

EU Forest watch. Special report: EU illegal timber products. July/August. http://www.fern.org/media/documents/document_851_862.pdf

Europa: Gateway to the European Union. http://europa.eu/abc/12lessons/lesson_4/index_en.htm

European Automobile Manufacturers Association (2009). http://www.acea.be/

European Union Biocides Directive 98/8/EC European Commission http://ec.europa.eu/environment/biocides/basic.htm

European Commission (2008). Environment, Consumers and Health. Is your city the green role model? http://ec.europa.eu/news/environment/080522_1_en.htm

European Commission, 2008 Environment, Consumers and Health. Green Production and shopping to be boosted http://ec.europa.eu/news/environment/080716_2_en.htm

European Commission, Directorate-General for Research (2005). Expanding knowledge to improve our well-being: EU research on environment and health. EUR 21458 2005 ftp://ftp.cordis.europa.eu/pub/fp7/environment/docs/env_health_en.pdf

European Commission. Public Opinion Eurobarometer surveys: http://ec.europa.eu/public_opinion/index_en.htm

European Commission. Public Opinion Eurobarometer surveys (2008). Qualitative study on the Image of Science and the Research Policy of the European Union. The study conducted among the citizens of 27 member states. October, 2008. http://ec.europa.eu/public_opinion/quali/ql_science_en.pdf

European Environment and Health Committee http://www.euro.who.int/eehc/implementation/20080208_3

European Environment and Health Information System EEHIS (2009) Environmental and Health Issues http://www.enhis.net/object_class/enhis_home_tab.html

European Environmental Agency (2008) Environmental Health http://ec.europa.eu/health-eu/my_environment/environmental_health/index_en.htm

European Environmental Bureau (1999) EEB Position on the Precautionary Principle. http://www.eeb.org/publication/1999/eeb_position_on_the_precautionar.html

European Environment and Health Action Plan 2004-2010 (The Public Health Portal of the European Union) http://ec.europa.eu/health-eu/my_environment/environmental_health/index_en.htm

European Environment and Health Committee (2009). Progress on environment and health http://www.euro.who.int/eehc/ctryinfo/ctryinfo

European Environment and Health Committee (2009). http://www.euro.who.int/eehc

European Investment Bank (2009). European Principles for the Environment http://www.eib.org/infocentre/epe/index.htm

Falconer, J. (2008). DG Development European Commission: EU Action Plan for Forest Law Enforcement, Governance and Trade—Chatham House Update. January 2008 http://www.illegallogging.info/presentations/17-180108/falconer.pdf

Forest Law Enforcement, Governance and Trade (FLEGT) http://www.dfid.gov.uk/eupresidency2005/flegt.asp

Goddard, V., Llobera, J. and C. Shore. Introduction. The anthropology of Europe. In *The anthropology of Europe*, ed. V. Goddard *et al.* Oxford, Berg, 1994.

Grammenos, S. (2006). Illness, disability and Social Inclusion (report). European Foundation for Improvement of Living and Working Conditions. Centre for European Social and Economic Policy (CESEP), Brussels. 17 February http://www.eurofound.europa.eu/publications/htmlfiles/ef0335.htm

Hansteen, I., Kjuus, H. and S. I. Fandrem (1998). Birth Weight and Environmental Pollution in the County of Telemark, Norway. In *International Journal of Occupational and Environmental Health*, vol. 4, no 2. http://www.ijoeh.com/index.php/ijoeh/article/view/720

Holmes, D.R. (2000). *Integral Europe. Fast-capitalism, multiculturalism, neo-fascism.* Princeton, Princeton University Press.

Hortus Botanicus [Botanical Gardens] Amsterdam. http://www.hortus-botanicus.nl/

IAEA International Atomic Energy Agency (1986, 1993). Environmental consequences of the Chernobyl accident and their remediation. IAEA, Vienna.

International Organization of Motor Vehicle Manufacturers http://www.oica.net/

Joint NGO Statement: How to ensure the EU FLEGT licensing scheme will contribute to controlling illegal timber and sustainable forest management. September 2005. http://www.fern.org/media/documents/document_3384_3385.pdf

Kolk, A. and Pinske, J. (2008). *International Business and Global Climate Change.* Routledge.

Kopnina, H. (2005). East to West Migration: Russian Migrants in Western Europe. Ashgate, Aldershot.

Kopnina, H. (2007). *Tourism and Migration: Formation of new social classes.* Cognizant, New York, Sydney, Tokyo.

Kosatsky, T. (2005). The European Heat Waves. Eurosurveillance;10:148-149.

Kovats, S., Patz, J. H., Dobbins, D. (1997). Global Climate Change and Environmental Health: Proceedings of the 1997 Annual Conference of the Society for Occupational and Environmental Health. International Journal of Environmental and Occupational Health. Vol. 4, No 1 (1998) http://www.ijoeh.com/index.php/ijoeh/article/view/715/596

Kroll-Smith, S., P. Brown, and V. J. Gunter (Eds). (2000). *Illness and the Environment: A Reader in Contested Illness*, New York University Press.

Lash, S., Szerszynski, B., Wynne, B. (1996). Risk, environment and modernity: towards a new ecology. London, SAGE.

Lucas, S. (2009). Press release 'Illegal timber trade: EU Parliament backs Green report to axe illegal deforestation and timber trade' The Greens. European Free Alliance of the European Parliament. http://www.greens-efa.org/cms/pressreleases/dok/280/280457.htm

McCormick, J. (2008). Understanding the European Union. A Concise Introduction. Palgrave Macmillan.

McDonald, M. (1996). "Unity in diversity." Some tensions in the construction of Europe. *Social Anthropology* vol. 4 no. 1, 47-60.

MacDonald, S. (1993) (Ed.) *Inside European Identities.* Oxford, Berg.

McDonough, W. and M. Braungart (2002). *Cradle to Cradle. Remaking the way we make things.* North Point Press, New York.

Medvedev, G. (1991). *The Truth About Chernobyl.* New York. Basic Books, Inc.

Milieudefensie (2009). Nederlandse overheid beloond voor jarenlang niets doen Uitstel Luchtkwaliteitsnormen; Milieudefensie gaat in beroep. [Dutch government rewarded for many years of inaction. Milieudefensie in appeal against the decision. http://www.milieudefensie.nl/verkeer/nieuws/uitstel-luchtkwaliteitsnormen-milieudefensie-gaat-in-beroep/view?searchterm=meta

Myers, N. (2001). *Environmental refugees: a growing phenomenon of the 21st century.* Phil Trans R Soc Lond B.;357:609-613.

Nielsen, J. (2009). The Killer Fog of '52: Thousands Died as Poisonous Air Smothered London News Report, NPR (National Public Radio) May 19, 2009. http://www .npr.org/templates/story/story.php?storyId=873954

Osborn, A. (2006). Fallout from Chernobyl Will Cause 100,000 Deaths, Says Greenpeace. *The Independent,* April 19.

Oussoren, G. (2009). Strijd Tegen Stoffen. Verkeer en veehouderijen zijn grootste luchvervuilers. [A fight against particles. Traffic and cattle farming are greatest air polluters]. Eigen Huis Magazine, 5 May.

ProClim (1997). Research on Sustainability and Global Change—Visions in Science Policy by Swiss Researchers, Bern, CASS/SANW: 39.

Raad van Verkeer en Waterstaat, VROM-raad, Algemene Energieraad (2008). Een prijs voor elke reis. Een beleidsstrategie voor CO2 reductie in verkeer en Vervoer [A price for single trip. A policy starategy for CO2 reduction in traffic and transport]. Gezaamenlijk advies. January.

Recuerda, M. A. (2006). Risk and Reason in the European Union Law, 5 European Food and Feed Law Review.

Raffensperger C. & J. Tickner (Eds.) (1999). *Protecting Public Health and the Environment: Implementing the Precautionary Principle*. Island Press, Washington, DC.

Rucht, D. (2006). Political Participation in Europe. In *Contemporary Europe*. Edited by R. Sakwa and A. Stevens. New York. Palgrave Foundations.

Santillo, D., Johnston P., Singhofen, A. (1999). The way forward: An Alternative, Precautionary Approach to the Regulation of the Manufacturing, Marketing and Use of Chemicals in Europe, Greenpeace. February 1999.

Schwarzenberger, A. (1999). Theoretical Concepts of the Precautionary Principle, Discussion Paper for the EEB, August.

Shore, C. (2000). *Building Europe. The cultural politics of European integration.* London, Routledge, 2000.

Shore, C. and D. Nugent, (Ed.) (2002). *Elite Cultures. Anthropological perspectives.* London, Routledge, 2002.

Smith, C. (2003). Human Rights, Environment, and Individual Action. In *International Journal of Occupational and Environmental Health*, vol. 9, no 1. http://www.ijoeh.com/index.php/ijoeh/article/view/20

Stacul, J., Moutsou, C. and Kopnina, H. (Eds.) (2005). Crossing European Boundaries: Beyond Traditional Geographical Categories, Berghahn Books, New York and Oxford.

Stirling, A. et al. (1999). On Science and Precaution in the Management of Technological Risk. Sussex.

Speth, J. G. (2008). The Bridge at the Edge of the World. Capitalism, the Environment, and Crossing from Crisis to Sustainability. Yale University Press.

Staatsbosbeheer (State Forest Management) http://www.staatsbosbeheer.nl/Over%20Staatsbosbeheer.aspx

Watson R, P. J, Gubler D, Parson E, Vincent J. (2005). Environmental health implications of global climate change. *Journal of Environmental Monit.* 7:834-843.

World Health Organization (WHO) Quantifying environmental health impacts. Preventing disease through healthy environments: Towards an estimate of the environmental burden of disease (2006) (http://www.who.int/quantifying_ehimpacts/publications/preventingdisease/en/index.html)

World Health Organization (WHO) Regional Office for Europe (2009). Health development action for south-eastern Europe (SEE) http://www.euro.who.int/stabilitypact

World Health Organization (WHO) Regional Office for Europe (2009). Environmental burden of disease (EBD) in the European Region http://www.euro.who.int/envhealth/data/20090108_1

World Health Organization (WHO) Regional Office for Europe (2009). Environment and Health. Themes. http://www.euro.who.int/envhealth/topics/20060220_1

World Health Organization (WHO) Regional Office for Europe (2004). Burden of disease attributable to selected environmental factors and injuries among Europe's children and adolescents (http://www.euro.who.int/childhealthenv/Monitoring/20040519_1)

World Health Organization (WHO) Regional Office for Europe (2002). Children's health and environment: a review of evidence (http://www.euro.who.int/childhealthenv/ Monitoring/20020725_4)

World Wildlife Fund for Nature (2008).

Illegal Logging and the EU http://assets.wnf.nl/downloads/eu_illegal_logging_april_2008.pdf

Wynne, R. (2008). Employment guidance services for people with disabilities/ Chronic illness and job retention http://www.introdm.be/idm/idm01.nsf/52b2da8b666e069080256 aaa002ab228/995022d8285bee6fc125727300383d77/$FILE/Presentation_Wynne.pdf

In: Health and Environment: Social Science Perspectives ISBN 978-1-60876-216-3
Editors: Helen Kopnina and Hans Keune

Chapter 11

ENVIRONMENTAL HEALTH ORGANIZING IN A GLOBALIZING WORLD: THE EMERGENCE OF A GLOBAL ANTI-TOXICS MOVEMENT AND ITS POLITICAL, LEGAL AND ECONOMIC CHALLENGES

Stephen Zavestoski

ABSTRACT

This chapter begins by arguing that although the chemical industry has been international since its birth, the last half of the 20th century saw an unprecedented spread of toxic chemicals around the world. The next section examines evidence that just as economic globalization was leading to the global spread of toxic hazards, a parallel process was occurring in which social movements to resist toxic threats were becoming increasingly global. Particular attention is paid to the evolution of the international organization Pesticide Action Network and the constellation of other organizations in its orbit. Despite certain successes of these organizations, there are significant challenges to making an impact on the global spread of toxics. Three types of challenges—political, legal, and economic—are discussed. Political obstacles include limited ability to mobilize constituents outside of affected areas and weak regulatory regimes at the domestic level. Legal obstacles include the challenge of acquiring expertise to engage the legal process and the absence of legal precedent for prosecuting multinational corporations. Economic obstacles range from the costs entailed in monitoring and documenting health impacts in a community to the lack of monetary resources needed to launch trans-national campaigns. The conclusion points out what social scientists interested in health and environment gain by understanding how environmental health organizing, and in particular the global anti-toxics movement, is shaped by the political, legal and economic challenges of the rapidly globalizing world and points to possible future challenges for the global anti-toxics movement and environmental health advocates.

INTRODUCTION

The Industrial Revolution uniquely transformed both natural environments and human health, as Hans Keune and Helen Kopnina note in the Introduction to this volume. Keune and Kopnina introduce McDonough and Braungart's cleverly conceived heuristic exercise that asks us to "Design a system of production that puts billions of pounds of toxic material into the air, water, and soil every year; produces some materials so dangerous they will require constant vigilance from future generations; … [and] requires thousands of complex regulations—not to keep people and natural systems safe, but rather to keep them from being poisoned too quickly …" (McDonough & Braungart, 2002: 18). McDonough and Braungart go on to note that the Industrial Revolution was not designed at all. Precisely for this reason, its consequences for the environment and human health have been profound.

At the same time, the Industrial Revolution brought with it scientific, technological, medical, and even public health advances that have greatly improved the quality of life for certain segments of the human population. Paradoxically, these developments sustain a logic in which environmental and human health risks introduced by industrial or technological advances typically are seen as outweighed by the benefits of the advances. As Keune and Kopnina point out, "proponents of globalization, industrialization and development argue that it is precisely through globalization and technological development that both health and environmental problems could be tackled." Do the contributions to human well-being brought about by industrialization, and later globalization, outweigh the environmental and human health consequences? To convince ourselves that they do depends on a short sightedness verging on blindness. After all, if our advances undermine the long-term ability of the earth's ecosystems to sustain healthy human lives, then what have we gained?

This chapter explores a significant human social response to the physical effects of the perverse logic that justifies a system "that puts billions of pounds of toxic material into the air, water, and soil every year [and that] requires thousands of complex regulations … to keep [people] from being poisoned too quickly" (McDonough & Braungart, 2002: 18). The response of which I speak is the formation of a global network of social movement organizations that works to counter the consequences of the global diffusion of toxic chemicals. I call this network the global anti-toxics movement, and it is a part of the broader environmental health movement discussed elsewhere in this volume. After providing evidence of the global spread of chemical hazards over the last half of the 20th century, I explain how the global anti-toxics movement emerged in response and describe its evolution over the last 25 years. The remainder of the chapter focuses on the main obstacles—political, legal and economic—that the movement faces. The conclusion points out what social scientists interested in health and environment gain by understanding how environmental health organizing, and in particular the global anti-toxics movement, is shaped by the political, legal and economic challenges of the rapidly globalizing world and points to possible future challenges for the global anti-toxics movement and environmental health advocates.

THE CHEMICAL INDUSTRY, GLOBALIZATION, AND THE SPREAD OF TOXICS

The industrial revolution's rapacious appetite for raw materials used in the production of glass, textiles, and metals gave birth to the chemical industry around the mid-1700s. Early developments ranged from experiments in England in the production of sulfuric acid needed in the manufacturing of dyes and metals, to the development of alkalis in France for glass-making, soap-making, dyestuffs, and bleach. In addition to England and France, the industry developed centers in Holland, Germany, and the United States.

Even before the 19th century, the industry had a global reach. By the time of the French Revolution, the French chemical industry relied on raw materials imported from Sicily, Spain, Scotland, India, Peru and the West Indies (Aftalion, 2005).[1] The industry's expansion across the 19th century followed the conventional path of the colonial powers. Raw materials were extracted from colonies and transported back to Western Europe or North America, where they were processed into the chemical products increasingly essential for industrial production of goods. But by the mid-1800s, as movements to reform the hazardous working conditions of industrial manufacturing gained momentum, a number of companies were beginning to set up production elsewhere around the world. Britain's United Alkali Company was operating in China and Japan by the mid-19th century. Later in the century, British and American companies began operating in India.

Although the industry was slowly developing a global network of chemical plants, most of its international activity up until the 1960s involved sourcing raw materials and sales of products manufactured elsewhere. According to a Dow Chemical Company historian, when Dow established an international division in the 1950s, its "employees were order-takers and pushers of surplus product" (Brandt, 1997:297). In some cases, Dow's international sales force discovered that industries in developing countries, like Indonesia, lacked foreign currency to buy from suppliers in their own regions. But with U.S. aid dollars, some industries could afford Dow's products.

More rapid expansion of production facilities began in the second half of the 20th century, when the forces of globalization created conditions conducive to the expansion and mobilization of a wide range of industries. In addition, population and economic growth in the developing world meant new markets were emerging for the industries' products. For example, by the 1960s, the German company BASF had built plants in Argentina, Australia, Belgium, Brazil, France, United Kingdom, India, Italy, Japan, Mexico, Spain and the United States. U.S.-based Du Pont, another major player in the international chemical industry, also expanded during this period. Increased demand, higher growth potential, and cheaper costs of production led the company to double foreign sales between 1960 and 1970, and triple investments in foreign subsidiaries (McNamee, 1983: 212).

---

[1] In fact, when the chemical industry in the U.S. surpassed Europe's chemical industry in production and sales by the mid-1900s, it was primarily due to the fact that American companies had access to domestic mineral deposits and to oil, coal, and natural gas.

Globalization's Link to Global Chemical Production and Consumption

The preceding section illustrates a pattern of global expansion by the chemical industry, first to seek out raw materials for chemical manufacturing processes, and later in the form of new manufacturing facilities abroad. Oddly, many of the most popular definitions and theories of globalization fail to take up systematically this material dimension of globalization. One example is Giddens' definition of globalization as "the intensification of worldwide social relations which link distant localities in such a way that local happenings are shaped by events occurring many miles away and vice versa" (Giddens, 1990: 64). Giddens emphasizes "how the emergence of instantaneous global communication and mass transportation, as well as the expansion of complex global systems of production and exchange, reduce the hold of local environments over people's lives" (Inda & Rosaldo, 2008: 11). This understanding of globalization fails to acknowledge the role played by the global spread of industrial hazards and toxic risks, for example, in determining the viability of local environments to sustain human health.

Many other globalization definitions and theories similarly focus on the social, cultural, political, and economic consequences of the intensification of global interconnectedness. For Harvey (1990), as with Giddens (1990), attention is paid to the compression of space and time. Other approaches similarly emphasize the increased "extensity, intensity, velocity and impact" reflected in the transformation of spatial organization of social relations and transactions (Held et al., 1999: 16).[2] Appadurai (1996), on the other hand, is primarily concerned with the tension between cultural homogenization and cultural heterogenization. Steger prefers that "globalization be used to refer to a set of social processes," at the core of which are "shifting forms of human contact" (2003: 8). Beck, though not working strictly on globalization, comes closest to offering a theory relevant to understanding the local environmental and health consequences of the global spread of toxics. Even for Beck (2009), however, "world risk society" is defined less by the local risks posed by the global spread of industrial, technological or other hazards, and more by the global risks that nurture global reflexivity.

What is unique about the global expansion of the chemical industry is that the environmental and human health impacts of this spread are primarily local. Although there is evidence of PCBs in the fat of arctic ringed seals and of dioxins in the breast milk of mothers in the most undeveloped parts of the world, where the chemical industry's global spread has triggered social responses is in those locations where a factory has contaminated a nearby village or workers have experienced some form of collective exposure. In other words, just as Marx believed that the organization of work in capitalist society, by placing exploited

[2] Held and colleagues (1999) include a chapter titled "Catastrophe in the Making: Globalization and the Environment," in which they attempt to theorize the transformation of environmental conditions using this definition. Their use of the example of acid rain illustrates the limits of their framework. They note, for example, that the greater the ratio of externally generated to internally generated pollution, the more tightly the problem of acid rain is linked to globalization. This chapter's focus on the global spread of the chemical industry's production capacity—as opposed to the global spread of toxics emitted from a geographically finite number of production points around the world—highlights the nuanced relationship between forms of economic globalization that locate inherently hazardous forms of economic activity in new spots around the world and the social movement responses, also global in nature, that emerge when newly contaminated local environments begin to compromise people's health. The same logic applies to the global spread of inherently hazardous products (as opposed to manufacturing processes), such as pesticides, that transform otherwise health-sustaining local environments into dangerous environments.

laborers in factories where they work side-by-side, would lead to the collective conscience necessary to mobilize protest, globalization's concentration of toxic hazards in workplaces and in communities where chemical plants operate creates the conditions for a collective conscience to form.

An example of this process can be found in the growth of the use of chemical fertilizers and pesticides in the agriculture sector. For the chemical industry to expand as rapidly as it did over the last half of the 20th century, new markets in the developing world were essential, as well as opportunities to cheaply manufacture chemicals and chemical products in the new places where they would be sold. Put simply, rather than economic development attracting the chemical industry, the industry saw development as a means to create new markets. Using foreign direct investment (FDI) as a proxy for economic globalization, defined here as an increase in economic integration, reveals the link between globalization and the global expansion of chemical production and consumption. This is precisely what Jorgenson's (2007) study of patterns of FDI in developing countries demonstrates. Jorgenson found that as FDI increased in the developing world between 1970 and 1990, use of pesticides also increased. Integrating agricultural systems in developing countries into the global capitalist system is typically done through the use of foreign capital. The infusion of foreign capital means that investors want cash crops to be planted, and at a scale that requires the industrial monoculture system in which pesticides play a vital role in order to ensure a return on investment. Put differently, foreign investment in developing countries transformed their agricultural sectors into new consumers of chemical industry products.

There is also evidence that growth in the use of pesticides was accompanied by the growth of facilities to produce these and other chemical products. In fact, throughout the decades of increasing FDI in countries of the developing world, growth in hazardous industries outpaced all other types of industrial growth. If pesticide use was being driven by internal economic growth, its rate of increase would have paralleled, not exceeded, growth in other industrial sectors. According to Frey, "The growth rate of hazardous production processes has been greater than that of the overall industrial growth rate in the periphery, and it has been far greater than the growth rate of hazardous industries in the core" (1998: 67). Frey also argues the case that stricter regulations drove hazardous industries to developing countries:

> regulatory actions undertaken by … agencies have further curtailed the production and marketing of hazardous pesticides … Chemical TNCs [trans-national corporations] responded to such controls (and an increasingly stagnant market) by turning to LDCs [less developed countries] as markets and formulation/manufacturing sites for banned and highly regulated pesticides because regulations have traditionally been nonexistent or weakly enforced in these countries. … TNCs often exaggerate the economic benefits of pesticide use and minimize (or avoid discussing) the health and environmental risks. Lax regulatory controls on the production, transport, storage, marketing, and use of pesticides in LDCs have been translated directly into reduced costs of doing business for TNCs and increased profits (Frey, 1995: 156).

Within the pesticides and agricultural chemicals segment of the larger chemical industry, there is another powerful example of globalization's link to the spread of chemicals: the Green Revolution. Linking Third World development to increased agricultural yields, the Green Revolution stimulated growth in the agricultural chemicals sector that had begun to expand following World War II. Union Carbide Corporation, the company later responsible for the 1984 chemical disaster in Bhopal, India that killed more than 4,000 people, was one of many companies that sought to profit from the demand for agricultural chemicals generated by the Green Revolution.

Triggered by Norman Borlaug's work at the Rockefeller Foundation-funded Centro Internacional de Mejoramiento de Maiz y Trigo (CIMMYT), or International Maize and Wheat Improvement Center, in Mexico, the Green Revolution was based on new high-yield seed varieties that required intensive chemical inputs to grow. Borlaug's work at the CIMMYT resulted in a high-yield, semi-dwarf wheat that led to Mexico's self-sufficiency in wheat production by 1951 (Hesser, 2006). Next, Borlaug convinced the Rockefeller Foundation to apply his work to famine-stricken India and Pakistan. Dealing with a series of famines related, in part, to inconsistent monsoons, the Government of India was looking for ways to become more self-sufficient in its food production. Borlaug topped his results in Mexico when Pakistan nearly doubled its yields of wheat in five years while becoming self-sufficient in wheat by 1968. India became self-sufficient in wheat, a grain not historically part of Indian agriculture, by 1974. Thus was born the Green Revolution, the basic components of which were increased crop yields through new crop cultivars, increased use of mechanical irrigation, systematic application of fertilizers and pesticides, and the mechanization of agricultural processes.

Companies in the agricultural chemicals sector had all the stars aligned for them when it came to the creation of new markets for their products in developing countries. Much of the R&D for the high-yield varieties had been done through funding from a private foundation. Furthermore, once there was evidence that the new cultivars increased yields, the World Bank, the U.S. Agency for International Development, and other development agencies around the world formulated policies that would make loans or other assistance dependent on farmers embracing the new agricultural methods. The World Bank had already been engaged in strategies, according to Brian Tokar, at least since the 1950s, aimed at shifting developing countries toward cash crop production:

> With the development of the so-called "high yielding varieties" of wheat and rice in the 1960s, Bank lending was often limited to governments that promoted the use of the new seeds and accepted their dependence on mechanization and costly chemical inputs. Seeds were often given away and enterprising farmers were offered attractive loan packages subsidized by the Bank. These policies came to define the so-called "Green Revolution" of the 1960s and 1970s. (Tokar, 2004)

The benefits to agricultural chemicals manufacturers were obvious: more farmers growing higher yield cultivars would mean greater demand for fertilizers to replenish the soil and for pesticides to fight off the pests in the pest-prone monocultures of the industrially managed farms. But it also meant more people around the world exposed to the risks of chemical hazards. In time, a movement would emerge to challenge the forces justifying these risks.

TRANS-NATIONALIZATION OF ENVIRONMENTAL HEALTH MOVEMENTS

When the global anti-toxics movement first took shape in the early 1980s, concern over the occupational hazards of work in an industrial society had existed for more than 100 years, the modern environmental movement had been around for about a decade, and local anti-toxics struggles had been going for about the same length of time. Yet, none of these were direct influences on the emergence of the global anti-toxics movement. In fact, the movement's roots are linked to the consumer's movement that arose in the late 1950s. More specifically, the global anti-toxics movement that formally arose in the early 1980s can be traced to 1971, when David Weir, working for the U.S. Peace Corps in Afghanistan, noticed that a packet of Kool-Aid he had bought listed cyclamates, a sweetener Weir knew had recently been banned in the U.S., among its ingredients. This experience triggered Weir's decade-long investigation into the problem of U.S. corporations dumping their banned products on overseas markets. His work culminated in 1981, in *The Circle of Poison*, a book focusing, in particular, on the dumping of pesticides. *The Circle of Poison* illustrated how chemical companies were protecting profits by selling, or flat out dumping, products that had been banned in their countries of production in countries without the regulatory infrastructure to evaluate the risks of agricultural chemical use.

Weir's book was followed in 1982, by *A Growing Problem: Pesticides and the Third World Poor*, by David Bull. That same year, Anwar Fazal, a Malaysian activist who at the time was the first person from a developing country to head the International Organization of Consumers Unions (IOCU; later known as Consumers International), organized a meeting in Penang, Malaysia to explore the possibility of an international network of activists aimed at halting "the indiscriminate sale and misuse of chemical pesticides throughout the world" (IOCU, 1982). Fazal, who had led IOCU's campaign for a code of conduct to regulate the marketing of infant formula, drew on his extensive network ties to pull together the May 1982 meeting where the Pesticide Action Network (PAN) was born. At its inaugural meeting, which included the authors Weir and Bull,[3] participants proposed a model that would be based on a communication network with three main components: regional nodes that would serve as action hotline centers; data hotline centers, and a newsletter. The idea was to link the communication network to Consumer Interpol, a global citizen alert system on hazardous products and processes created by IOCU.

Consumer Interpol was designed to alert citizens' groups around the world about products that had been identified as a hazard by a particular country's regulatory authorities. The underlying idea was to step in to provide information that could protect the safety and health of consumers when the regulatory agencies to do so either did not exist in certain countries or lacked the resources to do so. Linked to IOCU, Consumer Interpol was also designed to draw on IOCU's ability to launch global campaigns against products or companies. According to Fazal:

---

[3] Fazal, familiar with their work, invited Weir and Bull. Weir represented the Center for Investigative Reporting and Bull represented OXFAM. They were among the 14 participants from the developed world: six from the U.S., one from the UK, four from Switzerland, one from Japan, and two from Australia). The remaining 25 participants, 11 of whom were tapped through IOCU's network of consumer's unions, were from developing countries.

> If a specific company is involved, we will make sure that as much effort is made at the source of the problem as is made at chasing the particular products in different parts of the world. We will go to the factories where these hazardous goods are made, and to the company that manufactures them. With this kind of a Consumer Interpol, we can give a multinational response to the multinationals (Rothschild, 1982).

Why would the global environmental health movement grow out of the consumer's movement? Having begun as a conglomeration of the British, Dutch, Belgian and U.S. consumer movements, IOCU served as a clearinghouse of information about product testing procedures (Hilton, 2007). IOCU evolved and by 1982, according to Fazal, two-thirds of its membership was from developing countries, including 62 consumer organizations in Brazil, where a decade earlier there had been none (Rothschild, 1982). Fazal became the first president of IOCU from a developing country in 1978. Most importantly, IOCU had gained valuable knowledge of how to organize trans-nationally, and how to engage in international negotiations at the level of the United Nations. IOCU utilized its growing experience and influence to develop new sister networks like the International Baby Food Action Network, Health Action International, and, of course, Pesticide Action Network. These networks, each of which came into existence during Fazal's leadership, represented IOCU's new focus on challenging the power of multinational corporations.

Just as the consumer movement identified multinational corporations as the primary source of new risks related to the global spread of consumer products, so, too, did the nascent global anti-toxics movement see multinational corporations as the driving force behind new toxic risks. In fact, the two are quite intricately related. Many environmental health threats, after all, are the result of the by products of the manufacturing of consumer goods, or the processing of the raw materials to make those goods. Focusing strictly on the risks related to the goods themselves, as the consumers movement was doing, was insufficient. The pesticides that individuals were consuming when eating fresh fruit and produce were also contaminating the workers who applied the chemicals. Furthermore, through drift and absorption into groundwater, the chemicals used in agriculture and elsewhere were having a much broader impact. The realization of this, however, lagged behind the initial spread of the chemical industry during the Green Revolution.

So when PAN formed in 1982, it was very much a product of the work done by IOCU over the preceding two decades. IOCU had learned a great deal about how to challenge the power of multinationals, how to create global communication networks, how to negotiate at the level of the UN, and how to split off "subsidiaries," like PAN, that could take on new issues while IOCU remained focused on its core objectives. PAN quickly developed its own identity and developed new alliances and areas of focus that would not have been possible for IOCU. In particular, farmers became a natural ally.

PAN immediately called for an expansion of traditional, biological and integrated pest management; import and export controls on the movement of hazardous chemical pesticides; immediate notification from countries when they ban or restrict pesticides; the elimination of support from international financial institutions and development agencies for projects relying on pesticides that cannot be used safely; and "an end to the vicious cycle whereby hazardous pesticides applied in the Third World end up as residue in food products consumed by people all over the world" (IOCU, 1982). Finally, linking its efforts specifically to the growth in pesticide use resulting from the Green Revolution, PAN called for: "Reversal of the practice by nine international 'Green Revolution' research centres, including the International Rice Research Institute (IRRI) based in the Philippines, of developing and distributing seed varieties which are heavily dependent on expensive and hazardous inputs like chemical pesticides and fertilizers" (IOCU, 1982).

Within two years of its founding, PAN had already held several regional meetings in Latin America, Africa and Europe and engaged in negotiations with the UN's Food and Agriculture Organization on the development of the International Code of Conduct on the Distribution and Use of Pesticides that was approved in 1985. It had also organized and launched its international "Dirty Dozen" campaign. PAN's dirty dozen included aldicarb, toxaphene, chlordane and heptachlor, chlordimeform, chlorobenzilate, DBCP, DDT, the "drins" (aldrin, dieldrin and endrin), EDB, HCH and lindane, paraquat, parathion and methyl parathion, pentachlorophenol, and 2,4,5-T. At the worldwide launch of its campaign on June 5, 1985, actions included protests at plants manufacturing chemicals on the list such as the Dow plant in New Zealand that produced the herbicide 2,4,5-T.

PAN's Dirty Dozen campaign would eventually be transformed years later into the Stockholm Convention on Persistent Organic Pollutants. In fact, negotiations over the Stockholm Convention became so time-consuming that PAN spun off a new organization known as the International POPs Elimination Network (IPEN). IPEN became one of the most prominent nongovernmental organizations (NGOs) in negotiations over the Stockholm Convention and continued to play an important role in subsequent chemicals-related international negotiations.

Even prior to work on the Stockholm Convention, PAN had developed the notion of "prior informed consent," or PIC, which would become the basis of the Rotterdam Convention on the Prior Informed Consent Procedure for Certain Hazardous Chemicals and Pesticides in International Trade. The Rotterdam Convention, which includes among its objectives "to promote shared responsibility … in the international trade of certain hazardous chemicals in order to protect human health and the environment from potential harm," was shaped by the voluntary PIC procedure adopted in the U.N. Food and Agriculture Program's 1987 revision of the International Code of Conduct on the Distribution and Use of Pesticides. PAN's initial intent with PIC, which is very much captured in the Rotterdam Convention, was to give nations that might be importing hazardous chemicals or pesticides access to information on the regulatory status of the chemicals and pesticides in the exporting country. The real force behind PIC is in giving a country the right to refuse to accept imports based on assessments of their environmental and human health risks.

Structured around five autonomous regional centers, PAN's success in lobbying for the inclusion of a PIC standard in the United Nation's Food and Agriculture Organization's (FAO) code of conduct and later in the Rotterdam Convention depended on the mobilization of PAN partners in many countries around the world. Partner organizations gathered evidence of the harm done from the unregulated imports of chemicals and pesticides into their countries. They presented their findings to the FAO, and later to the U.N. Environment Program, and in 1998, saw the Rotterdam Convention adopted. In 2004, the Convention entered into force. PAN also played important roles, often through its progeny, IPEN, in ushering through the original chemicals management treaty, Basel Convention on the Control of Transboundary Movements of Hazardous Wastes and their Disposal, and the most recent chemicals management agreement, the Strategic Approach to International Chemicals Management as described in the "Dubai Declaration."

Despite these successes, chemical manufacturing and waste continue to grow. In the U.S. alone, revenues of the "chemicals and allied products" sub sector in 2008 totaled $715 billion. Furthermore, certain key countries have yet to ratify a number of the above treaties, and what appears to be a global commitment to supporting safe and sustainable use of chemicals by developing countries—the Dubai Declaration—is woefully under-funded. The next section discusses the major obstacles preventing the global anti-toxics movement from being more successful.

Obstacles to Global Environmental Health Organizing

To illustrate the types of obstacles faced by global anti-toxics organizers, this section will focus primarily on a campaign by the Pesticide Action Network–Asia & the Pacific (PAN-AP) and other organizations to ban the herbicide paraquat (1,1'-dimethyl-4,4'-bipyridylium dichloride) (also known by the trade name Gramoxone).[4] Paraquat, which disrupts photosynthesis in plants, has been in use since the 1960s. It is usually sold in liquid concentrate form and diluted by farm workers in the field before application. It is used widely to control weeds that affect oil palm, banana, coffee, plantain, sugarcane, corn, citrus fruit, oil palm, macadamia, mango, avocados, and other types of crops. It is also used as a pre-emergent herbicide where no-till agriculture is practiced, as a defoliant on cotton, for destruction of potato stems and tops, and in roadside weed control (Wesseling et al., 2001).

Global sales of paraquat were estimated at $396 million for 2001. Syngenta, a Swiss chemical company and the world's second largest agrochemical producer, which held the patent for paraquat until recently, accounts for 50% of annual sales (Fernandez & Bhattacharjee, 2006). The Asian market accounts for roughly 45% of all sales, or $178 million by 2001 figures. Latin America accounted for nearly $115 million of 2001 sales (29%) (Dinham, 2003).

A 2001 study meta-analyzing peer-reviewed research on the health effects of paraquat identified skin problems, eye irritation, and nosebleeds as the most common symptoms. Systematic exposure can lead to multi-organ failure (Wesseling et al., 2001). A 2009 study found that in combination with the fungicide Maneb, Paraquat exposure is associated with higher rates of Parkinson's Disease (Costello, 2009). Due to increasing concern about its health effects, paraquat is now banned in Austria, Cambodia, Finland, Denmark, Ivory Coast, Kuwait, Slovenia, Sri Lanka, Sweden, Syria. We now turn to a discussion of the obstacles faced by the global anti-toxics movement in achieving these outcomes and that have prevented an even more widespread ban.

Political Obstacles

The first political obstacle global anti-toxics organizers face is in mobilizing constituents outside of an affected area. In the case of paraquat's use in Malaysia, for example, for years, workers on plantations suspected that the chemicals they were being asked to apply to crops

[4] Most of the details in this section come either from Fernandez and Bhattacharjee (2006) or personal interviews and archival analysis conducted by the author at the PAN-AP offices in Penang, Malaysia.

were having ill effects on them. But plantation labor is organized such that one group of workers mixes the pesticides, another group sprays them, and yet other groups of workers have no direct interaction at all with the pesticides. Under the threat of losing their jobs from foremen or higher ranking plantation conductors, laborers seldom risk collective actions or even raising objections with their superiors. Furthermore, families living adjacent to a plantation and who may be exposed through the drift of pesticides often depend on the plantation jobs. Finally, many plantation workers are indentured immigrant workers from Indonesia, Sri Lanka, India, and Bangladesh who stand to lose their work permits if they insist on rights. Even if they understand the language, they are unlikely to challenge their superiors when they depend on continue employment to pay off the cost of their passage to Malaysia.

It is against these odds that an organization like PAN-AP attempts to organize to demand the environmental health rights of workers, their families, and anyone else exposed to toxic risks. In fact, as PAN-AP's partner organizations, like Consumers Association of Penang, Selangor Consumers Association, and Friends of the Earth Malaysia, began documenting cases of pesticide poisoning in the early 1980s, they found that plantation workers were being forbidden to meet with them. Nevertheless, they persisted in raising media attention and eventually forcing the government to respond. Apparent victory occurred in 2002, when Malaysia's Pesticides Board announced a ban on paraquat. But before the ban could take effect, Malaysia's government announced in 2005 that it would delay the ban in order to assess the implications for Malaysia's agriculture sector. In 2006, Malaysia notified the secretariat of the Rotterdam Convention that it was allowing the re-registration of five variations of paraquat. Despite the support of its international partners like the Berne Declaration (based in Basel, Switzerland, home of Syngenta's headquarters), the Swedish Society for Nature Conservation, PAN United Kingdom, and Foro Emaús, PAN-AP could not convince the Malaysian government to uphold its ban. Even if the ban had been upheld, the obstacle of ensuring enforcement of the ban would have posed an additional challenge.

Enforcing existing laws under weak regulatory regimes is a second type of political obstacle faced by global anti-toxics organizers. For example, Malaysia has had, since 1974, a Pesticides Act that, in theory, is intended to regulate the trade and use of pesticides in order to protect public health. Yet virtually no training in safety precautions takes place, and until PAN-AP's campaign demanded it, there was no system for reporting and monitoring cases of pesticide poisoning. According to Fernandez and Bhattacharjee, "a constant theme in the fight for a safe working environment is the dichotomy in the government's actions: good laws; but dismal enforcement; deep concern in one ministry, and a callous response from another" (2006: 137).

One strategy for working around this problem is to advocate for laws at the international level. PAN's international coordination of its regional centers allows each region to work on local issues while also contributing to, and benefiting from, international campaigns. In 2005, for example, with its Malaysian partner organization Tenaganita, PAN-AP brought Nagama Raman, a 48-year old plantation pesticide sprayer, to Rome for the second Conference of the Parties of the Rotterdam Convention. During the conference, Nagama shared her experiences at a side meeting organized by PAN-AP, PAN UK, and the Berne Declaration. Nagama had fought against what she believed were unjust practices on the plantation in relative isolation until she happened to encounter two Tenaganita activists in 2001, who were educating plantation workers on their rights. Nagama completed several training programs so that she could empower workers in the same way that she had become empowered. Nagama, who has also served as the voice of women sprayers at international meetings in Hong Kong and Bangkok, represents how global anti-toxics organizations like PAN-AP provide links from the grassroots level all the way to the highest levels of international politics.

Unfortunately, when organizations like PAN-AP try to work around local political obstacles like weak regulatory regimes by engaging in international negotiations on chemicals management, the outcomes are not always effective. Developing countries often lack the resources to carry out risk assessments. The Rotterdam Convention, for example, requires each country to appoint a Designated National Authority who is authorized to act on a country's behalf in the performance of the administrative functions required by the Convention. Countries that lack the technical expertise or the administrative infrastructure to support such a position are often unable to carry out the necessary responsibilities to bring the Convention's force to bear.

One response of NGOs to this obstacle is to push for funding mechanisms in international agreements. Funding mechanisms can take the form of dedicated funds like the World Bank's Global Environment Facility, or one-time contributions from developed nations. When such funds are not forthcoming, international agreements have a very low likelihood of being carried out in developing countries. In turn, trans-national NGOs like PAN dedicate themselves to training grassroots groups to carry out monitoring and enforcement in the absence of state-level commitments. They can also work with governments to strengthen their technical capacities and alert government agencies of their responsibilities under international agreements.

According to PAN-AP, some of the barriers to NGOs carrying out these roles are the lack of mechanisms to ensure participation and consultation, the lack of political will to ensure participation, the complex relationships states often have with the chemical industry, the lack of accountability, and the lack of important types of transparency such as community "right-to-know" laws. Of course, such laws, as has been argued above, might not even be enforced were they to exist. Nevertheless, as will be discussed next, overcoming legal obstacles is a part of the challenge faced by the global anti-toxics movement.

Legal Obstacles

The Industrial Revolution was facilitated by a number of profound social changes that had begun in the early 1700s. Three intertwined changes were the widespread transitions to democratic forms of governance occurring across the 18th century, the embrace of rational-legal forms of authority that reside in a legal order, and the rise of scientific knowledge as one type of rational-legal authority. Legal order is a system of law in which rules are based on rational grounds and created and enforced by individuals holding rational-legal authority. For global anti-toxics activists, however, the legal order and rational-legal forms of authority are often obstacles to more effective enforcement of chemical regulations.

At the local level, the challenge has to do with acquiring the necessary expertise to engage existing legal systems. For example, if workers are guaranteed certain rights under occupational safety legislation, or a pesticide act is intended to protect public health, a local struggle for the rights of workers or exposed citizens needs to understand how these laws function. At PAN-AP, a conscious decision has been made to avoid legal battles as a movement tactic. PAN-AP works with partners in at least ten difference countries that speak at least as many languages, each with its own unique legal system. PAN-AP has decided that it would be too great a drain on its resources to devote itself to acquiring the necessary legal expertise to challenge the laws of these countries.

Assuming an anti-toxics organization would want to engage in legal battles, a related obstacle has to do with the uncertainty of how to approach a case in which the responsible party is a multinational corporation. The movement for justice for the survivors of the 1984 Bhopal disaster confronted this problem when their case against Union Carbide was thrown out of a U.S. Court on a ruling of "forum non conveniens," or inconvenient forum. The judge ruled that because India had a functioning legal system, and because most of the witnesses who would need to testify were in India, the case should be tried there. Other famous environmental health suits brought against multinational corporations include: Bowoto v Chevron, in which a group of Nigerians sued Chevron for its role in aiding and abetting a brutal 1998 attack by Nigerian government on activists who had boarded an oil platform to discuss their concerns with Chevron officials; Carijano et al. v. Occidental Petroleum Corporation, accusing Occidental of harm to the indigenous Achuar people over a thirty-year period in the Corrientes River basin of Ecuador; and Aguinda v ChevronTexaco, in which 30,000 people from five indigenous groups in Ecuador allege massive oil contamination of their ancestral lands. Although there are signs of a possible victory in the Aguinda V ChevronTexaco case—a case that has extended over the last 16 years—justice in each of these cases has been hampered by the ability of multinational corporations to employ legal strategies that play one country's judicial system against another, or in other ways subvert legal systems that lack precedents for dealing with such cases. When they do engage in legal battles, movement organizations are often dependent on support from other advocacy groups, like EarthRights International, that specialize in global environmental justice cases. In short, without significant amounts of time, expertise, and monetary resources, legal challenges against multinational corporations have little chance of winning.

Economic Obstacles

The major economic obstacle that global anti-toxics organizations face is the obvious obstacle of raising the funds to carry out its work. With limited budgets, organizations must decide how to balance their efforts among activities such as monitoring and documenting health impacts in a community or across an entire country, participation in international negotiations, legal challenges, and media or other awareness-raising campaigns. These challenges are exacerbated for organizations in developing countries where not only is there less access to funding for the organization, but organizations often have even more work to do when under-funded governments fail in their responsibilities. Furthermore, economically weak countries become easily influenced by chemical companies and other industries that promise them economic growth if allowed to site a facility in their country, and then threaten them with pulling out if regulations become too strict.

Trans-national organizing has proven to be an effective strategy for dealing with the economic obstacles. When poorer NGOs from the developing world team up with wealthier NGOs from the developed world, resources can flow both ways. The wealthier organizations can provide funding to their counterparts in the developing world that might help bring them to an international meeting in Geneva, for example. They also often have the technical expertise to know how to deal with the U.N. or other international governmental organizations. This can help organizations from the developing countries gain seats at negotiating tables. Organizations in the developing countries bring to the relationship new ways for the wealthier organizations to go after old targets.

Although this type of mutually beneficial relationship happens throughout the Pesticide Action Network, another example is the Dow Accountability Network (of which PAN North America is a member):

> The Dow Accountability Network includes public health advocates, chemical disaster survivor groups, legal experts, investors, environmental health & justice organizations, corporate accountability groups and human rights organizations. The core of the Network includes survivors of the Bhopal chemical disaster, as well as activists in Vietnam, Nicaragua, and Michigan and the many other places faced with the public health and environmental fallout from Dow Chemical's production and use of dangerous chemicals. (Dow Accountability Network, "About Us")

Organizations working on Dow-related campaigns in the U.S. gain strength as they link up with activists in India, Nicaragua, and Vietnam. These relationships help identify new liabilities for the corporation that can lead to new strategies in the campaign to end the production of toxic products. Trans-national relationships such as these have become economically viable as the costs of telecommunications and transportation have gone down over the last decade or two.

CONCLUSION

If recent publications are any indication, there is a heightened concern about the impacts of toxics in our lives. A number of books have tapped into this concern with titles such as *The Body Toxic: How the Hazardous Chemistry of Everyday Things Threatens Our Health and Well-being* (Baker, 2008), *What's In This Stuff?: The Hidden Toxins in Everyday Products and What You Can Do About Them* (Thomas, 2008), *Poisoned Profits: The Toxic Assault on Our Children* (Shabecoff & Shabecoff, 2008), *Exposed: The Toxic Chemistry of Everyday Products and What's at Stake for American Power* (Schapiro, 2007), *The Secret History of the War on Cancer* (Davis, 2007), and *How Everyday Products Make People Sick: Toxins at Home and in the Workplace* (Blanc, 2007). While these titles tell us much about the effects of chemicals in our lives, they offer less in the way of analysis of collective social responses to chemical risks.

For such analysis, we have to turn to research from the 1980s and 90s that covered toxics activism in communities such as Love Canal, New York (Levine, 1982), Woburn, Massachusetts (Brown & Mikkelsen, 1997), Legler, New Jersey (Edelstein, 1989), and Times Beach, Missouri. Other works aimed to describe the emerging anti-toxics movement more broadly (Szasz, 1994). Although it made important contributions to the understanding of local cases of community contamination in a developed country like the U.S, this body of work needs to be updated to reflect community and worker contamination in developing countries where cultural, political, and economic contexts—such as globalization's introduction of multinational corporations to developing countries—shape different movement processes and outcomes. Social scientists need to understand better how social responses to the global spread of toxics are shaped by these forces. Fortunately, we can draw on work such as Cole & Foster's (2001) attempt to make conceptual linkages among the civil rights, anti-toxics, and environmental justice movements, or Gould, Schaniberg & Weinberg's (1996) effort to theorize the relationships between local and trans-national environmental justice struggles (Gould, Schnaiberg, & Weinberg, 1996). Focused specifically on the resistance to global toxics, Pellow's work identifies important connections among the human rights, indigenous rights, anti-toxics, environmental justice, and racial justice movements (Pellow, 2007). More squarely in the environmental health field, there are promising advances such as Brown's (2007) theory of embodied health movements that highlights the importance to movements of building alliances between lay activists and scientific experts.

This chapter has attempted to demonstrate that the globalization of the chemical industry has had significant consequences for environment and health, that these consequences have triggered a social response in the form of the global anti-toxics movement, and that this movement confronts certain political, legal and economic challenges in its efforts to reduce the threats of toxic hazards. NGOs are important in this effort for several reasons. First, since even in developed countries there is often insufficient government monitoring of environmental health threats, NGOs play the essential role of working with citizens who have identified potential threats. They help them to collect community and population health data that can be brought to bear in national and international policymaking arenas. Second, as Brown's (2007) research demonstrates, the institution of science does not necessarily ask the kinds of questions that are useful to anti-toxics struggles. Activists must find allies in the scientific community that are willing to work with them to collect evidence of toxic harm as well as conduct the basic science linking specific chemical exposures to biological markers. Future research should build on Brown's work on the formation of these "citizen-science alliances" in the U.S. to explain how or whether such alliances function in the developing world. Social scientists from science and technology studies traditions, medical sociologists and anthropologists, environmental sociologists and environmental and health policy scholars are well suited to lead these investigations. These and other scholars can contribute to an understanding of the conditions under which anti-toxics struggles emerge and the challenges that shape their effectiveness. Such an understanding will be crucial in shaping the broader human response to the parallel trends of increasing global economic integration and the global spread of toxics.

Achieving a less toxic future in the face of the political, legal, and economic obstacles discussed in this chapter causes one to wonder whether the global anti-toxics movement will be capable of handling future challenges. For example, no one in the movement is thinking ahead about how a changing climate will change patterns in when, where, how, and why we use chemicals. As the climate changes, for example, we will attempt to adapt our agricultural systems to the new climate. It is quite likely that we will hear arguments justifying the increased use of pesticides to assist us in adapting. A changing climate is also likely to expose dormant, contaminated sites that currently pose minimal threats. Many of the most contaminated sites around the world are underwater—in the sediment of waterways that were once dumping grounds for industry. Floods, droughts, or even less severe changes in precipitation patterns, all could expose currently submerged sources of contamination. Finally, as we already know, climate change will introduce diseases in new places, carried by disease vectors such as mosquitoes and rodents. Again, one can imagine public officials and chemical industry executives arguing for the aggressive use of pesticides and rodenticides to stem epidemics.

By understanding the nature of the obstacles the global anti-toxics movements faces today, we stand a better chance of assisting the movement to develop effective responses as new challenges arise. We can only hope that such knowledge will make for a more effective global anti-toxics movement—one that can protect health and environment for present and future generations.

REFERENCES

Aftalion, F. (2005). *History of the international chemical industry: From the "early days" to 2000* (2nd ed). Philadelphia, PA: Chemical Heritage Foundation.

Appadurai, A. (1996). *Modernity at large: Cultural dimensions of globalization*. Minneapolis, MN: University of Minnesota Press.

Baker, N. (2008). *The body toxic: How the hazardous chemistry of everyday things threatens our health and well-being*. New York: North Point Press.

Beck, U. (2009). *World at risk*. Malden, MA: Polity Press.

Blanc, P. (2007). *How everyday products make people sick: Toxins at home and in the workplace*. Berkeley, CA: University of California Press.

Brandt, E. N. (1997). *Growth company: Dow Chemical's first century*. East Lansing, MI: Michigan State University Press.

Brown, P. (2007). *Toxic exposures: Contested illnesses and the environmental health movement*. New York: Columbia University Press.

Brown, P., & Mikkelsen, E. J. (1997). *No safe place: Toxic waste, leukemia, and community action*. Berkeley, CA: University of California Press.

Bull, D. (1982). *A growing problem: Pesticides and the Third World poor*. London: Oxfam.

Cole, L.W., & Foster, S. R. (2001). *From the ground up: environmental racism and the rise of the environmental justice movement*. New York: NYU Press.

Costello, S., Cockburn, M., Bronstein, J., Zhang, X., & Ritz, B. (2009). Parkinson's Disease and residential exposure to Maneb and Paraquat from agricultural applications in the Central Valley of California. *American journal of epidemiology, 169*, 919-926.

Davis, D. (2007). *The secret history of the war on cancer*. New York: Basic Books.

Dinham, B. (2003). The perils of paraquat: Sales targeted at developing countries. *Pesticides News, 60(June)*, 4-7.

Dow Accountability Network. About us. http://www.thetruthaboutdow.org/article.php?list=type&type=10 (Accessed May 25, 2009).

Edelstein, M. R., (1989). *Contaminated communities: The social and psychological impact of residential toxic exposure*. Boulder, CO: Westview.

Fernandez, J. M., & Bhattacharjee, R. B. (2006). *The politics of paraquat*. Kuala Lumpur: Tenaganita/Pesticide Action Network Asia and the Pacific.

Frey, R. S. (1998). The export of hazardous industries to the peripheral zones of the world-system. *Journal of developing societies, 14(1)*, 66-81.

Frey, R. S. (1995). The international traffic in pesticides. *Technological forecasting and social change, 50*, 151-169.

Giddens, A. (1990). *The consequences of modernity*. Stanford, CA: Stanford University Press.

Gould, K., Schnaiberg, A., & Weinberg, A. S. (1996). *Local environmental struggles: Citizen activism in the treadmill of production*. New York: Cambridge University Press.

Harvey, D. (1990). *The condition of post-modernity*. Malden, MA: Blackwell.

Held, D., McGrew A., Goldblatt, D., & Perraton, J. (1999). *Global transformations: Politics, economics, and culture*. Stanford, CA: Stanford University Press.

Hesser, L. (2006). *The man who fed the world: Nobel Peace Prize Laureate Norman Borlaug and his battle to end world hunger*. Dallas, TX: Durban House.

Hilton, M. (2007). Social activism in an age of consumption: The organized consumer movement. *Social History, 32(2)*, 121 - 143.

Inda, J. X., & Rosaldo, R. (2008). Tracking global flows. In J. X. Inda & R. Rosaldo (Eds.), *The anthropology of globalization: A reader* (2nd ed., pp. 3-46). Malden, MA: Blackwell.

IOCU. (1982). PAN International calls for halt to global pesticide proliferation. IOCU Press Release, PAN archives.

Jorgenson, A. K. (2007). Foreign direct investment and pesticide use intensity in less-developed countries: A quantitative investigation. *Society and natural resources, 20*, 73–83.

Levine, A. (1982). *Love Canal: Science, politics and people*. Lexington, MA: Lexington Books.

McDonough, W., & Braungart, M. (2002). *Cradle to cradle: Remaking the way we make things*. New York: North Point Press.

McNamee, S. J. (1983). Capital accumulation and the Du Pont Company: An historical analysis. *Organization studies, 4(3)*, 201-218.

Pellow, D. N. 2007. *Resisting global toxics: Transnational movements for environmental justice*. Cambridge, MA: MIT Press.

Rothschild, M. (1982). Consumers take the offensive against multinationals. *The Multinational Monitor, 3(7)*. http://www.multinationalmonitor.org/hyper/issues/1982/07/interview-fazal.html (Accessed on May 25, 2009).

Schapiro, M. (2007). *Exposed: The toxic chemistry of everyday products and what's at stake for American power*. White River Junction, VT: Chelsea Green Publishing.

Shabecoff, P., & Shabecoff, A. (2008). *Poisoned profits: The toxic assault on our children*. New York: Random House.

Steger, M. (2003). *Globalization: A very short introduction*. New York: Oxford University Press.

Szasz, A. (1994). *Ecopopulism: Toxic waste and the movement for environmental justice*. Minneapolis, MN: University of Minnesota Press.

Thomas, P. (2008). *What's in this stuff?: The hidden toxins in everyday products—and what you can do about them*. New York. Penguin.

Tokar, B. (2004). The World Bank and the next green revolution. *Z Magazine*. http://www.zmag.org/zmag/viewArticle/14111 (Accessed May 25, 2009).

Weir, D., and Schapiro, M. (1981). *Circle of poison: Pesticides and people in a hungry world*. Oakland, CA: Institute for Food and Development Policy.

Wesseling, C., Van Wendel de Joode, B., Ruepert, C., León, C., Monge, P., Hermosillo, H., Partanen, T. (2001). Paraquat in developing countries. *International journal of occupational and environmental health, 7*, 275-286.

In: Health and Environment: Social Science Perspectives ISBN 978-1-60876-216-3
Editors: Helen Kopnina and Hans Keune

Chapter 12

Conclusions on Social Scientific Perspectives on Health and Environment

Hans Keune and Helen Kopnina

Abstract

We look back at the contributions to this volume by focusing on the problem-framing and problem-solving perspectives. A main critique of most contributors on the current state of the art of environment and health science is that important elements are not taken into account, thus resulting in a limited view on environment and health issues. First, we focus on how contributors want to broaden and deepen the health and environment science by taking into account complementary aspects of the complexity of environment and health issues. In achieving the objective of enriching the view on environment and health through this complementary social scientific lens, several obstacles are noticed. These obstacles arise from theoretical and social interpretations of the concepts of health and environment as well as from socio-political and economic aspects. From a problem-solving point of view, we highlight how contributors seek to overcome these obstacles and to what extent theoretical insights can be applied in practice. Finally, we reflect on the role of social science in relation to health and environment from the point of view of the contributors to this volume and discuss future research perspectives.

Introduction

This book introduced several social scientific perspectives on environment and health. We did not pretend to be either exhaustive by presenting a full picture of the vast diversity of social scientific issues in this field, nor to be conclusive about the role of social scientists in this dynamic and contested field. The collection of contributions displays diversity in topics and in viewpoints as well as in theoretical considerations and methodological approaches. We may characterize the contributions in many different ways, of which many supposedly are both relevant and interesting. What most contributions seem to have in common is a critical view on current environment and/or health-issue framing and problem-solving strategies. Contributors touch upon debates within the field of environment and health science, dominated by natural scientists (e.g., scientific proof and weight of evidence) Contributors also address debates within the social sciences involved in environment and/or health (e.g., the role of the environment in health issues), as well as social and policy debate on environment and health (e.g., the role of stakeholders in environment and health debate). Most of the contributors seem to propagate a broadening of the view on issues to be taken into consideration in that respect, supported by arguments ranging from scientific to normative. Some also give clues for practical application of these ideas: how can society benefit from the body of knowledge about health and environment from social scientific perspectives?

Picture 12. Conclusion.

We distinguished a problem-framing perspective and a problem-solving perspective. *Problem-framing* not only relates to the definition of the problem but also defines what type of knowledge about the complex nature of an environment and health issue is considered to be relevant: which information (data) about the issue do we need? Moreover, it defines our understanding of the problem: how do we interpret the data? Closely related is the *normative dimension*. Not only is our problem-framing coloured by normative aspects: what is important? We also have to envision what we consider to be desirable, which targets do we want to set for our approach of the problem? *Problem-solving* (or transformation knowledge) deals with practical application of solution strategies: how do we reach our goals? We will use this typology as a yardstick: where do the contributions to this book fit in? What is the contribution of social science to the field of environment and health? And which horizon for further and future social scientific involvement in this field seems appropriate?

Broadening the View on Environment and Health

How do contributors describe the state of the art of the environment and health science, and in which areas do they want to open up new perspectives? Generally speaking, we can distinguish two broad spectra: the natural-humanistic and the positivist-social constructivist. Several contributors bring the environment into the problem-framing domain. Bowden propagates to include ecology into the social scientific lens on the environment, thus opening up to the nature-side of the environment and health coin. He presents examples of how the ecological perspective introduces relevant aspects of environment and health issues from a broader, more systemic and more realistic approach. Baralt and Stingl propagate to broaden the view of diagnosis of health conditions (breast cancer and ADHD), by including the role of "the environment." Baralt shows how the biomedical paradigm obscures the important role of environmental pollution in causing a rise in cases of breast cancer, and thus limiting problem-solving strategies by not taking into account the environmental aspects. Stingl stresses the importance of taking into account the role that different natural, social, and cultural environmental factors play in diagnosis and treatment of what is conceived as ADHD. Like Bowden, Baralt and Stingl thus also propagate a whole-systems approach.

Other contributors focus more on the relevance of different social actor perspectives in framing the issues, thus challenging a traditional positivist or expert approach to environment and health and propagating a social constructivist and more open approach. Dépelteau highlights how risk assessment is limited and undemocratic when those who are responsible for creating the risks have a big influence on the analysis and interpretation of the issue: in the case, he represents how the polluting industry plays a major role and even has a kind of veto right with regard to conclusions. This results not only in a biased picture of the problem, but will also hinder broader support for risk management due to limited trust from the general public. Zavestoski introduces the perspective of environment and health activist groups and does so at the global level. Along with the globalization of the chemical industry and along its spread of toxic hazards around the world, also social movements to resist toxic threats are becoming increasingly global. Potrata and Akobirshoeva show how the Western Orientalist lens obscures a realistic picture on conditions in Central Asia. Local observations not only highlight much greater diversity in different local or regional settings, but also show that the main (Western) assumption supporting the World Health Organization's strategy for promoting traditional medicine and native knowledge is inadequate. The supposed beneficial potential of traditional medicine for environmental protection is questionable. In his contribution, Ney widens the social picture by discussing the narratives on environment and health from the diversity of social groups' perspectives. Ney ponders how stakeholders and policy actors navigate in the midst of complexity and uncertainty, so characteristic for environment and health issues. Different social perspectives on what at first sight may be seem as part of unambiguous "knowledge," "problem," or indeed as Ney puts it, "facts and objective data," may lead to totally different interpretations. Keune et al. discuss how the involvement of a variety of social groups (science, policy, stakeholders) with regard to policy interpretation of complex environment and health issues can be structured in policy practice. Finally, most contributors propagate more interdisciplinary involvement of social scientists in a domain that is dominated by natural sciences. Both Keune et al. and Kopnina address this topic in the *Introduction* to this volume and in their individual contributions. We come back to this issue in the section on *the role of social science.*

The whole-systems approach propagated by several contributors (e.g., Bowden, Baralt, Stingl, Keune et al.), flies on various wings of argumentation, but is mainly propagated for better and more realistic understanding of the environment and health issues under discussion. Here, the quality of the knowledge (science) is the main focus. Dépeltau, Potrata and Akobirshoeva, and Zavestoski seem to take a more normative stance in that they clearly choose sides. These contributors focus on the balance of power and influence that is part of the social construction of environment and health and aim for further emancipation of those who show to be less powerful in environment and health issues. Both Dépeltau and Potrata and Akobirshoeva focus on the involvement of local communities who do not have a strong voice in problem-framing or the definition of problem-solving strategies. Zavestoski does so at the global level, in focusing on global social movements that find many structural obstacles in their way when propagating public health in the face of the rising global contamination with toxics from the global chemical industry.

Obstacles to a Better View on Environment and Health Issues

Keune and Van Larebeke pointed out that knowledge about the environment and health problems does not easily result in problem-solving, due to a diversity of obstacles. We already discussed the limited view on environment and health as being an obstacle to adequate understanding according to all contributors. We will now turn to other hindering aspects that are specifically highlighted by several contributors, some case specific, and some more general.

Keune and Van Larebeke show how the social construction of environment and health partly are dominated by biased expertise, that is, expertise mainly voicing industry interests. This not only is a problem of hidden agendas and affiliations: several well-known cases show that some scientists obscure their linkages to industry. Even when industry experts are openly involved in assessing environment and health risks, problems may rise. When industry experts dominate expert groups, like was the case in the International Agency for Research on Cancer (part of the World Health Organization), the social balance of evidence is questionable. This imbalance of biased expertise is also highlighted on a local level by Dépelteau, resulting in the lack of social trust, which may be a stumble block for any problem-solving strategy. Kopnina discusses how in the Dutch cases of air pollution and illegal timber, the pleas for action from the side of environmental social and political movements to rebound unhealthy developments in the area of environment and health do not easily resonate in adequate policy actions. Zavestoski sketches the obstacles to a more balanced assessment with respect to the global social movement to resist toxic threats: both legally, politically and economically social movements are limited in their capabilities of engaging in discussions, negotiations and legal battles on the issues that are of importance to them. Thus, we may conclude here that some kind of imbalance of representation of relevant actors in the process of dealing with environment and health issues is an important issue for consideration. Relinquishing the obstacles to achieve a more balanced approach though is not straightforward at all. Gathering a balanced diversity of relevant social perspectives does not do the trick either, though. Ney sketches how the incommensurability of different social perspectives on environment and health issues may result in policy impasse: "dialogue of the deaf."

At a theoretical level, Bowden sketches three scientific barriers to the incorporation of the ecological perspective in environment and health sociology. One is the dichotomy between, on the one hand, the belief in human society capable of solving all problems including the scarcity of resources and, on the other, the idea of biophysical limits on resource supply. Another barrier is the everlasting debate between realists and constructivists: do we consider environment and health to be unambiguous and objective or do we consider environment and health to be socially constructed, subjective and dependent of perception and norms and values? Finally, the allergy of sociologists towards system thinking is mentioned. According to Bowden, for quite some time, sociology turned its back on the functionalist idea of self-regulation in social systems. This resulted in a blind eye for self-organizing principles developed both in systems thinking and complexity theory. Ecological sociology will only be possible, he argues, when these barriers are resolved.

PROBLEM-SOLVING

Good quality of problem definition and analysis is an important part and a good start to problem-solving. Most authors display interesting contributions in that respect, both in case studies and analytical approaches. The diversity in options for broadening the view on environment and health presented and the different analytical models enrich our view on and understanding of environment and health problems. Bowden points out that the theoretical obstacles to this broader view presented in his chapter may be overcome by building on important steps that already have been made within several scientific communities.

The resulting whole-systems approach, also propagated by other contributors, (e.g., Baralt, Stingl, Keune et al.) and the lessons learned from the perspective of complexity theory (e.g., Keune et al.) touch upon (at least) two important issues from a problem-solving point of view. First, the observation that the complexity of environment and health issues will per definition be hampered by limited understanding; both from a natural scientific and a social scientific point of view. Or as Stingl puts it: "*Therefore, good science means science that clearly understands the limits of its explanatory power. That leads to a pluralist approach, namely, the ability to enable the use of concepts where they are appropriate and enable critical relations between science, public and polity.*" Where several contributors point at blind spots that need to be and can be resolved by incorporating complementary aspects e.g., from an environmental point of view, several contributors also point at the limitations: next to the fruitfulness of a more diverse perspective on complex reality, we also have to be humble about it and acknowledge that collecting more diverse and richer data will not solve problems automatically. Due to several aspects of complexity waiting for traditional science to produce adequate answers is like waiting for Godot.[1] Keune and Van Larebeke propagate a more qualitative approach in that respect.

Second, the need for a diversity of critical mass; both a relevant diversity in (both natural and social) scientific disciplines (or other forms of expertise) and diversity in stakeholders. Several contributors plead for such diversity, but only few of them sketch a problem-solving perspective that deals with conflicts of interests and difference of opinion that will often occur. As already pointed out, Ney does not believe that merely putting different stakes at the negotiation table will unambiguously solve problems when problem-framing and definition of problem-solving strategies are characterized by difference of opinion. Nevertheless, he perceives difference and diversity of viewpoints as an important source of relevant information for policy learning: conflicting narratives act as vehicles for translating science to policy. Albeit from a totally different analytical perspective, in essence, Ney's view connects well to the way Keune et al. discuss social interpretation of science: "*the proof of the science is in the discussion*." Keune et al. describe how such rationale was applied and tried out in practice in a discursive setting.

This brings us to the practical side of problem-solving. Action- or transformation-oriented knowledge is not prominent in most contributions. There seems to remain a gap between academia and society. Opposed to the saying "the truth is out there," we might put it as "the truth is in there," in the ivory tower of science that is. Lessons from action research, transdisciplinary research and policy relevant/oriented research in translating knowledge to real life action may prove to be relevant in bridging this gap (see e.g., Keune et al.). Although most contributors promise that their analytical perspective will help resolve issues, and may feed into the process of problem-solving, most contributions do not clearly state how a practical process of change might be organized or how practical experience of such endeavours works out in practice. Thus, the translation of insightful social scientific fruits from the ivory tower of academia to society, to a large extent remains a mystery.

The Role of Social Science

What is the added value of social scientists' involvement in the field of environment and health? Do they have a different, perhaps complementary, role to that of natural scientists, stakeholders, policymakers or journalists? Based on the contributions to this volume, we may conclude that social scientists have a different view at least different from natural scientists: social scientists put issues in different perspectives, they ask different questions (see e.g., Baralt, Stingl, Kopnina).

[1] *Waiting for Godot* is the title of a play by Samuel Beckett (1948-1949). *Waiting for Godot* follows two days in the lives of a pair of men who divert themselves while they wait expectantly and unsuccessfully for someone named Godot to arrive. They claim him as an acquaintance but in fact hardly know him, admitting that they would not recognize him were they to see him. (http://en.wikipedia.org/wiki/Waiting_for_Godot http://www.samuel-beckett.net/Waiting_for_Godot_Part1.html)

Social scientists contextualise issues. Examples of contexts may be social contexts like the political, economic and cultural, or indeed the context of nature, or environment in general. Whatever the context, social scientists focus more on social aspects, social relations, and social dynamics. The social scientific lens focuses on human experience, perceptions, public opinion, and behaviour.

Do social scientists have a special social responsibility? Is it the responsibility of social scientists to choose sides, we may wonder when we look at the political and social dynamics of environment and health issues? Or should social scientists take an independent, neutral position? In broadening the assessment of environment and health issues, we may conclude that several contributors propagate the involvement of a variety of relevant actors in this process: potentially beneficial for several reasons of which knowledge input and support for the process are important features. The involvement of stakeholders, though, mostly propagated by social scientists it seems, may be more of a normative issue, especially when this concerns lay people, ordinary people, people who normally are not in charge, who perhaps benefit most from solving environment and health problems but lack the capacity to bring this about, due to lack of expertise, resources or power. Some contributors plead for a democratization of science, stress that public involvement is a democratic right; some propagate or focus on alliances between lay activists and scientists, on the emancipation of social movements. When does broadening the view on environment and health become political we may wonder? Is an independent neutral role or position, important for scientists in general and social scientists in particular? Is this possible at all? And if not, if biased expertise is part of the game anyhow, how should we deal with this? We all choose sides, one for industry, one for the local population, one for the general interest; and we all try to be transparent about it.

Some contributors attribute social science with a critical role. Here, also, we can distinguish two drives. First, the scientific or knowledge drive of bringing up critical issues from a radical perspective, radical in the sense of questioning (and investigating) the roots, the foundation of any knowledge. Second, the political drive of choosing sides: in this volume, the example of experts selling their scientific soul to industry was discussed as well as the example of more emancipatory science, supporting the weak and vulnerable. In general, it seems, be it from a political or scientific point of view, social scientists tend to focus more on the meaning of environment and health issues: what is the social meaning of knowledge, what does it mean for individuals and groups of people, what are policy implications, what is the social relevance in general? Also, the process of social change is highlighted as an important focus for social scientists. If problems are considered to be important for society, how, then, may societal change come about in order to solve these problems? Knowing about environment and health issues, as we saw, does not automatically or easily lead to adequate societal adaptation or change. These processes are of special interest to social scientists.

Next to focussing on the vast amount of relevant social aspects, perhaps best captured under the umbrella term of the social complexity of environment and health, social scientists also have a role in mapping and structuring complexity in general. Mapping complexity from a social scientific perspective may be qualified as gathering relevant diversity: diversity in aspects (content) and diversity in viewpoints (perspectives). This goes beyond different forms of specialized expertise and different stakes, beliefs, frames, narratives. Such holistic (or complexity, whole-systems, integrated) perspective, perhaps, is not unique for social scientists. Social scientists, though, tend to have a more open attitude with regard to environment and health issues and the relevance of a variety of viewpoints and contexts. Moreover, in coordinating this diversity, the interaction of which is a social process in itself, the expertise of social scientists regarding social dynamics is beneficial. Next to providing specialized content (social complexity), they tend to focus on social processes. Structuring this diverse view on complexity may thus also benefit from social scientific involvement. The variety of social scientific contributions in this book displays a diversity of social scientific expertise in this respect, of which qualitative methodology and interactive, participatory process are important examples.

The promise of social scientific involvement in environment and health issues may also have a downside, though. It will not wash all problems away. Including the social scientific lens does not necessarily make life easier for those involved in the environment and health issues under discussion. Opening up the black box of complexity perhaps sheds a more realistic light, but also adds complexity to a field that is characterized as complex even when only the natural scientific aspects are considered.

Future Perspectives

In general, more involvement of social scientists in environment and health research may prove to be relevant from a problem-solving point of view. Generally, we may distinguish a specialized disciplinary contribution with regard to social aspects of environment and health (social complexity) and a more general both holistic and structuring contribution (complexity in general). Both seem to be undervalued and underrepresented in current environment and health research, from a scientific, a policy and a public opinion point of view.

With regard to specific social scientific contribution, especially the traditions of qualitative research and social processes, they may turn out to be relevant. The complexity of environment and health issues more or less condemns society to a more qualitative approach, it seems, in which discussion between relevant actors will prove to be of main importance. Social scientists will need more opportunity for more intense (interdisciplinary) involvement and will need to produce more proof of the value of their expertise applied in real life practice.

ABOUT THE EDITORS AND CONTRIBUTORS

THE EDITORS

Helen Kopnina is an anthropologist (Ph.D. Cambridge University, 2002) currently researching the relationship between occupational health and policy at the Amsterdam Institute of Advanced Labour Studies (faculty of the University of Amsterdam). Kopnina is the author of three other books, *East to West Migration* (2005); *Crossing European Boundaries: Beyond Conventional Geographical Categories* (2006) and *Migration and Tourism: Formation of New Social Classes* (2007). On the personal level, Kopnina considers herself a type of environmentalist who views non-manmade nature as an entity having a "right" to exist independently of human interest. The "right" in this case refers to intrinsic value quite apart from human legal or moral connotations of it. Being a mother of three children, the health of future generations is also one of Kopnina's great personal concerns.

Hans Keune is a political scientist working at the Faculty of Political and Social Sciences, University of Antwerp-Belgium. His expertise is mainly related to environment & health (science, social issues and policymaking), risk (perception, communication and assessment), knowledge (production, integration, interpretation and application), complexity, inter- and trans-disciplinary (boundary work science—government—society and participation), action research, group assessment/decision support methods (e.g., multi-criteria decision analysis, Delphi method, focus group). He works for the Flemish Centre of expertise for Environment and Health and has experience in several international projects related to his work in the field of environment & health.

THE CONTRIBUTORS

Anzurat Akobirshoeva is one of the rare researchers in the region of post-Soviet Union who works on ethno(veterinary) medicine in Central Asia. After completing her MPhil on the use and monitoring of medicinal plants in Rushan in Gorno Badakhshan, she joined academic staff at Pamir Biological Institute Yusufbekov in Khorog in Tajikistan. She is involved in research on, monitoring of and development of good practices in sustainable use of medicinal plants in this region and promotion of bio-diversity. She is also a deputy director of the

Institute of Professional Development of Gorno Badakhshan Autonomous Oblast in Khorog, Tajikistan, responsible for the educational programs for biology teachers in this region.

Lori Baralt is an Assistant Professor of Feminist Health Studies in the Women's Studies Department at California State University-Long Beach. She received her PhD in Sociology from Michigan State University. During the PhD program, she earned an interdisciplinary certificate in Gender, Justice and Environmental Change. Her primary interests include the sociology of health and illness, environmental health, social movements, and gender. Her research thus far has focused on both environmental and biomedical breast cancer advocacy in the U.S. and globally.

Gary Bowden is an Assistant Professor of Sociology at the University of New Brunswick. A past President of the Environmental Studies Association of Canada, his environmental work has ranged across a diverse range of topics, among them: social factors affecting geologist's estimates of how much oil is left in the ground, rural-urban differences on the value of wilderness, understanding hurricane Katrina as a "system accident," and campus sustainability. His present work involves a series of historical case studies examining the long-term interplay of energy, nature and society in the rise or collapse of those societies. This work uses concepts from complex adaptive systems theory and the concept of a socio-ecological system to empirically exemplify the theoretical idea of an ecological sociology.

François Dépelteau is an assistant professor (Sociology, Laurentian University, Canada) who is specialized in social theory, social movements and environmental sociology. He graduated at Laval University (PhD, 1995), did two post doctorate researches in Montreal and Paris (CADIS, Écoles des Hautes Études en Sciences Sociales), and has taught sociology in Azerbaijan, Belarus, Canada and United States. He has started to work on the Sudbury Soils Study in 2008-09, in cooperation with the Community Committee. He published various books and articles on research methods in human science, Belarus, social movements, social classes and relational sociology.

Ilse Loots is Professor at the Faculty of Political and Social Sciences and the Institute for Environment and Sustainable Development, University of Antwerp, Belgium. She received her PhD in urbanity and settlement patterns from the University of Antwerp, Belgium. Her current research focuses on environmental sociology and policy sciences, environmental health perception and risk communication, social impact analysis, spatial planning and location issues. Part of her work concerns the Centre of Expertise for Health and Environment in Flanders (Belgium).

Bert Morrens is Master in Sociology at the Faculty of Political and Social Sciences, University of Antwerp, Belgium. He works as a research assistant for the Centre of Expertise for Health and Environment in Flanders. His current research focuses on the policy interpretation of human biomonitoring data, risk communication, and environmental health disparities.

Steven Ney completed his doctorate in political science at the Department of Comparative Politics in the University of Bergen. In July 2005, Steven took up his appointment as Assistant Professor of Political Science at the Singapore Management University. Trained as a policy analyst at the University of London, Steven Ney has worked on a wide range of policy issues in a number of research institutes including the LOS Center in Bergen, the Interdisciplinary Centre for Comparative Research in the Social Sciences (ICCR) in Vienna and, not least, the International Institute for Applied Systems Analysis in Luxenburg. Starting as a researcher of environmental issues, particularly climate change,

Steven Ney has increasingly focused on issues surrounding societal and policy responses to demographic ageing in Europe and South-East Asia. He is currently exploring institutional change in the field of health care policy.

Barbara Potrata is a research fellow in qualitative methods as applied to health studies at Leeds Institute of Health Studies in UK. She obtained her doctorate at Cambridge University on the practitioners of spiritual practices and complementary and alternative medicine (CAM) in her home country, Slovenia. She was a senior fellow at Harvard University, in the U.S., working on examination of how CAM practitioners, working with breast cancer patients in Boston area self-define efficacy of their practices. Her other research includes the study of issues involved in the use of CAM among the nurses in Campinas in Brazil, and various mixed-methods projects in the area of cancer studies. As a fellow of Cambridge Central Asia Forum, she undertook a research on traditional herbal medicine in Uzbekistan and Tajikistan.

Alexander Stingl (PhD University of Erlangen-Nuremburg, Germany 2008). His research combines history of science, medical anthropology, and sociology of knowledge. He is interested in the conceptual construction of normalcies and pathologies in the discourse of the life and cognitive sciences in the nineteenth and twentieth century. His present research project focuses on the normalcies, pathologies, and interventions regarding "attention." He is interested in the use of internal martial arts as preventive medicine and complementary therapy. In a second line of research, he focuses on conceptions of justice and tolerance in public discourses since 1945. He is currently working on a manuscript about the production of theoretical knowledge in the nineteenth-century life sciences, a substantial essay regarding the development of early American medical sociology, and several essays on ADHD in regard to the concept of the neuro-chemical self and the whole-systems frame of reference.

Nik Van Larebeke is a Medical Doctor (University of Ghent, 1972) who obtained a PhD in 1977 (University of Ghent) on the induction of plant tumours by Agrobacterium tumefaciens. He worked on the mechanisms of cancer induction and on the prevention of cancer since 1970. In 1978, he founded a genotoxicity laboratorium in the Pasteur Institute in Brussels. Between 1984 and 1994, he worked on cell motility and invasion. Since 1994, he is mainly active in cancer prevention and human biomonitoring. He is spokesman of the Flemish Centre of expertise for Environment and Health.

Stephen Zavestoski is Associate Professor in the Department of Sociology and the Environmental Studies Program at the University of San Francisco. His research areas include environmental sociology, social movements, and sociology of health and illness. His past research has explored the use of Internet technology as a means of increasing public participation in environmental decision-making. Currently, Dr. Zavestoski is engaged in research examining the mobilization of trans-national social movement networks in response to the increased risk of industrial hazards brought on by globalization.

AKNOWLEDGMENTS

Special thanks to Riley Dunlap for reading and commenting on the early draft of Introduction and Chapter 1. Thanks to Carla Krooze, Claire Moore, and Melanie Hughie-Williams for proofreading an early draft. Thanks to Engelbert Fellinger (www.engelbert.org) for providing his photographs for this volume.

INDEX

B

C

D

E

F

G

H

K

L

M

Q

R

S

T

U

V

W

X

Y